CONTEMPORARY SOCIAL SCIENCE RESEARCH

(An Evaluation of National and Non-National Contributions)

By the same author

— Violence and Religion : Cross Cultural Opinions and Consequences

— Social Behaviour of Children

About the Author

Ralph Tanner has a B.Sc. and Diploma in Social Anthropology from Oxford University and a D.Phil in Law from Stokholm University. He has done fieldwork in Thailand the Philippines, Guyana, Zambia, Zimbabwe, Tanzania, Uganda and Kenya as well as in Britain and Eire. He was Chairman of the East African Institute of Social Research and Lecturer in Comparative Religion in the University of London, He has published books on Murder in Uganda, Witchcraft killings and religions and successful change and a cross cultural study of the Roman Catholic Mass. He has co-authored books on the social ecology of religion, the recreation of tribal identity and religious change and on religion and the environment. He has written numerous articles on the social aspects of religious change, translation and language use in religion, as well as on behavioural theories in the Journal of Modern African Studies, Africa, Journal of the Social Sciences, Nordic Journal of African Studies, Anthropos, Studia Missionalia and others. He is currently working on a cross cultural study of spirituality, well-being and health.

CONTEMPORARY SOCIAL SCIENCE RESEARCH

An Evaluation of National and Non-National Contributions

R.E.S. Tanner

CONCEPT PUBLISHING COMPANY, NEW DELHI-110059

Cataloging in Publication Data—DK
Courtesy: D.K. Agencies (P) Ltd. <docinfo@dkagencies.com>

Tanner, R. E. S. (Ralph E. S.), 1921-
Contemporary social science research : an evaluation of national and non-national contributions / R.E.S. Tanner.
p. cm.
With particular reference to developing countries.
Includes bibliographical references (p.)
Includes index.
ISBN 13: 9788180695162
ISBN 10: 8180695166

1. Social sciences—Research. 2. Social sciences—Research—Developing countries. 3. Social sciences—Methodology. I. Title.

DDC 300.723 22

ISBN-13: 978-81-8069-516-2
ISBN-10: 81-8069-516-6

First Published 2008

Published and Printed by
Ashok Kumar Mittal
Concept Publishing Company
A/15-16, Commercial Block, Mohan Garden
New Delhi-110059 (India)

Phones : 25351460, 25351794
Fax : 091-11-25357103
Email : publishing@conceptpub.com

Dedication

To

The Memory

of

My Father and Mother

Acknowledgements

In a work on social science methodology by a writer who has undertaken many forms of research, my thanks must first go to the many hundreds of informants in four continents who have participated in my work and who have tolerated my intrusions into their privacies. I am deeply grateful for their role in making possible the publication of so much of my fieldwork.

More specifically I am grateful to Ali Farahani of Kilwa, Tanzania who taught me to understand the intellectual significance of non-literacy and to Juma Ndamulo of Mwanza, Tanzania whose personal social 'radar' of what might happen and where in the religious rituals of the Sukuma, enabled me to see much that would have usually been hidden from an outsider.

Lastly, my thanks must go to David Tanner MA (School of Oriental and African Studies, University of London) who cast a critical and time consuming eye over what I had written and suggested many valuable improvements.

CONTENTS

Chapter 1

THE DIMENSION OF SOCIAL SCIENCE FIELDWORK

1. Post modernism and social science research in developing societies

The unknown as it relates to human behaviour is unlimited in scope and more importantly in the factors that may be involved. A basic problem for the social sciences must be this professional title — social science. The definition of science refers to trained skill, a branch of knowledge and a branch of study concerned with a connected body of demonstrated truths or with observed facts systematically classified or more or less collegiated by being brought under general laws and which includes trustworthy methods for the discovery of a new truth within its own domain (Oxford English Dictionary,1956). It is certainly taken to cover a certain hardness of facts and conclusions that can be verified by others. In reducing this unknown there has always been an uneven approach, with some areas of behaviour receiving continuous or intermittent attention from researchers and some aspects none at all. Most of the research in the social sciences has been highly specialised while tending to become the foundations of more general conclusions. So it is always a question of who does, what, when, where and how so that while the whole research process is theoretically open, it is in fact always severely restricted by personal, professional and place restrictions and variations. It may well be the fact that academic considerations have to be subordinated to factors that have nothing to do with what is chosen and indeed what can be researched.

Very little of human behaviour can be classified as exact enough to parallel inorganic chemistry or DNA analysis which would have effects independent of human variation. But even if we accept the accuracy of a DNA analysis, a medical diagnosis or a marriage certificate, the 'social' aspect of the social science label comes into the equation. The fact may exist but how it is treated or come into social existence is immensely varied in how it is observed, understood, and acted upon by those involved; a pool of potential knowledge with inexact boundaries and innumerable invasive factors. The social sciences are faced with the problems of human variability at any stated time and this variability is ongoing and continuous. Any research in its conclusions is tied to circumstances which have passed into history. Professor Evans-Pritchard has suggested that anthropologists are in fact historians as whatever they have studied was in the process of change, but those who are observing are also changing and are as varied as their research subjects. So it would seem that social science strains after a scientific exactness which is denied to them by the circumstances in which people are studied. Certainly there are facts, influential or not, round which there are not exact behavioural events but brief periods of coincidence which may have correlations. However the key variability in all research is the researchers themselves in relation to the country in which they work.

The citizen as a researcher involves two categories of people. The citizen category in law is a general one in which the individual is no different to several million others. The value for researchers to have this legal status will involve positive assets which will enable them to do certain things, while at the same time will restrict them in other just as significant ways. However the citizen as a researcher is entirely individual. Any group affiliation of researchers will be circumstantial. Their personal contributions to any research situation will be a combination of what they bring into it as individuals and the particular circumstances in which they are involved.

The importance of citizenship is only in theory a static value and much will depend on the researchers awareness of what is involved nationally in not only their ability to do research but in their ability to manipulate the national social and political environment; the factors which officials and politicians want to employ at that time. Citizenship is a card in the pack which has legally defined values but its value at any particular time and place depends on the value of the other cards which are being played then. The researcher on the other hand has to be evaluated in entirely individual terms. Any groupings into categories are bound to be largely artificial because the numbers of researchers in any country at any time is not likely to be large. Their personalities and approach to their research work are differentiated by any social, psychological and physiological categories employed. In addition the research situations are by definition different and changing all the time while the research continues.

It is the contention of this author that those who undertake research can be divided between those who are nationals of the country in which research is undertaken and those who are non-nationals with a third category of nationals who have left their country of origin and undertake research in or about that country from a position in a citizenship in a third country whom he classifies as "half and half" researchers. He feels that this division is more important in influencing the nature of research than the more conventional divisions based on ethnicity, sex and age. It is also his contention that since the social sciences originated in European thought, it is Westerners who have continued to dominate research not only from a theoretical perspective but through their greater access to research funding. So it would seem that in their Asian and African intellectual and political assessments there is an element of unconscious exploitation.

This study looks particularly at the advantages and disadvantages of citizenship for research in Asia and Africa comparing the methods and enablements of the three categories which we have defined. It will be assumed that citizens with their superior insider knowledge of the societies in which they work would have significant advantages over the non-national, while imposing balancing disadvantages which outsiders may be able to avoid. As far as is possible we look at social science research from the perspective of African and Asian academic and political realities and at the published work of nationals as much as non-nationals to illustrate these issues.

2. The insider-outsider division between social science researchers

This may be an unwise dichotomy however attractive it may seem to those who see researchers as fitting into such a classification, dividing nationals from non-nationals. It is obvious enough that an Asian Indian anthropologist researching American Indians is an outsider and that in his comparative work with Indians in India he was an insider (Pandey,1979), just as a Saudi Arabian woman with high American educational qualifications researching Saudi-Arabian family life is an insider (Altorki 1982). But these are extremes and between them there are many gradations between the legal distinction of being a national or a non-national which have to be discussed. The vast majority of social science researchers are insiders and outsiders in varying degrees at the same time according to the aspects of the topic on which they are working. An Afro-American studying Nigerian social life is both a quasi-insider and outsider at the same time just as a European studying European social life in East Africa (Tanner,1964,1966), made himself into a specialised outsider among other outsiders in terms of their nationality and the surrounding African majority.

It has always to be remembered that those undertaking a research project in the social sciences are by self-definition making themselves into outsiders to those that they are studying. How far this 'outsiderness' goes is very much a matter for their personalised methodologies, a combination of personality, topic and accidental circumstances, but outsiders they remain from the moment that they start their research and inevitably when they put pen to paper. Cross that boundary-line of professional detachment and they have to consider whether they are any longer social scientists. It seems likely therefore that these are analytical classifications of limited usefulness except in Merton's comment (Merton,1972:36) that 'the problem is not to ask whether it is the Insider or the Outsider who has monopolistic or privileged access to social truth but to consider their distinctive or interactive roles in the process of truth seeking'.

The researchers are insiders when there is some mutual shared interest which at least to some extent overlays their obvious differences as researchers; for example, ethnic similarities, belonging to the same family, lineage or clan, shared birthplace, shared experience of understandable misfortune, shared interest in football or religion, and a shared dislike of certain categories of outsiders.

In their fieldwork do the researchers let the information come to them as at least partial insiders but never completely so or do they seek for information which makes them more distinctively outsiders; what are the dimensions of social acceptability in the roles of being accepted as curious outsiders to being outsiders who are curious? Researchers are outsiders when they are known to be investigating matters which are not seen by those being researched as commonly accessible to outsiders; their dividing line between public and private curiosity. The question 'why' does not sit easily into any inter-relationship in any culture as it is intrusive. Once researchers take an active part in obtaining information they are automatically outsiders and this becomes almost always a potentially hostile act since it has no part in the research subject's social life and is related to the bureaucratic entanglements of modernising societies. The actively seeking researchers whether they are nationals or non-nationals are following on these peoples' experiences in schools, government offices and the presence of inspectors from a wide range of departments. The cards are marked against researchers when they go beyond the passive role of letting themselves be slowly absorbed as far as they possibly can into the research environment and it is this which we will be discussing.

3. The aims of research

We must assume that the aim of social science research is the betterment of the lives of the particular parts of society which are being investigated by particular researchers and that they have some particular insights in this and that almost inevitably they have some personal concern; they are biased by their personal and professional commitment to this field. Further, it is clear that the aims of social science researchers should not in the first instance be professional success and career fulfilment. Of course, this exists but surely this should be mainly secondary and should be kept severely subordinated to the essential aim of the betterment of the lives of those researched. If we accept that the aims of social science research is this betterment of the lives of those researched, then we are immediately concerned with how this is to be brought about since we, the researchers, are not administrators nor political and social power holders. At whom are we directing the results of our research and who are we trying to make aware of how the people we have researched live?

The aim of this study is to illustrate the problems of fieldwork by the well-meaning and socially aware researchers who devote much time, money and effort, and their social capital, to finding out about the social life of people whom they research. The inarticulate people who are researched know their own ways of behaving well enough but about whom the local power holders and initiators of change may know very little. Thus social science research has a threefold aim beyond the moral aim of enabling betterment. To find out what they want to know without disrupting the lives of those researched. To enable better understandings about these parts of society by those seeking to change it. Finally, to persuade outsiders about these understandings which have been acquired through social science research.

Of course in all this there is the need for foundations based on appropriate and acceptable methodologies and through this the ability of researchers to portray and persuade. The basis should be the possibilities enabled by methodology, not the methodology enabling the possibilities. The study of human behaviour can never be an exact science and forcing exactness on to data by using the methodological approaches appropriate to working on inorganic material, is self-defeating; the data cannot cover the ranges of variation and is out of date almost as soon as it has been recorded. The aim of methodology should be to produce data that has some logical coherence between the views of the observed and the observer and not just one created by the intellectual imagination of the researchers.

4. The dimensions of the unknown

Let us start by referring to this author's own experiences of trying to find out about the people with whom he was involved as an administrative officer. In north Burma almost all records had been destroyed either by accidents or deliberately during the Japanese invasion and the subsequent reinvasion by the British and Indian armies and United States Army Air Force bombers. Offices were bare except for the recent productions of the Military Civil Affairs Directorate. Even if he had known that there had been a study of the Palaungs (Milne,1923) but there were no locally owned copies.

When he came to work in the Kilwa coastal district of Tanganyika also as an administrative officer, there was nothing in the office except recent records related to cases that had been heard in court and tax registers. The so-called District Book contained no more than a few unsystematic idiosyncratic reminiscences of previous administrators. There was some data on the Arab rule over the island of Kilwa Kisiwani and some published descriptions of the 1905 Maji Maji rebellion against the German colonial occupation which had centred on Liwale and Kibata over a hundred miles inland (Bell,1950). Thus it seemed to him as it must have occurred to almost any social science researcher that whatever fieldwork project he started it would be a very small item in an enormous field of the unknown and however thorough he might be, it would be no more than a pebble thrown into a pool of ignorance.

In both the two areas mentioned above there were small populations in enormous areas so that when this situation is seen in societies numbering many millions such as in China and India, this amount of the unknown must be multiplied. In contemporary terms the situation must be much the same except for small communities in which individuals have specialised such as the

Manus islanders, the Coorgs in India, and some American Indian communities in the United States. The theories behind much contemporary social science fieldwork will have been based on the sporadic earlier coverage of a limited range of communities, may seem important in retrospect, particularly for those who do not have any personal knowledge of cultures or parts of cultures other than their own, but not nearly so important as new data which for most of the world still continue to be more or less blank slates.

5. Developing societies and research

There is a basic difference between the research environment in poor or at least not wealthy societies and economically successful ones in which higher education is able to develop its own ideas of how research should be conducted because they have the resources to be independent. A developing nation with its own limited revenues is often dependent on the major Western nations for its own plans for social and economic development. They would agree to non-nationals researching inside their boundaries on subjects and producing results for which they have no immediate or indeed foreseeable use. They know that these researchers are furthering their own careers but it is not worth the political difficulties in trying to restrict such work to areas which they consider important. Of course, this happens to some degree in national committees vetting research topics before visas are granted, but they cannot be too bureaucratic about this because they need the goodwill of outsider universities for exchange students and support for their own universities.

Most developing societies have immediate needs as do their universities and they want to maintain control of their own institutions so that they provide value for money. Of course, there is a theory that the education is a 'good thing' for all but in practice the cost of education beyond the primary level is so high that it is in fact always elitist and limited, even though they understand that education is the means to achieving higher standards of living for all or at least in keeping economic development in time with population growth. They would see little need to educate engineers and agriculturalists when national opportunities are limited and they would be financing a 'brain drain' in reverse. In so far as they have the means to control and promote social science research it would be in applied research in support of their immediate problems such as the control of AIDS, urban planning, agricultural improvements and low level selective change. Overall research has sometimes been seen to support a particular regime and its agenda which is usually controlled by a corrupt national socio-political oligarchy.

6. The factor of credibility

At any time in the professional world there is a framework of acceptability for ideas, methodology and results which is the basis for research work in the social sciences. This is not static and changes occur slowly as new research work is carried out on new issues and in new environments but there can be violent and radical changes which require an immediate reorientation of methods. The fact that the 'tsunami' in Indonesia killed far more women than men has meant that men there have been forced to do their own cooking and laundry so that the whole structure of those coastal societies will have been reorganised in days in ways that could not have been envisaged in any of the existing scenarios for rapid social change and within the dominances of Islamic ideology. Such enforced social change provides important opportunities for original and acceptable research. Then there is the work of those such as Castenada (1968), who go beyond these boundaries of existing methodology and however 'right' they may be in the long run certainly find their acceptability limited to those on the fringes of academic and mainstream social science. Of course, we can say that his works come from his imaginative reconstructions of what Yaqui Indian intellectuals may have been thinking. But on the other hand professional parameters do exist and restrict acceptability of ideas which challenge existing non-national understandings.

Demographic data is usually based on documented birth data and movements tied to maps so that accurate figures can be produced on the movements of populations. But for most populations birth dates are not known except in relation to key events for which dates are known and even if maps exist there is not much use of them in recording movement to and from places which have no relation to village names or streets which have no recorded names. So the recording of movement by direction is as valid a method as any for studying migration but not one acceptable to the professional rules of demography.

However radical the individual researcher may be in their personal views and political affiliations, their professional standing and future careers rest within the parameters of acceptability which is correlated to specialised forms of credibility. It has little if anything to do with whether these conclusions fit into the understandings of the people about whom the research has been done, indeed in most cases there is no backward consultation or the passing on of their conclusions. Perhaps there is an understanding that it is difficult enough to get professional acceptance without the complication of putting these conclusions into a more popular genre often in another language.

In the developed nations the social science parameters of acceptability allow a much wider range of latitude than exist for the harder sciences but they are still laid down by those in senior positions in these fields of enquiry. In what could be described as a profession seeking wider acceptance for its work not much has been done to offset Karl Popper's comments that 'the triumph of social anthropology is the triumph of pseudo-observation, pseudo-descriptive and pseudo-inductive generalising methodology and above all marks the triumph of a pretended objectivity and hence an imitation of the methods of natural science' (Banton,1964).

However, creditability has no general acceptability according to predetermined standards and is always dependent on what matters and particularly for whom. In professional circles credibility is likely to be related to on-going understandings of evidence and how the whole discourse hangs together and often as sponsoring organisation will be looking for some assurance from referees, that they are getting a reasonable return on their expenditure. The political power holders will want to be assured that what has been researched and the results available for public use are not contributing to unrest and indeed not over-critical of their regime and of course show some concern for the people who have been researched who may not have been told what has been written about them

Social scientists learn from their university education but if they are to enlighten those who read their publications, their writings have to be more than and indeed qualitatively different from the use of the techniques and ideas which they have received from their academic superiors. Human behaviour is so complex that the understandings of most people are tied to commonsense pragmatism rather than to those of academic analysis. The Tanzanian Sukuma when asked the old English riddle as to why does the chicken cross the road, answer that it wants to get to the other side. Similarly, if they are asked why they propitiate their ancestors, the answer will be just as it will be to such difficult questions in all societies and at all levels, this is what we have always done; analytical thought is always a rarity. Too often commonsense is replaced by the satisfying crudities of quantitative science and theoretical complexities. The social scientists if they want to have influence outside their own profession have to use their imaginations to link what they see and hear with what those they are observing are thinking (Miller,1976). An intelligent Sukuma may understand perfectly well the anthropological theories relating to mother-in-law avoidance and the unity of alternate generations but moving onto more complex ideas such as the Oedipus complex they would think that the enquirer is trying to be funny. It seems likely that the national may be less likely to waste time on issues which both they and their people already know.

7. The inaccuracy of accuracy

Whatever approach is used in discussing social science research beyond our own cultural understandings rather than just their behaviour, there are enduring dilemmas. We are seeking to impose standards of scientifically acceptable accuracy on human behaviour when we already know that it is intrinsically and permanently variable. It seems likely that too much attention and effort has been put into the scientific aspirations of the social sciences and the consequent feelings that many researchers have not so much failed but that they could not have succeeded in the first place to achieve any such overall standards of exactness and as accuracy in anything as changeable as human behaviour.

If we take the opposite aspect we can see how much the exact sciences have to use social science qualitative data in support of their exactnesses. A study of non-smoker mortality from living with smokers (Hill *et al*,2004) accepts the data provided by the sample that they had never smoked as believed, apart from the fact that not much is known about what is involved in second-hand exposure to smoking. Another study of the consequences of a public smoking ban on hospital admissions for heart attacks (Sargent *et al*,2004) lists among the weaknesses of the study that it did not and indeed could not measure exposure to second hand-smoking. Whether the conclusions will prove to be valid is not the point, only that these researchers had to depend on qualitative assumptions, even if we do not include in these cases that the public opinion on the unpleasant aspects of tobacco smoking may have motivated both the research and its conclusions. Such doubts will always exist unless the hard sciences confine their research work to environments which are scientifically controlled. Any current number of the British Medical Journal or the Journal of the American Medical Association will contain articles on medical research which combine factual data from laboratory analysis, standard quantitative information, correlated with qualitative information to which the same principles of pragmatism cannot be applied which are nevertheless treated as important enough to be tied to scientific data. A high proportion of research articles rely in part on the honesty of informants supplying information and there are no adequate ways of checking on the correctness of varied answers in varied situations. This is not to suggest that replies are deliberately dishonest although they may well be so, but the assumption that much information is correct is much on the lines of an act of faith.

While there are identical twins that are biologically similar, they are nevertheless subject to different environmental pressures. For the rest of society there can never be environmental situations even those of 'tsunami' sea wave disasters which will have similar effects on all the individuals involved as they are all basically dissimilar. Many social scientists continue to act and research on the assumption that in some way it is possible to group these innumerable and fluctuating differences and to halt their constantly changing human and environmental interrelationships. This is done so that the data becomes potentially usable and transferable to other situations in which it is assumed to have affinities. There is the assumption that human behaviour is intrinsically malleable by others but perhaps not the social scientists themselves.

While it is not necessary to go as far as the Buddhist framework of thinking that it is only humans who create the illusion of individual permanency out of constant and varying microscopic units, perhaps we are still nearer to Malinowski's approach that viewed the Trobriand Island society as a human laboratory. We are usually, it seems, nearer to the approach that treats those whom we research as 'them', rather than refractions of ourselves. There are few social scientists who approach their understandings of others as 'but for the accident of birth' I might have been there.

Since so much social science in both the developed and developing worlds is applied to sorting out immediate problems already defined by political and known social needs, there is at the back of the social scientist's minds not so much an unconscious wish to know more, but a desire to predict and control. This is not so much because of political pressures but because social scientists may be to some extent mesmerized by the successes of inorganic science. Social scientists are also not happy with the suggestion that they are not only biased in the ways that they collect data as individuals but that they are subject to fashion in their fields of research and the more social than scientific boundaries that this creates.

This study is an examination of some of the contemporary factors involved in carrying out social science fieldwork 'with' rather than 'on' people who are disadvantaged in terms of what they are and what the power holders hope to achieve within their life-times. There is immediacy in what social scientists aim to describe and relate to in their conclusions. There is the need to examine the predominance of Western methodology and understandings in social science research and further the range of influence enabled by the professional predominance of English. Although Mandarin Chinese is spoken and read by far more people than any other language and in which we can assume much valuable research has been written, it is not a language of global influence. There are of course Mandarin readers who are bilingual in English, but they aim to inform the power holders in their own society and do not have much interest or are restricted in how far they can draw attention to themselves by translating their work for extra-territorial readers. This examination of social science fieldwork is divided between factors restricting and enabling non-national social science researchers and the extent to which these are shared by national ones. We cannot assume that the national researchers have an overall advantage since there are internal expatriates, sex and social distance factors which are also important.

8. Culture shock

The social science fieldworker, even national ones working in their own country unless it is in their own social environment and amongst those with similar social understandings of themselves, are social and psychological migrants. The only difference to those who migrate in the usual sense of the term, is that they are committed to getting on with the people surrounding them on whom they are dependent without being employed by them. Culture shock can occur as the result of very personal experiences. For example, the middle class Brahmin researcher collapsing in the street and coming to in the slum hut of a sympathetic Untouchable family which resulted in him choosing this area for his fieldwork

(Varadachar,1979). This is a peculiar form of dependency which is much more inclusive than that of other forms of migration in which migrants keep or are kept to themselves except for those marrying into another culture or who move there for some form of missionary commitment.

However well-equipped and intellectually ready researchers may be for the experience of a particular piece of field work, it is always a personal experience. Whether the researchers move from Belgaum to Banaras or from Berlin to Birmingham it is inevitably filled with surprises that they can accept intellectually but nevertheless impinges on all researchers in ways that cannot be avoided; they are professionally committed to include their own reactions in balanced perceptions of what they have set out to do and record. In the fieldwork situation there is not really anyone who is so open-minded as never to be shocked by the behaviour of those they encounter. It is theoretically possible to take every experience as paralleling writing on a blank slate but in reality this is impossible even for the most imaginative and sensitive researchers. In such contexts even imagination may be a form of bias because it leaps over the boundaries of the available and perceived data.

Culture shock may be dyadic as was the case with Zigua adolescents in Tanzania who were shocked by the way that Europeans cleaned their teeth with water which they were thought to drink. Personal hygiene is inevitably a matter of disturbing realities whether it concerns bathing, excreting or menstruating. Some fieldworkers find their situation so personally difficult or distasteful that they give up and retire to do their fieldwork from within the social safety of a university campus or hotel. Most make some sort of a showing in at least attempting what they set out to do. In this they will be benignly shocked by some aspects of their experiences as their host's acceptance of strangers which they interpret in a personal way or malignly so boycotting others because of the noise, lack of privacy and dirt. No researchers feel totally at home because they have always been socialised into other patterns of behaviour to which they are socially and psychologically committed.

Obviously the non-national researcher may feel the greatest range of culture shock because they know themselves to be cultural aliens and they are always ready to be surprised ideologically by matters which are in practice on the fringes of everyday life. They are much less prepared for the intricacies of inter-personal experience in how they use their hands, eyes and mouth for which there are no internationally acceptable ways of behaviour. To be left-handed for example, may be a grave disadvantage since it is too deep set in the mind to be turned overnight into right-handedness and the Western habit of shaking hands may be polluting. In much the same way non-nationals may be delighted and indeed overbalanced by the freshness of their experience; blue skies and starlight can be confusing after the miseries of the European or North American winter until they have experienced the monsoon and the extreme heat of a northern Indian summer.

Of course, culture shock can be a matter of extremes, an agnostic non-national can be revolted by the practices involved in the worship of Kali, in the overwhelming poverty of a Dakar ~~or a~~. Lima urban slum and the treatment of women whom the male researcher may never meet. The culture shocks are certainly more intrusive and much more continuous in emotional matters. The uplifting excitement of seeing Sonjo men dancing in honour of and becoming possessed by the spirit of Hambagau the water related God, their tutelary spirit, of being in a Kachin long house for the dancing and drinking in honour of the dead house-holder and appreciating the many ways in which the Sukuma, like all societies, circumvent their inevitable miseries. The overall shock that so many people in other cultures seem better able by design to cope with their difficulties without intellectuals and bureaucracies telling them what they should do.

To what extent is the national researcher liable to experience culture shock and to be affected by it? Almost certainly they are in the sense that they cannot explain it away quite so easily as the non-national who is able by definition to explain these issues in terms of 'us' and 'them'. The national cannot so easily bracket off such issues because they are clearly by socialisation not so divided-off from 'them'. From socialisation they are committed to the intricacies of their own societies and are less able to circumvent them by design or social ignorance. Since these national social scientists have become a self-recruited intellectual elite, it may also be more difficult for them to excuse activities which in terms of their national image or social progress may shock their sensibilities.

The stimulus to research and the excitement which comes from this work will usually lie in the differences that researchers find between their own life styles and understandings and what they encounter in their fieldwork. Whatever form it takes there is bound to be some culture shock of 'the personal disorientation that may accompany entry into unfamiliar social worlds' (Macionis,1989:62).

This disorientation is obvious enough when anyone moves so far out of their own social environment that they are deeply shocked by what they encounter for which they have had no previous social warning. Few social science fieldworkers are likely to have anything approaching Changnon's first encounter with the Yanomamo 'I looked up and gasped when I saw a dozen burly, naked, sweaty, hideous men staring down at us down the shafts of their drawn arrows. Immense wads of green tobacco were stuck between their lower teeth and lips making them look even more hideous and strands of dark green slime dropped or hung from their nostrils—strands so long that they clung to their chests or drizzled down their chins. Then the stench of decaying vegetables and filth hit me and I almost got sick. I was horrified. What kind of welcome was this for the person who came here to live with you and learn your way of life, to become friends with you' (Chagnon,1983:10). Indeed we must also notice the use of the term 'friend' which the non-national often uses in relationships which has no parallels in their societies. Of course, this is an extreme case with a North American academic going to undertake fieldwork with an Amazonian tribe which had had little contact with the outside world and did not want to increase any such relationships. Any individual going to a locality with which they have not been in regular relationship before must experience some mental stimulation because these people are behaving in ways which are strange and at the first glance inexplicable. A woman fieldworker going from a university in

Mumbai to work in another part of India will be struck at once by innumerable differences, just as a graduate man whose research takes him away from the cultural environment of his education.

It is one of the truisms of fieldwork that national researchers have a social nearness which enables them to have particular insights which are not automatically available to non-nationals. It is certainly true for example, that a Hindu socialised with their particular social environments will have many insights into another Hindu community. Such socialisation enables national researchers to avoid the long working-in period when non-nationals attempt to become acculturated to the new communities in which they come to do research on some previously unknown aspect of their lives. The snag here is that such people may feel that they already have some natal insights into this community and thus they are lulled into not recognising the importance of the stimulus which must come to them from culture shock. However this shock is bilateral in that it is experienced both by the researchers and by the researched.

All researchers whether nationals or non-nationals go through a process of being liminalised in ways that parallel the patient entering a hospital; the individual is defined in terms of the receiving institution and community. In the research situation much of the field-workers social features are striped away and redefined. Shock of course is a loaded word and to change it to 'a confrontations with the unexpected' may diminish its realities. Bob Geldorf who started the global charity Band Aid has spoken of being anaesthetized by mass misery. The abrupt realisation that 'the other' is different and this is so obvious that we are inclined to neglect its significance; not the odd strangeness but the totality of 'otherness'. Hence, the importance of the researchers accepting their liminal status.

9. The roles of the researcher in fieldwork

To the researchers, fieldwork is a perfectly legitimate way of furthering the aims of their profession quite justified by the ethical need to know more about human behaviour for the ultimate betterment of mankind. This is an abstract ideal which is rarely going to be shared by those into whose lives these outsiders intrude.

The closer those being researched feel that their lives or the institutions in which they work or administer are being examined on logical and indeed to their minds scientific grounds which they personally value, the less likely they are to accept wholeheartedly these intrusions. They are even less likely to accept conclusions which challenge the ways in which they have become used to accepting as best. There is thus a difference between what can be done in fieldwork which has no more than the aim of general understandings and work which is done in applying their enquiries and understandings to particular objectives. There is almost invariably the feeling in the former case that these researchers will not be able to provide any information of which they think that they are not already aware. The researchers will be faced with various forms of obstruction as they disentangle what these people say that they do as opposed to what they actually do. In these situations national researchers may be better able to negotiate themselves round these successive obstacles (Minocha.1979), although non-national researchers with previous experience of working on institutions may be less impeded by these difficulties.

When researchers enter fieldwork situations such as a village, school, or institution, it is likely that the respondents there will label them as government spy, plain clothes policeman, inquisitive tourist or social worker. The respondents will always have some idea of what outsiders are doing and certainly there are no longer social environments in which outsiders are unknown. They ignore the general way in which traders come into the community from outside but can see researchers as agents of the government which in their experience and folklore are more likely to interfere than help them. But once within such a community a researcher will be given different roles according to the particular interrelationships in which they are engaged. A researcher on the island of Montserrat (Skinner,1997) with a population of eight thousand people, writes of being a person of ambiguity moving through different communities and social groups in the course of which he related in different ways to tourists, poets, firemen, librarians, boxers, government employees and Rastafarians because he had no single role and was in a constant state of adjustment. These constant adjustments which a fieldworker is required to make or are forced on them by day- to- day circumstances within which they have to work is wider and more comprehensive than that which an average person would have to undergo in their daily lives even in the most varied or developed communities.

No community at whatever level of development is homogeneous and it will have herbalists, diviners, young and old, men and women and those who are influential and those who are not. The approach and how each will respond to the stranger within their social environments will always be different. With the particular roles which these researchers may be able to establish they may well develop some affinities with the people with whom they are in contact; they may have similar familial relationships and the patrilineal connection makes mother-in-law avoidance a shared interest. They may have shared interests in pets, wild animals, fashions and food. The attempt to interrelate in the detached sense required by hard science is clearly impossible. The youthfulness of most researchers makes them, for some respondents, the social equivalents of grandchildren, siblings and classificatory children. There is also the fact that in some societies they cannot come within the scope of endogamy because they have married outside that particular social framework.

For some of the people in the researched community these intrusive strangers represent opportunities. Wanting to learn about divining or how to play a particular game turns the researcher into not so much a client but into a status enhancing pupil. Even outsiders come to consult these local specialists so he must be considered good at their job. For others these outsiders create opportunities for links with the outside world. So all the people with whom researchers relate see them in different roles and the researchers themselves would have to develop chameleon-like characteristics in order to get what they want from real or potential informants.

10. The roles available in social science researchers

These are divided between ascribed roles related to physiology and achieved roles to the extent that the researchers are accepted. The physiological ascriptive roles from which there can be no escape, such as sex, age and ethnic appearance; and however, much researchers convince themselves that they have detached investigative understandings; the roles related to physiology cannot be escaped. They cannot assume that their fieldwork has anything more than rudimentary affinities with experimental work under quasi-laboratory conditions. All researchers with their physiological characteristics are working in social environments which have well established rules and roles as to what is to be expected from people with these visible signs of identity. Researchers cannot expect to be able to assume roles for which they are not physiologically entitled. A fecund but not fertile woman is just that and a post-menopausal one has some marginal acceptability into some male dominated social groups. A young man cannot expect social equality with socially mature men however much that this may be disguised, on their part, by culturally required politeness to strangers.

While ageing may be useful in getting additional accesses to information and as does marital status and children, ethnic appearance does not change. This is unique in categorized reactions within a range of sight, an immediate categorizing of the seen individual as someone who is one of 'us' rather than being one of 'them'. The fact that such categorizations may or may not be socially or politically acceptable, does not alter it being the basic reactions to the appearance of any researchers from their first day in the fieldwork area; this perhaps has some ethological groundings. Of course a Sikh with his beard and turban is physiologically outstanding but even if he shaves he still has his visible Asian ethnicity.

11. No known role for the researcher

Unless researchers are working in a community which has already had substantial experience of social science fieldworkers with peoples such as the Coorgs of the Nilgiri Hills, they are social misfits, and the researched have a socially unclassifiable stranger in their midst. All the roles which he or she may want to have had in their own communities no longer apply. There is a loss of status because of their age as they expect to be able to talk to all on a basis of equality when in most cases they are too young to claim such a right. They are usually unmarried in communities which give little status to the unmarried whether male or female. Materially they are obviously better off than most of those with whom they have come to live and yet they are not prepared to share with those with whom they claim friendship. Finally, they are usually linguistically incompetent in the local dialect and generally behave accidentally in ways that are at best on the fringe of good manners according to local understandings of how visitors should behave. Few researchers entering fieldwork realise much less accept that much of their public persona is dictated by the succession of social environments in which they have lived prior to entering these communities strange to them. They are usually only marginally aware of their own social strangeness to those whom they are now approaching often expecting their cooperation without understanding the reluctance of all communities to the abrupt intrusion of inexplicable strangers. It is not surprising therefore that both national and non-national researchers find themselves confronted with a succession of new social situations for which they are psychologically and socially ill-prepared even if they have already some intellectual expectation of differences. Egoism may often make them feel that these differences will be overcome by their own self-evident goodwill towards the community and its members on whom they are imposing themselves.

In the past sociology has always been the field of the national 'native'; Americans and Europeans who have studied people who have at least nominally been members of their own cultures American life and so anthropology involved non-nationals in foreign environments. Now that African and Asian universities have departments of sociology and anthropology some of which have long-standing and distinguished reputations, we are investigating whether their researchers are experiencing the same problems that sociologists have long experienced in studying their own cultures or do their difficulties reflect those of anthropologists studying a foreign culture; a reduplication of the largely Western problems of the past.

12. The researchers initial ideas as to their role

We have seen that the roles of fieldworkers which are ascribed to them or to which they may achieve, are a combination of several different aspects; the roles to which the researchers aspire, those which the accepting community impose on them and the roles which eventually develop as a compromise between these two contrasting viewpoints.

How will researchers approach their fieldwork? It seems very unlikely that they will approach their work with anything that even approaches an open and detached mind. There is the academic requirement of completing the research project, a break from the materialist routines of their own social environments, wanting to contribute to a better understanding of other peoples and cultures, curiosity about how other peoples lead their lives, putting their own feelings of intellectual superiority to use and possibly having a breathing space in their personal lives. Whatever the hoped-for roles may be, they are made to a large extent in ignorance of the social conditions under which they will be working.

13. The final compromises over the researcher's role?

If the researchers stay for the full period which they have set aside for their research this must mean that they have achieved roles which enabled them to work adequately on their projects and that the researched people or community have not actively obstructed the researchers' work and their presence there; they have perhaps become unobtrusive and that the research has been passively conducted. What is certain is that the researchers' presence has been to the advantage of certain individuals there who have provided sufficient cooperation to enable each piece of research to be started and to continue but of course this is often not what these researchers may have set out too in the first place. What roles have researchers taken on which has enabled their work to be continued?

Researchers have become paying guests in some households which has provided their hosts with an unexpected income; the role here would be that they are seen as committed to the viewpoints of that particular family with which they are staying. Other researchers set-up household on their own with local domestic servants who do the cooking, water fetching and laundry. They are thus employers of locals who profit from being able to retail information into and out of the household.

Then there are those with spare-time either because of age or seasonal or semi-permanent unemployment who find in the researchers a status enhancing way of spending their time over a shared interest in what is going on in their community, a process of professionalizing gossip. The researchers have an interest in the community and what the respondents do so that they become for their hosts a useful repository of what goes to making them special which they are in their own eyes; they become as it were the community's official historian (Jorgensen1981). There are also situations in which the respondents become sufficiently used to the researchers presence so that they feel and are felt too be sufficiently at home to have more intimate insights than other strangers.

Finally, the political tool role. The researchers are seen as useful channels into the outside world from which their informants individually and the community collectively may get some advantages; it is for them an investment of their more or less usual goodwill to strangers who might well have connections since they have obviously been paid to do this work by their employers and their presence there anyway means that they must have at least the tacit approval of powerful outsiders.

The aim of fieldwork in the social sciences is to take a detached view of what is being observed although it is almost certain that the researchers however well trained and professionally minded, are going to be biased in their approaches. An American national undertaking long-term research in a community in which he and his wife were not ostensibly different to those amongst whom they were living, became his own primary informant. Perhaps it could not be otherwise since he was required to research the educational system in which he could have no undercover role in the five years set aside for this research project; there were just no 'us' and 'them' possibilities in this situation. Here was an altogether pleasant merging of roles but not any practical possibilities of detachment (Hennigh,1981).

There are other occasions when a researcher is forcefully deprived of this hypothetical distinction between 'us' and 'them'. An American social scientist was researching the cooperative movement in a Peruvian community (Brass,2003). He was involved in what the government considered to be a suspect organisation as a result of which he and the leaders were arrested with no distinctions between them as to how they were treated and in how they were interrogated before he was expelled. Although initially he may have assumed something in the nature of 'us' and 'them' towards the people he was researching, these experiences made him one of 'them' (de Kadt,1970:288). Whether they were guilty of subversion or not is not the point as here he was now part of a system, the cooperative movement which the government wanted to destroy as opposing to its rule.

14. The advantages of culture shock

Social scientist must usually approach their fieldwork with some conception of the philosophical unity of mankind and yet this is contradicted by their injection into communities which are not only different but prefer their differences to their inclusion in the larger world beyond their social and psychological horizons. Of course, we are always surrounded by the 'other' but researchers remain ego-centred throughout their lives and as much as they can they avoid adjusting this to the relationships which they encounter. The more striking the contrast with the 'other' in terms of culture shock, the more likely they are to resist these influences and to avoid the nibbling away of their personal identities by enforced change. So culture shock can be an essential element in stimulating the researchers to approach their fieldwork with some appreciation of their own ignorance. The nearness in terms of social distance to the 'other', the greater the researchers' difficulties in practising detachment. The importance of culture shock is the recognition by fieldworkers that they cannot be detached, even in experimental situations, from those whom they are working on. In practice there is little hope of human social life being successfully studied under the canons of Western scientific detachment initiated by the European Enlightenment.

So accepting that social science fieldwork in all its aspects, from start to finish, is a social relationship between the researcher and the researched, its most important component is culture shock because of its unrelatedness to previous experience. The immediate and important reactions to total differences show that the value to the potentially enquiring mind at its most obvious would be to those who move over the East-West divide such as Nirad Chaudhuri (1959) and Lafcadio Hearn. The shocks of difference and the immediacies of personal experience; the shock which we can anticipate in the mind of the African Christian who stated to this author that she was looking forward to visiting 'your great Christian London' lay the foundations for all subsequent experiences with the receiving community; a reaching out of the mind to understand why so much is so obviously different within this unity of mankind. Thus there is the cumulative importance of culture shock.

15. The citizen as researcher

It would seem obvious that social scientists working within their own country and having legal status as citizens there, would automatically have advantages over non-nationals ones but their overall position as researchers is divided over a number of complex issues.

The researcher working in the field is always at a disadvantage; however well-prepared they are and committed to their projects, they are outsiders by almost every social and demographic category used. Except under very specific, predefined and onerous condition, can they ever become participants rather than observers; in Western experience the results of research involvement have often been personally disastrous. In the main they are outsiders for reasons which have little to do with their personalities and professional competence. A major and growing distinction is between researchers who are working in their own social environments and those who come from elsewhere, as there are internal as well as external expatriates in countries as culturally diverse as China and India.

We need to examine their ability to work effectively according to a number of factors; the value and extent of legal status, academic relationships, grant support, self-imposed and legal censorship, and political issues, shared language, religious background and social status in terms of national and sub-national identities. It may well be doubtful whether we can accept that social scientists in these sub-divided groups are dedicated to finding out and using the same range of facts even when working in the same fields of enquiry. Their viewpoints, even their biases are likely to be different as reflections from a range of direct and indirect factors which may have little to do with personal choices.

16. Citizen and foreign researchers as a legal distinction

The distinction between the citizen and non-citizen is a legal one and although it has many social consequences, it is an abstract classification which is not aimed at any particular person beyond some idealistic conception of the impersonal obligation of citizens to the state and the state to its citizens; in terms of Orwell all the animals are equal but some are more equal than others. The value of this legal status will allow researchers to do certain things while at the same time restrict them in others. The citizen as a researcher involves a further category as an individual. Any group element is likely to be circumstantial so that their personal contribution to any situation will be a combination of what they bring to the situation as individuals and the particular circumstances of the research with which they are involved. Outside this legal definition, the social constituents of being a citizen even in an ostensibly homogeneous state can never be uniform. An individual doing research may state that he or she is a citizen of Zimbabwe, but this person turns out to be a member of the white minority whose influence and indeed ability to be heard is both marginal and probably suspect in the minds of the dominant black majority. This is not an extreme case since examples of such people being accepted can be found in many contemporary states, such as Verrier Elwin with his work among Indian tribal peoples and his having an Indian family. The importance of citizenship is only in theory a static value and much will depend on the circumstances in which individual researchers are involved and what politicians and officials want to utilise at any particular time. Citizenship is just a card in the pack which has a defined value but its power will always depend on what other cards are being played.

Leaving aside issues of academic intellectual immunity there is a somewhat instinctive assumption on both sides of this national divide that the citizen has research advantages over the noncitizen. The latter feels that the citizen has numerous automatic advantages in the sympathetic biases of those in political and social authority, language, social understandings, ease of acceptance and access and less likely to be the subject of police surveillance. We will see that this in fact is no more than an idealistic hope. The citizen researcher in every state and not just in nonwestern ones with a long history of academic independence, may have a somewhat instinctive assumption that they should have some sorts of priority based on various considerations; a refraction of the modern political obsession with nationalism which those in authority expect to be supported by state supported intellectuals. There is the need to support the intellectual aspirations of these newer states whose academics may feel that these should have priority since they as citizens have a more natural and automatic understanding of their own people, and the need to oppose the supremacy of the outsider academic establishment dominating grants and publications in matters to which they do not have proprietary rights.

This creates the formalisation of an almost ethological 'us' and 'them', and the need to exclude outsiders from using national data for their own advancement. Finally, some conception that specialised knowledge is both private and individual and thus marketabie and as part of the national stock which should not be used and not taken away by non-citizens; a popular and widespread feeling. By the mid, 1960s post-graduate foreign social scientists were popularly seen as 'sucking the blood of Uganda' for their personal advantage.

17. The intra and extra national fieldworkers

Has there been any change in the field-workers experience when they are of the same nationality, religion and ethnicity as the people with whom they work and into whose communities they are injecting themselves? The word 'injecting' is used advisedly as even if the researcher is studying her or his own community or work situation, there are always methodological problems. Few undertake such introspective research because of the personal and moral conflicts involved given that they have to live socially with their results.

In the earlier situation of Westerners researching non-westerners, the dichotomous situation was so obvious and all-embracing that no one is under any illusions as to the degree to which they might become participant observers apart from the fact that they were seen as operating under colonial sponsorship at one time. The personal relationship aspects were more or less ignored and only gradually did this become a factor with the expressions of gratitude to those researched and reference to key informants as 'friends'.

However well researchers get on with the researched community, they are and will always remain social outsiders; they probably look different for a start in how they dress, talk differently, eat different food whenever they can and in their treatment by the receiving community, there is always the initial ethologically based hostility. However, when the national researcher comes onto the scene to what extent can they expect to be seen and treated as people who have by definition reduced the social distance between the researcher and the researched. The national researcher working with nationals may well be experiencing a much more refined and difficult situation than that of their non-national professional colleagues. The first difficulty must be the overcoming of any feeling of 'spiritual affinity' and that the 'us' and 'them' environment of the colonial era has been eliminated by political independence; the experiences of Indian social scientists such as Srinivas are illustrative of this. Further that their dominance in and control of national institutions of higher education and research endorsing committees insures that there are shared national interests.

18. The internal expatriate?

The expatriate 'white' classification starting in the colonial era and now applied to a much larger number of foreigners working in nationally independent states, is based on an obvious ethnic division. Now it is a much more complex division with Chinese, Egyptian, Japanese, Pakistani, Palestinian and Philippine professionals working 'abroad' in large numbers. In terms of what nationals understood to be their community, there was a further classification from the earlier period of internal expatriate, when nationals entered the colonial civil services. Individuals serving in areas where the people regarded then as social foreigners such as a northerner working in the southern Sudan, a Shona in an Ndebele area of Zimbabwe, a Burmese in a Shan or Kachin area of Myanmar, or a Han in south-western or north-eastern China. The complexity of the cultural situation is perhaps at its most detailed in India where except in terms of nationality and national pride, there is little in common between Assamese, Bengalis, Rajputs and Tamils; they are distinct by whatever social category is used in addition to which there is the complication of caste over a much wider social arena than Hinduism. The situation of the internal expatriate is compounded by the fact that they may consider themselves to be working under handicaps which they feel should not occur from their shared national identities. In fact this classification augments the other distinctions between nationals and non-nationals in their ability to function as efficiently as they would like as social scientists.

19. Foreign researchers and his social competence

Foreign researchers are often accepted into their field of research by those involved as socially incompetent however well intentioned they may be and however much their good intentions are accepted. Basically they do not know practically how to behave according to their perceived status in the research environment according to their age and sex and whatever role is attributed to them, despite their theoretical and carefully learnt sensitivities. Most Westerners shake hands with strangers almost compulsively which is disconcerting for many in other cultures. Almost certainly the biggest area of social gaffes has to be in the use of language in which they will have either learnt scholastically probably as a national state supported language or taught to them by locals in terms of their own social roles. Every language is sophisticated in its own structure and even more so in its semantics. The lack of missionary success in China has been attributed in part to the lower class nature of their spoken and written Mandarin; no doubt a feature of all foreign leaning of non-alphabetical languages.

Then there are the permutations of age and sex correlated behaviour which is so much a part of socialisation so that he actions of the researchers may be triggered off by quite appropriate responses in their own cultures, but quite wrong elsewhere, particularly in relationships between the sexes; eating behaviour, joking relationships, mother-in-law avoidances, body language and particularly clothing. An Indian woman researching in Goa used four different styles of clothing according to the situations in which she found herself (Mascarenhas-Keyes,1988).

One suspects that foreign researchers not only provide a source of gossip and anecdotes in the communities involved but also provide these people with their first personalised expansion of scale; close contact with someone from the outside world. Much more than objects on sale in markets, mass media advertisements and the irregular presence of government officials and soliciting politicians, their role as resident outsiders provides their hosts with the opportunity for the continuous observance of the overall accidental rather deliberate weirdness of social outsiders. These foreign researchers are perhaps tolerated because they provide entertainment, commercial opportunities and mental stimulus for a few and a feeling that it would be politically unwise to reject their presence as pawns or spies for the government which has allowed them to come there in the first place.

20. National social distance factors

The instituting of democratic voting procedures has increased the importance in every state of the numbers of people who have shared interests. It is not just a question of whether 'tribalism' is a new phenomenon attributable to colonial machinations but its combination with ethnicity, religion and caste has insured that national researchers will have particular difficulties because of who they are before any professional status comes into play. Thus there are very strong antipathies made worse by the creation of permanent minorities of people who see themselves primarily as belonging to a certain type of communalism. Sometimes these are clear dichotomies as in India between Hindus and Muslims or between native Pakistanis and refugee Pakistanis from India. But almost everywhere researchers will be out of place in most of their national communities; they are not readily acceptable because they represent outsider classifications as with Outcastes researching Brahmins and vice versa and a Bengali researching Biharis.

21. The consequences of researching

The basic difference between the two categories of researchers is that one remains in the country in which the research is done and the other leaves and produces their results outside the political jurisdiction of that country and away from the national academic establishment. In these difficulties there are differences between transparent research into such matters as the social aspects of disease and work practices involving soil fertility in which facts are clear as stemming from the project as defined in the application for grant support or opaque research. In opaque research in which researchers are as likely as not going to find material which has discreditable implications for those in authority concerning political, social and economic aspects of internal change and international reputation. It might in fact be difficult to study any community without becoming aware of corruption, mal-administration, law and order illegalities and voting fraud. There are complex moral and legal issues which are more personal and difficult for the national than the foreign researcher for whom these matters have a more detached and theoretical aspect; they can always leave the country in which they have done research and publish in safety there.

Let us now examine the citizen and non-citizen researcher dichotomy, firstly in terms of basic issues and then for place of research, person researching, projects researched and finally, in terms of problems of contemporary research in both categories.

22. Basic split between the ideal and the actual

The split between the ideal and the actual may be accentuated. The model for research is western and in a published form, a particular aspect of supposed excellence. Failure in researching or failure to complete research is never brought into evaluations of relative success. Non-citizen researchers in the late 60s in Uganda had over 30% not writing up anything; a failure rate making them into academic tourists while being associates of the East African Institute of Social Research.

The model is of the individualistic and intellectually aggressive researcher working through a topic to the satisfaction of his or her peers and not one of an ongoing relationship between researcher and the researched. The Mead-Manus islanders long-term interrelationship may well have been considered professionally suspect as Margaret Mead's involvement lacked detachment. The national researcher may have a pattern of social relationships which are more laid back, group sensitive for a variety of reasons as well as being more self-effacing, and less socially pushy since they may have a longer time perspective.

Does the researcher have a dominant role or are there different roles at different times which elide or are there self-adjusting priorities. Researchers do have personal lives and the introduction to published work often contains laudatory remarks about how helpful their partners were. Has this always been so or have there been failures or handicaps in the doing of research which are not mentioned? The primary role of the researcher may not be clear either to those with whom they are working or in their own minds and when does citizenship arise as an issue. What are their roles as researchers in relation to their own countries or their host country, the host community researched as regards references in introductions to friends rather than acquaintances or work associates? (Tanner,2001) What are the roles of researchers in relation to grant providers, potential users of the research and the connection with the researcher as formal or informal teacher as well as being a long-term investment in outside contacts?

How does the state view research in the abstract and the researcher in particular? Supposedly no control at all as it is in the hands of sponsoring academic or other institutions, but nevertheless access to official information or bodies will always be restricted in covet if not overt ways as for example in the research into British gypsies which was seen by the British Home Office as going against official policy and the use of gathered research materials restricted (Okely,1987). Opportunities for research may be terminated when sponsors see a draft report (Tanner, 1970 a) which disclosed nothing derogatory other than the unsurprising conclusion that despite the overall 'good' conditions of prison care in Kenya which the prisoners may not have had on the outside, the prisoners did not like being imprisoned. The researcher has always to be careful and the non-national particularly so and thereby the conscious need for a degree of duplicity in what questions are being asked and how the information is stored. In China each Peace Corps volunteer can be assumed to have their own police file (Hessler,2001,). For example, research tapes of interviews conducted by this author in 1969 with citizen and non-citizen Roman Catholic priests in Zambia could have been stolen or supposedly lost, but they may have been taken by the police at the airport.

Whereas in a western state the amount of research available may be enormous and on virtually every conceivable topic, this is not so in Third World countries where the conclusions of research may have political uses coming from no more than the reading of the brief conclusions of complex research or indeed not read at all beyond the title carrying with it implied biases. This author's lectures in Swahili to Roman Catholic seminarians on the contemporary problems of the priesthood can be received with some hostility as these men said these were not problems but the very nature of priestly life. It is not only a choice of words but the sense to which audiences or readers attribute to key words; Kenyans objecting to being called 'chaps' in English and the problems entailed in translation are numerous and probably insoluble (Tanner,1993,1996,1998). There is always likely to be a filtering and controlling process for non-citizen researchers as in the case of the American, who was researching the contents of Islamic sermons given in mosques in middle Egypt. This is a theological topic unrelated to any political connections which was blocked off by shuttling him with great politeness and much coffee drinking between various departments in Cairo and Assyiut. Islamic social science research should be examined within the terms of *'tawhid'* wholeness and *'nasiha'* advice. It appears that so far texts on social science fieldwork have been largely written by Westerners for the use of western researchers and it is only in India that the long tradition of research has received recognition as an important aspect of national intellectualism in the work of such researchers as

Baviskar and Srinivas. In other countries such as Pakistan, China, Indonesia, Russia (Dragadze, 1987) and Saudi Arabia (Asad.1993) research work needs sympathetic understanding and examination of the built-in restrictions.

The researcher has always to be assessed in entirely individual terms and their groupings into categories must be largely artificial because their known numbers in any country at any time is not high; their personalities are disparate by every social, psychological and physiological category and the research situations are different by definition and changing all the time as their research progresses.

23. The impossibility of social neutrality

No fieldworker has ever approached their research in a neutral and scientific frame of mind. Of course, many can recognise and counteract their own biases which may be both personal and cultural but much may remain hidden in the subconscious or be the result of both professional and personal socialisation. There are many examples of the ways in which researchers may almost instinctively behave in ways that highlight their outsider status; even a semi-Christian who does not practice may have difficulties in accepting the social and religious implications of Hindu polytheism and blood sacrifices to the Goddess Kali. Most Westerners have a framework of thinking in terms of 'please' and 'thank you' as a form of necessary reciprocity while many societies do not have such semantic usages seeing their actions in a wider framework of continuing real and potential obligation.

Scientific research in which the object of investigation is inorganic and delimited can only be free of bias in so far as it keeps free of human uses. Involving psychological, social and biological effects and the factor of choice in these research situations bias can be more or less satisfactorily accounted for. In fieldwork, however, value free research is probably not so much a theoretical ambition but a practical impossibility. Researchers are always working within the personal attributes which they already have and socialisation along with their biological compositions has always and already provided them with likes, dislikes and indifferences which will inevitably guide or restrict them in what is seen and how it is interpreted. It is important to accept that whatever the project and the methodology the data obtained and its interpretation cannot be value free in the scientific sense and indeed it is a form of bias to even think that it can be. It is not usually the deliberate fault of the researcher but a by-product of the fact that they are individuals outside the topic which they are researching. Even researchers of their own communities can never be free of the tunnelling of vision which comes from particular social roles and their demographic status which provides them with concepts of normality and deviance.

It is difficult to see how individuality can ever be removed from a social activity such as fieldwork or even allowed beyond a limited extent. Even if bias is suspected it is more likely to be attributed to informants than to the perspective of the researchers. The mentality with which or through which research is conducted is always tunnelled by biological and social perceptions. There is always the tunnelling of natal language and the fact that the researchers are men or women, of a certain age and ethnicity who use it in specialised ways. They are socialised into caste or class status and specialised in their perceptions by professional training. All have likes and dislikes and all react in their individual ways to the unexpected. It is from this biased and biasing diversity that social research whether it is done by nationals or non-nationals is an important part of the dialectic by which human behaviour and sociality is always in the process of changing and being changed.

24. The ideology of development

In the past there have been numerous examples of societies which have been changed or changed themselves but there has been no universality in this so that much change could be classified as internal. The push to change now comes into societies from the almost overwhelming understanding of how the Western societies have reached high standards of living and the desire to be the same. This ideology of development has come largely from the media.

It is not so much that new states or old states which are underdeveloped think within patterns of development and that this is part of all discussions on national policy based on a belief, indeed a hope, that there is a necessary and known pattern of progress to which all societies aspire and which is feasible. This is in contrast to developed states which relate their policies on feasibility and on multi-party variations. This is no doubt in part because the leaders of these new and old states are often doctrinaire politicians who have seen their colonial experience and past western dominance in somewhat simplistic terms (Wallerstein,1977); they wish to see the processes of change in their countries follow some fixed alignments of change at which their policies are directed.

There is a framework of optimism which is often reflected in the thinking of both national and non-national researchers who imagine that changes for the betterment of these people however defined, can be engineered relatively easily using the information which they have discovered and their intellectualised understandings as outsiders. It might be that this materialistic optimism comes from the Western framework of thinking that man can always change their social environment for the better, a refraction of the European Enlightenment. On the other hand it may be that non-Western social science thinkers may be less optimistic within their mental paradigms as both their languages and beliefs are less determined by definitions and the supposed exactness of Western instituted change.

Nevertheless at most levels of these societies the underlying pattern of thought is envy of the better off Western societies and the belief that they have the right to such higher standard levels of living constantly refreshed by the electronic images which are flashing into their minds and rubbed in by mass tourism of well-to-do Westerners; it is the ideology of developmentalism.

25. The dominant need for applied social science

Whatever the value of theoretical rather than applied social science, research is an economically expensive business and takes up much individual time during which the researchers have to be housed and fed. This accounts in part for the fact of the predominance of religious thinkers coming from endowed establishments in which costs are not much of a factor and particularly from within the Eastern religions with their commitment to passivity. Theoretical thinking might also be correlated to the growth of universities and the plain fact that commitments therein handicap fieldwork possibilities. There is also the availability of what might be called spare money which could be used for matters not connected with economic survival or development. Certainly it correlates to endowments which allow monasteries and colleges to house thinkers without there being any obligations for physical work, apart from the small number of individuals who are able to research what they like as they have private resources.

Developing nations have always given priority to the establishment of their own universities but find themselves caught between the acquisition of knowledge as a 'good thing' in itself and the realities of employment for graduates once the expatriate civil servants had been replaced. Makerere University College in Uganda was graduating twenty sociologists per year in the 1960s and yet the Manpower Survey estimated that there would only be employment for two in the next twenty years. During this period there was a preoccupation with status and an intellectual brain-drain as a function of political instability in many countries (Kashoki,1982:39-48) as well as the fact that salaries never kept pace with inflation so that cultivating their own farms and moonlighting at other jobs became inevitable. Once this brief period of accelerated promotion has been passed, the employment of graduates declines as does the status levels of employment. Thus the social scientists who wish to use their training may have little opportunity of academic employment. So the governments of developing nations are faced both with the dogma of social and political development and see the social sciences as having an obligation to apply their skills to developmental issues. In this they would prefer to employ citizens rather than imported experts over whom they have only peripheral control although their cost may be covered by donor governments or non-governmental international agencies.

The role of the social scientist in the political and social environment which demands rather than just hopes for development must be more than to describe what they see in a detached way and leave it to others to make parts of their work useful in their projects. They should attempt to diagnose what the future might be and in this "the ideology should be 'value accommodating' rather than 'value accepting' or 'value neutrality' (Mukherjee,1976). Other national researchers are quite explicit that they have to relate to matters of social consequence. 'Social science cannot be and should not be practised in isolation from what is happening in the society and from what is sought to be done in it' (Dube,1973:21-22). In general developing nations with their urgent social and political issues cannot afford the kind of disciplinary specialisations which are common in developed ones. Social scientists should develop the approach of general practitioners in the issues of change in support of development (Fahim,1982).

Chapter 2

THE GLOBAL AND NATIONAL ENVIRONMENTS FOR SOCIAL SCIENCE RESEARCH

1. Social science research in the colonial era

A significant case has been made that in the colonial era social scientists were deliberately used by those governments to study societies and produce results which would enable their administrators to govern more smoothly. The Sudanese Government wanted Nadel to study the Nuba and Evans-Pritchard the Nuer because these 'headless' societies gave their administration constant problems and as they saw it blocked social and political development. This approach has been rigorously criticised (Faris,1998). Some of the issues need to be examined not so much to argue against any such correlation, but to establish whether and in what ways the contemporary situations are different.

In the colonial period social scientists working in colonial territories were few and far between and were seen as politically harmless but in a vague sort of way good to have around. They provided some local knowledge for administrators should they have the time and the personal interest to read their productions should the administrator work in the area of their research. They were not required to have their topics vetted by any national research committee before entry visas were granted. Social science research in developing societies is not only part of the intellectual aspirations of the scientific communities initiating and individuals carrying out such work, but is and was also rooted in the unequal power encounters of the West and the developing societies of which colonialism was just one aspect. Knowledge will always be used by those who exploit others of which applied social science is just one aspect. The colonial power structure made areas of anthropological study accessible and safe so that physical proximity to tribal communities over long periods became a practical possibility for those who were not Christian missionaries, some of whom wrote monographs on the customs of the people with whom they worked. For those who did study these peoples it was no part of their professional remit to study the colonial system in which these pre-literate underdeveloped societies were of necessity living. The work of Hans Cory as a Tanganyikan Government Anthropologist involved providing senior officials in the Lake Province with his opinions from his long association with the area and its peoples about matters of official concern; he did what he was paid to do and did not to act as a branch of the police.

It seems unlikely that anthropologists as social scientists then or indeed now promised anything to the governments of the countries in which they worked (Pels,1999:102). The fact is that the Tanganyikan Government thought it both useful and progressive to have the customary law of the Haya (Cory and Hartnoll,1945) and of the Sukuma (Cory,1953) recorded as codes involved no issues of colonialism. The fact that such codifying of custom proved to be a fruitless exercise since custom was basically fluid and creating a code destroyed this ability to react to each case in situational terms was a contemporary evaluation of what was then thought to be a valuable exercise. For the researchers involved then it was interesting employment under moral conditions which they may well have partly shared with the scarce administrators given limited funding.

There are also the possible parallels between traditional custom being the exercise of situational ethics and how these social scientists may have seen what they were doing similarly bounded by time and space. The duplicity characteristic of Occidental arguments over morality and politics and in the underlying idea that social sciences could contribute hard conclusions not just opinions valid for current affairs. This is not just to throw away the idea of abstract morality of which there is some latent necessity prodding away at the back of the human mind but the recognition that for all practical purposes morality is situational whether for the British or the Barotse. Its creation of something permanent and abstract is probably an offshoot of literacy combined with the consistent failure of Occidental philosophers to provide this fiction of universal morality with any ontological foundations (Dworkin,1976; Macintyre,1984:69-70O.

The work of social scientists should perhaps be more reasonably seen in terms of the career opportunities of these men and women rather than moralising over what they may not have done as more easily carried out from the safety of distance and assured employment. For these obvious reasons national social scientists do not comment on the contemporary coercive, corrupt and inefficient regimes in which they are resident when the issues confronting them are only distantly moral in the face of issues of personal safety. It seems more likely that these social scientists who worked within colonial regime shared with administrators an understanding that this form of government was in the process of disappearing anyway and that they were not even by implication supporting or indeed associating with a group of ideological reactionaries. But Nadel possibly more than Evans-Pritchard, was an apologist for the regime which was forcing some of the Nuba to move, using the police to summon and bully informants (Nadel,1947:322-3, 335, 347). Evans-Pritchard after a police raid on his camp stated that he could no longer work there profitably and moved back to the Azande. Anthropologists during the colonial period shared a framework of political acceptance of the existing regimes within which they worked. This was a comfortable enough way of working without necessarily requiring any written public endorsements of the existing government. When Gulliver was working among the Ndendeuli in the Songea district of Tanganyika, it seems unlikely that the administrators gave anything more than passing thought to his existence there during his long period of research.

No colonial government would have allowed access by a social scientist with a political evaluation topic and probably no university grants would have been made available for European researcher to go to colonial mini-states to study a system run by Europeans (Asad,1998). These early anthropologists were part of the colonial administration by the accident of occasionally needing social contacts with people speaking their own language, cultural rest and familiarity. Evans-Pritchard wrote with some affection of the long term administrator of the Azande with whom he stayed from time to time as a break from his fieldwork. Certainly these men were not uniform in their approaches to fieldwork and in its last years they kept a much greater social distance between themselves and government officials because they did not want the rapport with their informants to be spoilt by an association with expatriate political power. The anthropologists as expatriates worked under and benefited from the reflected protection of the colonial authorities of whom they may well have disapproved; it was a cosy one-sided relationship. They were protected without admitting that this was a major factor in the success of their fieldwork. The point surely to be made here is that the colonial governments allowed anthropological work to be carried out which produced a series of definitive studies societies which no longer exist in the forms they describe whether in Assam, the Sudan or Eastern Africa. These studies had the failing of omitting any detailed elaboration of their relationships with their governments such as the ever present threat of military policing. These studies also ignored the feelings of inferiority which the colonial system created among those of ability and ambition which were stultified by the closure of politics to their hopes until the post-World War II era of political Independence. But Victor Turner who worked in Zambia makes the case for the significance of anthropological studies in an altogether different light. The social scientists in their works came to represent the structural perspectives of their informants and thus impeded the efforts of the administration to govern more efficiently. They were unconsciously allies in what the new political powers saw as the divide and rule policies of colonialism.

It is flattering to suggest that anthropologists were influential and did more than support what have been assessed historically as oppressive regimes, but the reality must have been quite different. That a book may exist on a particular society does not mean that it would be read by those working with those people; such books may not have been available at that level even if those in charge knew that they existed. Those working in the Shan States had no knowledge of a detailed study of the Palaungs (Milne,1923) or of the Shans (Milne and Cochrane,1910) and copies were not in the houses of the local intellectual elite.

Colonial trainees in their year at Oxford, London or Cambridge before they were posted overseas, studied some anthropology as an aid to the understanding of the peoples with whom they were to work. But even if they knew the country to which they were going, they would not know in advance the areas in which they might be posted; there is an enormous difference between Baluchistan, Banaras and Assam. In their first tour they would probably work in two or three different districts. A trainee administrator doing his first tour of three years and posted to the Southern Province of Tanganyika would have had dealings with the Muslim Swahili, the matrilineal Yao and the demilitarised Angoni. For serving administrative officers it was certainly correct that they considered anthropology to be useful although they were perhaps never quite certain for what it was useful. For non-national social scientists this was a period of great freedom to pursue whatever interested them in an environment in which their safety was guaranteed and expatriate officials provided some social back-up against their loneliness; both Malinowski and Evans-Pritchard comment on how personally valuable they found these connections.

2. Methodological issues and the post-colonial divide

The end of colonialism and the start of political independence regardless of the extent of extra-territorial economic and political dominances, is often seen as a divide in how fieldwork in the social sciences should be conducted and in the accuracy of what is reported. A critical view is often taken of so-called classic studies completed during the colonial period by national researchers who are able to assess them by new standards free from the dominances of the earlier period. The colonial contexts, loose methodologies, the tendency to rely on a few key informants and the inability of these researchers to understand or communicate in the vernacular (Owusu,1978) are certainly valid criticisms when applied to certain studies as over-simplifications or being just wrong. But as a general criticism this has to be modified by the fact that studies of the Nagas, Azande and Bemba for example, provided the base lines on which further studies could be assessed and on which subsequent researchers would now be able to accumulate new data. There were two divisions then between those who were permanently associated with the government, salaried government anthropologists, serving officers and missionaries and those who came for shorter periods in an overseas dependency and whose criticisms were muted by the nature of academic life and the need to keep open opportunities for further research in the same locations. But surely the modern situation is just as confining and indeed more so. The national researcher does not have to get a visa or get any approval from a national committee on research planned by foreigners, but they certainly have to be careful as to what they do and what they publicly conclude. The new governments may be much less tolerant of the social science researcher and see their results as reflections on national dignity and much more influential than they may actually be. One can conclude that while research done during the colonial period can be subject to valid criticisms as indeed all research is likely to be, for some a golden age of new opportunities may have come, but there is no clear divide except for non-nationals.

It is perhaps suspect that the pragmatic possibilities of health care in developing societies except in the field of mass immunizations, will never catch up with public need in expanding populations which have severe budgetary restrictions. If this is so then the social sciences have an ongoing field of research relating to how people are coping with their health problems and the effectiveness of these methods whether they are psychosomatic or not. A study in rural Rajasthan, India (Bannerjee *et al*,2004) found that the main sources of health care were private practitioners and traditional faith healers. 41 per cent of those calling themselves doctors had no medical degree, 18 per cent had no medical training and 17 per cent had not even graduated from high school. The latter would indicate the possible role of psychomatics.

Further it cannot be assumed that if there is anything like an automatic and superior professional ability for nationals to research their own societies primarily it is because it can be assumed that in order to become this they have left their natal communities and their understandings and because of their prolonged processes of higher education and employment away from home. It is a long process to learn and experience enough to be able to write objectively about their own societies (Yang,1972). Possibly also to be outside that society during any periods of traumatic change as Yang himself who was away from China in the United States from 1939.

A characteristic of developing societies is that many of those in power are examples of very obvious upward mobility; the political process has made it possible for those with relatively humble origins to attain high political office without much education and without any long experience of administration. It is not surprising therefore that they are suspicious of both national and non-national intellectuals who have achieved their present status through prolonged subsidised study and who produce research results which they suspect gives them superiority over their own work. 'Politicians in their recognition that intellectuals carry the mantle of alternative legitimacy even while they themselves wield power and that independent ideas are more dangerous than physical violence. Part of the larger intellectual community where constructions of reality may be unacceptable to political controllers' (Cheater,1987).

3. The need for non-Westernised data from developing societies

Social science was born out of the European Enlightenment movement and it remains dominated by Westerners and researchers trained in Westernised methods. A high proportion of their fieldwork can be characterised as generalisations from white middle class experience (J. Scott,1973). Indeed much human psychology is based on the many studies of readily available white middle class American undergraduates (Serpell,1982:10). Much has been assumed from the work of white people as in the example of the American psychologist working in Uganda who demonstrated that Ugandan children working with a western child's modelling material were less well developed than a comparable group of New York children. When it was suggested by an Ugandan academic that if the tests had been done with red natural clay, the results might well have been reversed, he replied that this visiting academic would not wish to know.

It may well be that any new theoretical patterning can only be from the perspectives of non-Westerners who have been socialised academically in Western theoretical thinking as well as in the thinking of their own cultures. Of course, national social science researchers have been active in Asia and North Africa since the early years of the last century but only much more recently in Africa south of the Sahara. While their contributions to knowledge remain significant if only because of their rarity, the balance of social science information available has always come from Western fieldworkers who have had both the institutional and financial support to work in societies other than their own. What is important here is the enormity of the potential fields of enquiry. The academics and students of the University of the West Indies working in the finite area of the island of Trinidad with its small population, could keep a number of research projects going relatively cheaply; the field is on the doorstep.

Universities in the major cities of China, India, South Africa and many other countries have many millions of nationals just outside their doors and many more within a bus ride. It seems likely that little is known of these societies in human much less social science terms except perhaps in relation to political ideologies despite the fact that in China Mao Tse Tung stated that there was a need to know how the poor people lived. 'The basic method for knowing conditions at the lower levels is to concentrate on a few cities and villages according to a plan...and to make a number of thorough investigations. To do this first direct your eyes downwards...Unless a person is interested in turning his eyes downwards and is determined to do so, he will never in his whole life really understand things' (Mao Tse Tung,1969:11).

There is perhaps a need to rethink the processes of acquiring research data at a lower level of detail integrated into formal educational programmes rather than relying so much on the work of professional fieldworkers who will always be in short supply because of funding and career opportunities. Foreign universities can only process and provide a limited number of post-graduates either to return to their own countries or remain there as immigrant academics. Whatever the future of social science in developing societies there is an enormous need for not only more data but that this data should be collected by nationals inputting their own patterns of thinking.

4. The research problem of objective precision

Detachment is an acceptable ideal in working with inorganic objectives and to a lesser extent with those that are both definable and isolatable; there can be very few of the latter such as the situation had a social scientist studied Easter Island when it was first visited accidentally by Westerners. This is obviously impossible in researching contemporary human behaviour in which every facet has connections in some way with what is going on outside the defined area of research. The social scientists who detach themselves as far as they can from the fullest possible involvement with the lives of those they are researching are depriving themselves of much information which can only be obtained from participation. 'As long as I remained uninvolved, I did not get the information I wanted. My interaction was with people who were not my social equals as I was ascribed a high social status because of my foreign background and education. The more trusted I became the more I had to fulfil a role resulting from my much wider social contacts'

(Bellwinkell,1979). This quote covers in essence the major difficulties and indeed the impossibilities of being detached in the scientific sense from the people who are being researched whether they are partially reformed head-hunters in northern Luzon in the Philippines, the Managing Director of Chemical conglomerate in Singapore or Muslim women doctors in Delhi. It is impossible to be detached in any human interactions whether towards liking or towards being revolted as indeed indifference is not detachment.

There is also the overriding personal factor that researchers need company. They cannot exist for months without relationships which are not solely dictated by their research needs; they may want to turn their key informants into temporary friends which is perhaps easier now than with earlier generations of fieldworkers. Malinowski did not do this and went off to stay with planters of whom he may not have had a very high opinion and would not have had them as friends at home but he needed a break from the Trobriand islanders. If we look at this issue from the point of view of informants or the community or institution in general within which they are researching, researchers must see these efforts to be detached as showing at the very least their social and psychological abnormality; who indeed would want to detach themselves in this way from the people with whom they are living? This would be a difficult state to put into colloquial use in many languages without involving concepts of mental ill-health.

In Western societies the interpretations of detachment reflect a lack of common humanity whether in religious or psychological terms. In developing quasi-traditional communities any such isolation would bring the possibility of witchcraft in sub-Saharan Africa into their minds. To be involved does not necessarily mean that one becomes one of them because whether one is a national or non-national the researchers are always outsiders however much they may think that they have become classificatory relatives or clan members, no one can be a Rotarian and not a Rotarian at the same time. There are advantages in partial involvements even if it is no more than sitting in a tea-shop or beside the path, are that researchers are collecting information without having to do anything very much in order to collect it. Kumar sitting in the same tea-shop day after day (Kumar,1992) was in an excellent position to collect information as although no one but the owner spoke to her, everyone there were very much aware of her presence; her role there was detached but still involved.

There is then question of what to research when the social world, particularly in developing societies, is virtually unknown in any detail anywhere and for any period. The only general restriction would be for any potential researcher to check as far as possible whether there are any close associates who are working on the same theme and have a head start; it is prudent, as far as possible, to avoid topics already occupied particularly by those who already have made a name for themselves. Freeman's lengthy attack (Freeman,1983) for that is what it was, on Margaret Mead's first overseas fieldwork in Samoa (Mead,1928) seems both petty and unnecessary since most social scientists have very mixed feelings and uncertainties about their own earlier research efforts.

There are predefined research opportunities in which an individual is offered a topic. A senior academic may require an assistant to continue a project on which they are already working and a government department or charity may want a project to be investigated; it is prudent to consider why a national or non-national is being offered this work as it may be that those involved want to sideline the topic or have results that can be dismissed as the 'short-sighted' work of an outsider or 'biased' insider. Others will sense that certain topics are currently popular in donor and political circles and thus apply for funds to investigate a topic for which previously they will only have a marginal interest; it may pay to be a generalist rather than a specialist in a narrowed field.

Should an individual be working towards a higher qualification they may be pushed in a certain direction by their supervisors with whom they have personal as well as long-term professional relations whose enthusiasms may be infectious. This push may also come for more individual reasons such as a woman researcher wanting to work on the alleged depressed social position of women in a particular society. Others may want to work on long standing intellectual puzzles such as relating ethnology to the issue of natured or nurture for which they may be able to receive a grant. There are also the attractions of researching behaviour which seems to be so totally different to that of their own society and their own personal investigations; how can a polyandrous woman have two husbands and how can it possibly work that a married woman's children in matrilineal societies are the responsibility of her brother rather than her husband or the Hindu *saddhus* who abandon their families and who destroy their own bodies for religious reasons. No intelligent person thinking over the textbooks with which they are taught, can fail to be intrigued by the strange behaviour and understandings of seemingly intelligent persons outside their own social group. Whether these persons are national or non-nationals these same intellectual excitements arise and stimulate enquiries.

Then there are personal situations which provide targets of opportunity for research which no one could have anticipated but will not happen again such as the social consequences of the 2004 Indonesian tsunami leading to a shortage of women, and the researcher who took advantage of a *coup d'etat* on his doorstep to investigate rumour (Tanner,1978) or finding a maternity clinic in Mombasa, Kenya in which the records had been kept by the same nursing superintendent for seventeen years (Roberts and Tanner,1963). Creative researchers will always divert their interests and efforts to events which will never reappear. Perhaps the best position to be in is that of a researcher who has independent means or is already situated in a potential area of research so that they do not have to seek any support. Such people can undertake 'fishing trip' community wide research for which there are no predefined objectives and interesting leads can be followed up. So in some cases the problem has been defined in advance as existing and needs an outsider to work on simplifying the issues for political decision makers, while in other cases the more interesting issues arise after the researcher has entered an area and they may feel the compulsion to change direction in mid-term. Researchers can be certain that what is known about the subject of their choice will never be enough to suggest that the topic has been adequately covered and most of what is known is on the verge of becoming out of date.

Social research is both time consuming and expensive so each individual involved has to decide for themselves what the justification can be for such expenditures of personal effort. The achievement of success in a competitive world raising the market value of the researcher while at the same time providing significant information on matters in which there are gaps so that developing societies may be better able to take advantage of the abilities of their own people. Research for its own sake which involves financial support may be hard to justify in conditions of widespread poverty and physical distress.

In studying inorganic materials precision is always both possible and necessary but once the objects of research are both organic and human then the need for and possibility of precision is more ideological than real; something verging on a bureaucratic necessity which probably cannot be fulfilled. Some degree of precision is inevitable as an automatic response to what the researchers perceive as reality and an orderliness of presentation as a logical necessity but of varying exactness in relation to the material. There is always the incubus of quantitative evidence hovering over the approaches of any fieldworker which is assumed to have greater validity than qualitative impressions which may have come from an equally long-term and thorough involvement in a project.

The difficulties over the possibilities of accuracy and objectivity are multiple, starting with the choice of what they are going to research which may be as much the result of circumstances as of whim and rationality. The wider the field of enquiry, the greater the impossibility of objectivity and conversely the smaller the field there is the loss of the wider understandings within which individuals and communities as well as institutions must carry on their affairs. The researchers themselves are restricted in their ability to be precise by the limited range of what they see, in how they see it and in how they interpret it. The human objects of research are constantly changing and objectivity as much as accuracy requires its data to be relatively static. Lastly, the act of writing is a selective and individual process.

5. The privileged position of the non-national researcher

In all fieldwork except when the research is carried out covertly as with Goffman's fieldwork for Stigma, the non-national researcher is in a privileged position whether they admit it or not. They can leave whenever they wish should their local situation become too difficult for them personally or professionally. They are economically independent of the researched community or institution and their status as an independent 'investigator' is superior to that of the researched however much they endeavour to disguise this. Since they do not just inject themselves into a research situation they have the implied or accepted backing of a university both locally and in their own countries, and their presence and topic has been agreed to by a national research committee. They have the diplomatic support of their own local representatives and they are accepted by the local political authorities. As far as possible researchers are protected to a higher degree than might be available to ordinary citizens, particularly as in most states, researchers will be under police surveillance. For example non-national researchers were brought back to the safety of Kampala when there were rural disturbances in Uganda. The researched community will usually see that the researcher is reasonably protected as they do not want to be involved in the complications if they should come into trouble.

6. The researcher's behaviour in fieldwork

Although it would be easy to assume that national researchers would abide by locally sensitive codes of behaviour and that this would come to them more easily than to non-nationals for whom the whole social environment might be strange, this may not be so. Every society has its social divisions and superiority and inferiority are as likely to be expressed in fieldwork as it may be in any other inter-personal relationships. Professor Evans-Pritchard headed his list of norms for fieldwork behaviour by stating that the researcher should behave like a gentleman. Although this sounds ridiculous as an anachronistic phrase for a professional Englishman abroad, it should be remembered that he wrote this as a result of his prolonged fieldwork experiences with the Azande, the Nuer and the Arabs of Cyrenaica. He felt that his own behaviour should be guided by what would be required of him generally in a wide variety of situations within his own culture. Although it might be difficult to translate 'gentleman' into languages other than the major Western ones, the principle behind the term should be applicable to the behaviour of any researcher in any fieldwork situation. There are examples in which national researchers have failed to behave in an appropriate way. In an Indian graduate training exercise among the Oraong living near to Ranchi 'everyone was asking the people about their most intimate relationships and fondest beliefs without any regard for their feelings or convenience. I had the uncomfortable feeling that there was something indecent about such a field trip' (Madan,1975). He makes the particular point that in his field trips, he was expected to abide by the social, moral and ritual conduct prevalent among the people with whom he was working even though he was regarded as a somewhat wayward Hindu. It might seem likely that this type of insensitive behaviour among students who should have known better within a related culture rather than among non-nationals who might just not be aware of what is required of them if they live in close proximity to strangers for a long time. There is a need for researchers to be both professionally and personally sensitive to the effect of what they are doing and asking as if it was being done to them in their own culture. This is particularly so when the researchers come from a culturally superior class or caste whether from birth, education or the business success of their families.

7. Political uncertainties and the researcher

The uncertainties of non-national social scientists are usually concerning professional and employment opportunities and their domestic problems. They rarely have politically connected anxieties except for those researching extreme groups or who are themselves members of such groups. The McCarthy period in the United States was an exception and affected a small number of academics with alleged affiliation to the Communist Party or who had particular understandings of China such as Owen Latimore (Latimore. 1950).

Outside the Western democratic countries citizens as social scientists in a number of countries have been both restricted and persecuted. At its most extreme even marginally unorthodox thinkers were killed or imprisoned in the Russian Stalinist purges of the 1930s and more recently in South America there has been assassination by right wing groups or governments. The most persistent persecution has been of Chinese intellectuals first in the Anti-Rightist campaign of the 1960s and then the Cultural Revolution. Professor Fei (Pasternak,1989) said that he lost twenty years of his life in being forced into prolonged work as an untrained and distrusted citizen and being prevented from either teaching or doing field research; in fact he was made the lavatory cleaner for a block of apartments and forced to separate from his wife. He referred to this last period in which all universities were shut as one in which the Chinese nation produced no educated people at all.

Certainly widespread violent disturbances have occurred in a number of developing societies and non-nationals have either not gone there at all as their governments' have advised against being there or the host governments have refused permission for social science researchers to come there at all. Whole countries or major parts of them have been bared from fieldwork for years such as Chad, southern Sudan, Myanmar, Zimbabwe, Afghanistan, Iraq and Vietnam. Other areas are blocked off by ideological and religious difficulties such as Iran and Saudi-Arabia. At all times substantial areas and enormous populations are blocked off from fieldwork by not only non-nationals but nationals as well. Their roles as critiques of particular situations and regimes can only occur when they are safely out of the country when there is no longer any danger of their assassination for quite impersonal ideological reasons. In this the country's authorities were not necessarily acting in order to protect what might have been assessed by outsiders as unjust and preventing its activities from being detailed and publicised outside. The death of a non-national researcher perhaps involved in work which has only humane and possibly abstract connections, is an embarrassment and a matter for diplomatic exchanges; it is better for local authorities to be safe than sorry. Most administrators and police would act on the side of caution.

The national researchers are in a much more difficult position and they have every reason to be cautious over what they are seen to be or thought to be doing. India is perhaps the only exception to this with its long history of intellectual independence both inside and outside the domain of social science research. There are powerful restrictions on research in Islamic states and in states which have opposed federalist solutions to their internal problems such as Iraq, Turkey, the Sudan and Sri Lanka. In many cases social scientists' fieldwork is tied into political situations in ways which must affect their work. In many fields of enquiry there are tensions between central and local governments and their communities. By the nature of fieldwork researchers are tied into the communities in which they live and work and yet they are only too well aware of the broader national and international society in which the people they are working with are entangled whether in Egypt or Norway. In the case of many developing societies central governments are new creations often of previous colonial regimes and in others the government, whatever the reality, is out of sight and is very generally not thought to be all that interested in the ordinary concerns of citizens except at election times. While this is certainly likely to be the situation in working for government or an agency which is dependent on governments such as a hospital or a development project, it is unlikely to overshadow any research except perhaps in the purely personal or in limited social distance interrelationships. If there is not hostility to the central government there may well be suspicious that the researchers are acting against local interests because it is dominated in its decision making by elements even those elected by democratic and constitutional processes in which there are antipathetic factors; a dominant party, ethnic interests or a particular religion.

There are always macro-considerations bounding research.

During the time that fieldwork is being undertaken the national or international scene may have changed because of an electoral turn-over, assassination as in the recent case of the killing of the leading Lebanese politician, the failure of a harvest and the increase in people who have AIDS; the popular knowledge in Uganda that of the 77 doctors who had graduated from the national medical school in 1984, eleven had died from AIDS and a further five had committed suicide because of AIDS connected worries (Dambisya,2004), leading to an even wider feeling of despair amongst those who have already been infected.

So in any country social science research might be seen as an asset or an embarrassment and in such an eventuality the alliance with a local university dependent on government support for its annual budget is unlikely to provide more than nominal. Researchers whether national or non-national are unlikely to have sufficient status to become the subject of serious negotiations with any central government which decides that research is too intrusive, particularly, if they are non-nationals or of the wrong tribe or religion.

We can thus appreciate that social science researchers are always at the mercy of matters outside the boundaries of what they want to work on. Unexpected riots between Sinhalese and Indians prevented the projected study of the latter as it related to their assimilation and the disclosure of the discriminating legislation involving them and after a year an alternative proposal to study caste was approved (Sayaranan,1979). An application to study an Indian farming community in Australia was opposed because a national academic was already studying them and there was perhaps a nervousness of the linguistic advantages which an Indian researcher would have in studying such immigrants; this may have been a valid reason for the postponement of such research but not on the grounds that the topic was already committed (Jayaraman,1979). In view of such uncertainties it would seem more advisable to put up research proposals in such general terms so as to allow not only a switching of locality but one which would provide a range of alternative approaches should one be blocked. Bureaucratic hurdles have to be taken into account.

8. Information stakeholders in developing states

In practical terms there are at least four parties involved in the processes and results of social science research; the researchers, those who are researched, those who provide the grants or institutional back up which enables research to take place and the political authorities in the researched area. Each of these groups and their social environments will view the research results in different ways as providing both opportunities and restraints. Most will see the results of research in terms of possible immediate benefits to those involved and not in terms of what the results contribute to abstract theories. Whatever views each of these parties have will have consequences for the conduct of social science research and for fieldwork in particular limiting the opportunities for covert as well as overt research. Overall the pursuit of knowledge for its own sake would be considered a luxury by the power holders in most of the contemporary world. Abstract research with its affinities to hard science aims at some aspects of universality whose validity is independent of particular scientists and whose data can be checked by its reduplication by others, and into this comes the concept of usefulness. Come and go well-intentioned non-nationals do not usually share this need for usefulness and thus go their own way; they do no harm and they are useful in demonstrating to the outside world the international acceptance of the current government. These researchers also bring in hard currency.

It is not going too far to suggest that most developing societies see abstract research as not only useless in terms of their current national needs but also in its framework of thinking as just another aspect of Western dominance. It is often difficult for such researchers to avoid local suspicions that they are in some way agents working for the continuance of Western cultural and economic dominance. While the smaller nations may have no alternative to this, many Eastern nations or Pacific nations such as China, India and Japan have a sufficient number of intellectuals as well as a history of intellectual endeavour to question this dominance and to seek their own patterns of thought without the socialisation which comes from Westernised education. These nations have developed their own approaches to the social sciences which have appeared from time to time in the publications of Western societies. They may have been relegated there to the side-lines but they have political significance in non-Western countries (Saran,1958).

Just as in the developed countries with their own social and political anxieties, there are social scientists who work from the assumption that the past was more harmonious than the present and that this has been ruined by industrialisation of which Mahatma Gandhi was a famous proponent. Marxism even with the collapse of the Soviet Union as the founder state for its implementation, still retains for some its power of providing a theoretical framework which is adequate for the analysis of any society and freeing it from involvement with capitalism. Then there are the social science nationalists who consider that the Eastern nations have long standing and indeed long lasting ways of thinking and doing which have outlasted a long series of physical and cultural invasions. These are adequate for all future needs and social science researchers should work within this paradigm which is quite a rigid one with substantial political support such as the Ujamaa and Harambee movements in Tanzania and Kenya. One cannot imagine social scientists in Muslim states researching the possibility that their economic, social and political position in the world is due to the failure of Islam itself to modernise rather than being due to the failure of Muslims to follow Islamic precepts formulated for conditions in mediaeval Arabia.

Some nationalist politicians have seen research and researchers in terms of their subordination to outside intellectual values and indeed see it as a common situation in poor states (Uberoi,1968). While this may be partially correct this subordination is more likely to come from their use of Western style education and its syllabi than from any deliberate pattern that there is not much difference in the thinking between the leaderships of any type of nation. The information gathered in research is much more likely to be felt as something individual and valuable allied to personal property and that its value rests in the existing system of knowledge which is firmly based on thinking created in the Western world; only long established thinkers can question the systems in which they themselves have achieved status.

Where status has a firm rather than just pervasive control over social science researchers and their research topics, there may be long periods in which no independent research at all is accomplished and indeed no research at all which is not part of the state's political processes. A Chinese anthropologist who had done post-graduate work under Firth and Malinowski has described in some detail (Pasternak,1989) how he became a non-person outside normal society but not really classified as an enemy. Eventually, he came to feel that he was guilty so he became suspicious of himself and so participated in accusing others and in the process losing his field notes and translations. It is important to recognise that for such periods in the histories of states on the verge of totalitarianism, the information stake-holders are the state which may well use the information in reports in ways that the original researchers feel has no relation to what they actually tried to do and report on. For such intellectuals it is likely to be a period of stagnation quite alien to Western patterns of unsupervised and wide ranging thinking.

There is a more material aspect to the accumulation of knowledge from people who have always regarded information as something which is exchanged as a negotiable asset rather than given to others for free. So information which is needed in social research is not some abstract matter that can be treated quite easily from a neutral stand point such as a computer memory or contained in bureaucratic records. It is always the property of someone and as such can always be traded; it is not free. In a closed situation such as a prison, a convict may have information but it will not be just given away but traded for cigarettes. Much the same applies to the information required by social science researchers who have either to find it out for themselves or trade for it.

This, the provision of information to strangers, for that is what researchers are, researchers should remember that the word itself entails reciprocities, not free gifts. It is not a relationship that stands in isolation without obligations. The reciprocities in social relationships are usually negotiable within known social alignments. Relationships on the other hand with researchers are usually unbalanced so that for informants it is like playing in an unknown game in which even the system of the cards themselves is not known. Thus the persons who are being asked questions by a stranger to whom they have no obligations whatsoever, is always going to be reluctant to give direct answers even to the most obvious ones. Age and where they live for example, are not questions which a respondent may find easy to answer without implying some ulterior motive to the questioner. To go beyond that in order to more specific questions makes for issues of disclosure that are not part of their normal social experience and make for suspicion. These people have no automatic hold over the person to whom information is being given which comes from affinity, kinship or even distant classificatory relationships. From the informants point of view it might well be better to trade with non-nationals who by definition are not involved in local affairs and are known to be well-off and have potentially useful outside connections. The national on the other hand may be locally connected by ethnicity, class, caste or family. In these situations there is no certainty that these researchers would be discrete in how they use their data since it is written down, photographed or taped and everyone knows that what can be written down can be used by anyone who has access.

9. Power exerted during and after the research process

Social scientists in any fieldwork other than being in unnoticed observation roles, are effectively submerged into their primary role of researchers, and as such they are very obvious and therefore exercise some forms of conscious or unconscious power. They are dealers in ideas which they either voice or represent and however careful they may be, they are injecting into the minds of their research assistants, informants and anyone who hears gossip about what they are doing, concepts and practices which may not have been there before; a question implies alternative answers as is detailed in the book on the Mugwe (Bernardi,1959) which suggested that the question about belief introduced the concept of unbelief where none had existed before. These questions almost always come from someone who is overwhelmingly different to themselves and obviously doing well as a result of being different. An idea which is visibly tied to material success is a form of power for which the social science fieldworker is a mobile advertisement.

It seems unlikely that there can be any overt research situation in which the social scientist's research is entirely passive; there can be no such situation as just sitting there and observing as their known presence must disturb the social environment; four people and a researcher are not just five people. Thus researchers have the latent power to disturb the social environment in which they are working. The differences which they have to those researched from education, age, sex and status are powerful agents for change. Of course in economically deprived communities and this is in practice a comparison to the imagined or real standards of living of university staff members or post-graduate students, they inject economic influence by eating regularly, having more material equipment even just in a plentiful supply of paper and ball-point pens, making payments for services rendered, the distribution of gifts if only cigarettes and the very real feeling among some that they are well connected and can help those who help them. Indeed they have the power to provide help in many situations where those involved do not have the means to do so and the researchers feel themselves obligated to do something. Some of those researched may have been of so much use to the researchers that they have forged a bond of reciprocity to which those outsiders feel obliged to respond. In such situations the non-national may be under greater personal pressure to respond than a national one who is perhaps better able to avoid entanglement in a network of reciprocities which the former may have unknowingly created for themselves. It seems likely that the non-national may be at least partially unaware of their powers in the research situation if only because they wish to see their own personal status as irrelevant to the research in which they are involved.

Those researched know without being told that researchers have connections in the world outside the communities in which they are involved. To the researched they represent an undefined outside power which will always have some relevance to them; if properly managed this can be turned to their advantage after all they have directly or indirectly helped the researchers who have lived with them which implies some return from the surrounding social-cum-political world. Researchers also have power because they are in the business of obtaining information which is in practice a commodity. Of course, there may be some who provide information out of goodwill but in practice the gathering of information and its use on anything more than a casual basis should be seen in buying and selling terms; the phrase that 'there is no free lunch' is relevant. Thus the providers and receivers of information which is not a matter of public record, are mutually manipulating aspects of power in which process both will be adept in their own spheres of influence.

Since no community is homogeneous in anything other than in uniting in the few matters which affect them all such as witchcraft, poor water supplies, police interference or a communal tax. Informants are interested parties in the provision of information to outsiders who keep records and send them away to unknown others. Informants who are often first on the scene for getting the attention of researchers will provide data from their point of view and into their field notes goes this information. So researchers are given this particular information which is translated into power as this or that side of community disagreements are put on record. Researchers may not realise until too late that they have been used by people who are often far more skilled in the manipulation of influence than they could ever be. So the researcher who is more than just a passing visitor and particularly those who even if they do not return to the area in which they first did fieldwork, can be and often are of considerable influence which goes beyond the provision of medicines from a private stock, small loans and being useful as a taxi driver.

Much fieldwork never goes beyond the notebook, tapes and

photographs which remain in the files of researchers so for majority of those researched the period in which the researchers were present can only have been a short event about which gossip only remained subject to folk tale elaborations. However with publication the power of print becomes a factor which may affect the community which has been researched which is perhaps out of the control of either the writers or those who have been written about. Firstly, the data is there in print with the elaborations of the author, usually in journals with small circulations and in languages unknown to the researched so that for them it is a non-event. It may be published as a book which is available to certain individuals and the community for a lot of money even if they did know that it had been published. But it would be a professional publication and hard reading. It is not likely to be known to or the subject of local comment. The leading character in Redfield's 'The village that chose progress' concerning the village of Chan Kom near Chichen Itza (Redfield,1950) had not seen the book which contained his picture, as this author discovered during his own fieldwork. Finally, the work of researchers are translated into the local language and becomes widely available and subject to local comment and those who read these works comment that they are one sided and do not consider all the issues and influences known to them as locals. It probably could not be otherwise as research is not usually published with the researched being seen as potential readers.

However researchers who become interested in the people they have researched as people rather than as objects of depersonalised research, can make a direct influence on their lives. Margaret Mead was able to get substantial help for the Manus people and they regretted her death. 'An American Dilemma' written by a Swedish researcher (Myrdal,1944) must have had considerable influence on the power holding elite in the United States when the book showed the pragmatic effects of legal and social discrimination against black Americans.

10. The foundation factor of superiority

Much social science research involves members of a highly trained intellectually disciplined professional elite imposing themselves on people who are their inferiors according to a range of categories with which they evaluate others. Such barriers can usually be surmounted by friendly persistence, but there is always the background fact that these researchers come from the high culture of the surrounding or outsider major societies and the experiences of those who are researched with that part of their own societies or connected foreign ones are not usually benign. So researchers whether they are nationals or non-nationals and whatever their official connections are seen as coming from these major societies which taxes them, enforces laws over which they have not been consulted or which controls their trade and makes them indebted without seeing the benefits of these loans. Those researched do not accept that these outsiders should intrude and indeed behave in intrusive ways which would not be acceptable amongst themselves. It is not surprising that researchers are inevitably given guarded welcomes (Mead,1972).

Scientific research is increasingly governed by sets of ethical rules promulgated by professional bodies that are able to control the acceptance of research results by their control over membership. When human beings are the subjects of research their rights are considered to be of primary importance whatever their status may be. The social scientists cannot invade the social world of the school, hospital or factory without a considerable amount of preliminary agreement or some deliberate deception. Of course this involves the subjects being made aware of the nature of the research and their part in it which may well undercut the main object of the research in the first place. Being in either situation involves the researchers in a considerable amount of personal arrogance (Agar,1996:91),whether the individual researcher is aware of it or not. Here the issues are more distinct with non-nationals who are less concerned with the immediacies of moral issues since both mentally and physically they come and go. They may justify their research by suggesting that the pursuit of knowledge by strangers and people who are strangers to them, is valuable in itself; the pursuit of the unknown is a law unto itself.

For the national researchers there are more difficult issues as to a greater extent they are holding up to public view and publishing what they see in a mirror which they themselves are holding up; it is a more personally involved situation. The general scenario for both national and non-national researchers is for young persons in their twenties from a different social background to impose themselves with varying degrees of success onto a social situation to which in practice they have no unavoidable obligations. It is also unconsciously arrogant as their professional training and career needs have a profoundly alienating affect on some aspects of their understandings. They may be unaware of the biases which they bring into the field with them but it may mean that it channels their research away from what the researched may consider to be the realities with which they live.

The short-term involvement of most field workers who have to complete their research within a set period before their visas, money or university vacations run out, often compels them to be more actively intrusive than the host culture can accept which can lead to passive evasiveness (Agar,1996:95). This may mean that the researchers have to scale down the scope of their work to what they can observe rather than what they originally wanted to do through questioning which most people in all cultures regard as intrusive. It would be difficult to think of a fieldwork situation in which researchers do not feel that they are at least the intellectual superiors of the objects of their work. It is not a relationship which has the potential for social equality even though the individual may feel some humility for his or her ignorance of the topic being researched as it involves learning. It is their general assumed capacity as intellectuals to overview the situations in which they are involved which gives them this feeling of superiority regardless of any sense of quasi-friendship and reciprocities.

On occasions social science researchers from developed societies together with their allies in other fields have an overconfident approach to the solution of social issues in which they imagine that the difficulties can be swept away in much the same ways that mass diseases can be approached; they confine social problems by seeing them in finite terms of proximate optimism. Reaching conclusions that suggest finite solutions is much in line with attempts to evaluate human behaviour in scientific terms. Research work on the international trafficking in children and women tends to underestimate the issues involved (Busza *et al*,2004) by recommending legislative solutions despite its voluntary nature and the female need to earn apart from issues of corruption, and ignoring the inappropriate nature of short-term solutions. It seems likely that national researchers may be less likely to see such issues in such emotive terms except as a way of getting funds for further research. In developed societies social scientists do their research work on well-known topics and in the framework of acceptance by those in power and administration that they can contribute useful information; there is not the feeling that they are in the centre of important events but probably an acceptance that they are not all that personally important.

In the colonial situation no one in power took much notice of researchers and they were not involved in researching major issues. However in the contemporary world with its wider publicities through the media, social scientists, non-nationals in the first place and then nationals, are faced with major situations of change and disruption which are matters of international and national knowledge. Violence and mass diseases spill over national borders, refugees pile up in unexpected places, cities outrun their capacities to organise themselves and ethnicity and religion emerge again as nearly uncontrollable factors in national lives of nations. So social scientists are often involved in working on matters of major importance in situations in which the administration is weakly structured and the state is a legal entity but not necessarily a social and political one. Whether they are researching the capacity of communal farms to raise their productivity or whether individual farms provide more food or in the pilgrimage situation of the millions attending the Kumbha Mela on the Ganges or at Mecca for the Hadj, they are much more in the centre of affairs than previously.

There is also the feeling common to researchers of whatever national origins that beneficial changes can be made and that they from their detached and intellectualised viewpoints are in a better position to see what is wrong and advice on how it might be put right rather than just adjusted. They may have little influence on those in power and their decision making but there is the feeling that they are at the centre of major changes and important events. They may be humbled by the situations which they come across whether in the spread of AIDS, kidnapping of children in northern Uganda and perennial flooding in Bangladesh, but not in their individual feelings that they can contribute understandings of which those involved may not be capable. Coming as they did originally from more highly developed societies and subsequently from the better educated and more economically secure sectors of their own societies, it is not surprising that social scientists have had and continue to have a sense of their own superiority. Their own ability to assess their projects comes from their positions as outsiders from better organised parts of their own societies with a superior culture which is usually disguised as detachment as well as a much wider ranging understanding of how societies work and indeed perhaps should work. There has always been this sense of superiority even when there is involvement in moral issues as in attempts to preserve less developed if not straightforwardly primitive tribes such as the Aborigines Protection Society founded in London in1837 for work in Australia.

It seems unlikely that contemporary national social scientists approach their research in a radically different frame of mind from their colonial era predecessors or indeed non-national co-workers. Certainly there is the element of recording information for posterity but much research work is slotted into change which is either already happening or is expected to happen. It is not often that the researchers and the communities in which they are working share their approach as working as equals for their mutual benefit. Participant observation is an assumed pattern of behaviour but does not involve any change in attitudes of mind.

11. The resolution of role conflicts in the lives of researchers

Everyone in every society has roles which conflict and their socialisation teaches them ways in which conflict can be avoided or at least reduced with varying degrees of success. We may not even be aware that there are such role conflicts, as the ways of coping with them are so much part of what we are. If role conflicts are a constant feature of all societies in what might be called their normal activities, then we must expect them to be an additional feature complicating fieldwork. Most of the obvious ones are related to the researcher's demographic status where the obvious ones of age and sex conflict with their investigative needs. This is further complicated by the fact that in all societies even those with a long experience of social science research, there are no properly accepted roles for a free ranging investigator. They have to be doing something and that something has to be related to some existing social realities; a trainee catechist investigating the religious practices of his congregation, a social worker looking into the marital backgrounds of the couples she encounters and so on (Minocha,1979). These are important gate opening roles which do not tie these researchers to any particular duties and obligations.

Every fieldworker has these same role conflicts built into their lives wherever they are working; they are still mothers and fathers, sons and daughters as well as their own emotional involvements. The fact that they are doing field work away from their normal social environments complicates the lives of both national and non-national researchers (Agar,1996:101). It is difficult to be a friend or family member when researchers have deliberately geographically and socially shut themselves off from their more basic functioning relationships. Thus it is easier to fail in the roles of parent or partner in the execution of fieldwork because there are no predefined ways of coping which their society has worked out for itself; fieldworking is not a 'normal' way of earning ones living. This is therefore one aspect of the lives of researchers that has to be juggled with various degrees of success.

The second group of role conflicts which have to be juggled are the relationships with members of the communities or institutions being researched. The fieldworker enters the arena with the roles which those people attach to age, sex, social outsiders, education, and marital status since in many cultures the unmarried adult has inferior status, and then not practising the locally dominant religion. These involve difficulties which occur in every society and they are made more complicated by the researchers own socialisation and the particular one of being a graduate; they cannot avoid being what they are and yet the ideology of research requires that they should take a neutral view of the work and social environment on which they are working. Of course, they are cut-off from their own cultural commitments, but not the ones which are attributed to them by the host community.

Some conflicts are avoided by the host community providing for them a role which is both explanatory and convenient for both of them and the researchers. One of the commonest is for the researchers to be considered to be entertaining oddities. Or as outsiders with potentially valuable connections and providers of employment for a few who are otherwise unemployed or perhaps unemployable. The fact that they make so many social mistakes and behave in ways that would perhaps be inappropriate for an eight year old as well as and their ability to appear foolish, makes much of their behaviour excusable as well as providing diverting gossip which is free from dangerous in-house social handicaps. Then there are the role conflicts connected to the professional life of researchers which often carry within them the necessary discipline of living in small closed communities. In both social environments of field or university, acquaintances are not friends but rivals in the use of ideas and information as commodities. Much role conflict is avoided by reticence; the more that the researchers disclose of their experiences they become subject to personalised envy rather than just being part of professional dialectics.

When social scientists study their own community or profession they have much more complex relationships with their informants than if they were straightforward outsiders coming into a social situation for which they have no previous connections. An American social scientist as a non-national studying Irish culture and the ethnography of Irish intellectuals (Sheehan,1993) was involved in role conflicts which were perhaps impossible to reconcile. This may be the case in all research situations because no society has had researchers involved in them for long enough to find acceptable roles for them and ways in which their role conflicts can be satisfactorily avoided or reconciled. In this case the social scientist was involved in three conflicting and irreconcilable roles; a colleague studying affairs at her own level, as a student studying upwards in order to climb her own academic career ladder and as a social scientist studying down to others as objects of study. Such conflicting roles may be particularly difficult to solve because the issues have no exact definitions of what might be acceptable or not acceptable within the ambiguities of academic freedom which can cover a wide variety of personal failures and successes. 'Too many potential rich contradictions get lost in the politeness rules of academic rhetoric' (Agar,1996:99), perhaps driving them underground into the permutations of inter- and intra-disciplinary academic politics. It is not so much that these role conflicts exist and occur at the same time as they would in anyone social setting but that they occur in situations which are socially disparate and have to be sorted out within the research lives of each fieldworker as individuals. There must be some personal distress in this which can only be modified by blocking off one social environment at the expense of another which can be done by the commonplace fact that research takes place in distinct, geographical and often quite distant social environments.

The more obvious role conflicts are no different to that of other employees which are related to their indefinite hours of work, commuting to and from their research and the perennial problems of the urban-industrial societies in which almost all researchers live. Fieldwork is different as it almost always involves some measure of deception in which researchers for a time have to pretend they are what they are not, while at the same time have attributed to them roles which vary from government spy to foreign idiot.

So researchers have to resolve for themselves how their work role is to be integrated into what they see as their own ongoing and permanent persona. There are perhaps two extremes in this. The individual researcher who makes no attempt to adjust to any role other that that of someone doing their own thing in which they come and go as their work requires, Then at the other end of the scale there are those researchers who abandon their attempts at social science detachment and endeavour to become one with the community which they are studying or at least some part of it which they feel needs their involvement and support. Of course it is theoretically possible to become someone else as happens to some extent in Witness Protection Programmes in either Britain or the United States in which new external persona are created for some individuals to keep them out of danger, but it is doubtful whether such people think that they have become different people. Becoming a person who merges with a community which is not their own, must involve serious strains on behaviour both in the field and afterwards even assuming that there were some initial biological similarities.

In most cases researchers as far as they are able and their work permits attempt to assume some aspects of the people with whom they are living, taking part in communal work or harvesting even if this shows up their comparative incompetence. As far as possible they create for themselves a place in which they can feel at least the beginnings of being at home as a good neighbour, lodger, pupil, tourist, temporary adopted classificatory relative or just a friend of the family. But of course this feeling of being at home is not the same as their acceptance by the researched community's members. There is always some measure of illusion in such attempts to be included unless they are carefully planned probably for limited periods. Fieldworkers may think that they are being initiated into some group because of their appreciated personalities but their sponsors are only doing this for perhaps some personal purposes which cannot be identical with those of the researchers. So because of the role conflicts built into the work of social science fieldwork researchers have to accept considerable occupational stress.

12. Social science and contemporary political involvements

Much justified comments have been made about the work of social scientists during the colonial period when they were protected by the governments which had permitted them to work there in the first place, indeed often subsidising the research; a protected species of outsider. But this connection is perhaps glossed over by the self-righteousness of contemporary commentators who have not realised the extent to which these researchers distanced themselves from the administration. This was shown by Evans-Pritchard in his first field trip to the Nuer 'a government force surrounded our camp one morning at sunrise, searching for two prophets who had been leaders in a recent revolt, took hostages and threatened to take more if the prophets were not handed over. I felt that I was in an equivocal position since such incidents might recur and shortly afterwards I returned to my home in Zandeland' (Evans-Pritchard,1940:11). There are two aspects to consider in this since he stopped attempting to do fieldwork in one tribe while considering himself better able to research the Azande where he had a 'home' but still under the same colonial rule. But in this case even before the army arrived the Nuer would have already concluded that the anthropologist as a type of European but who was not a missionary or trader, must have been related to government anyway according to their understanding of political realities. It is more likely that he moved because he sensed he would be in danger from attack by some Nuer warrior once the punitive expedition had moved on.

More recently in a post- colonial political environment Geertz records running away when the Balinese police raided a cock-fight which he was observing (Geertz,1973:412-453), but we cannot assume that the sophisticated Balinese were going to conclude from this one incident that he was on their side and not on that of the government. The contemporary researcher whether national or non-national may well be a commoner type of person but their roles in the researched community may not be any more understandable when looked at from the point of view of the observed. When government officials visit the area of research they may well visit the social scientist or at least enquire about them, but do the researchers avoid them in return? Is it not likely that they are thought to be in someway connected with at least the local government in its many bureaucratic forms and thus thought to be of some potential influence there or at least best not to boycott much less quarrel with them.

13. Political relationships

As we have already stated colonial governments usually took a tolerant view of the rare social scientists who came to do field-work in their countries unless their activities brought them to the attention of their local civil servants. They were generally felt to be a good thing to have about and that more knowledge about the nationals which was available the better they would be able to administer them peacefully. While such people may have had a substratum of criticism of authoritarian governments in general, they rarely found matters of substance about which they felt obliged to be critical and about which they felt compelled to publicize in the outside world; they were researching a defined subject and not the larger political structures.

Independent states are far more sensitive to real or implied criticisms which they can attribute to a non-national who can be expelled without having to give any reasons. Nationals are in a much more difficult position particularly in states with dominant political parties, single party systems or military governments. Such researchers may well feel well advised to keep away from any research topics which are even remotely tied to politics. Politicians may be forced to tolerate political opposition which has become institutionalised and an accepted part of national life but not the academic producing seemingly hostile evaluations of what they have been doing. It has no part in the politics of most developing societies who have no understanding of the proper work of their academic elite.

Academic freedom as a social reality also has no part in the political landscape of many countries both historically as in the McCarthy period in the United States and in China during the Cultural Revolution in which there was systematic and wholesale humiliation of intellectuals and their condemnation to years of manual labour. These difficulties were typified by the experiences of Zhong Wei Chen, an innovative micro-surgeon who performed the first reported hand transplant, who was restricted in visiting the West where he was much in demand and in contacts with Western surgeons visiting China (British Medical J, 2004). The national researcher not only finds it wise to keep clear of political issues but often finds that any association with government officials as possibly aiding them in their work is likely to be counter-productive. Civil servants do not have the ethos of helping the public and the national researcher is caught by a personal contempt for systems of bureaucratic government. Kumar, the daughter of an Inspector General of Police stated, 'After my experiences in Lucknow I no longer needed convincing that distaste and fear were the appropriate feelings for all government institutions' (Kumar,1992:218).

14. Ethics and contemporary social science research

Any discussion of the ethics of research is usually around the major issues of complicity in crimes and the disclosure of information. Mehta's study of Mumbai involved long interviews with professional murderers and police who have done the same thing in attempting to control crime outside a paralysed legal system (Mehta, 2004) and much the same would apply to any work with criminals (Tanner,1970). In failing to provide the police with information about crimes which have been committed, they have committed the crime of concealment, abetment.

There is also the ethical issue of using the ideas of others for personal advancement. This may be clear enough with the ideas of colleagues, but fieldworkers certainly benefit from the ideas which they have received from informants, research assistants and the researched community in general which have only got a return in such rare cases as Margaret Mead's relationship with the Manus islanders (Mead. 1996). There are two practical aspects to the consideration of the ethical issues in social science research work which are outside the boundaries of morality. Firstly, if ethical considerations are to be more than a matter of dealing with contingencies, how can they be assessed in open ended research in which little or nothing is known about what may or may not be found in situations which are as yet undefined and indeed indefinable. To consider ethics in such situations would seem to be an exercise in legal imagination. Ethical issues might arise in the course of fieldwork. Just as the researcher might get ill or break a leg but such hypothetical dimensions make for anxieties in the researchers which they do not necessarily need for making the best of the opportunities confronting them. Risk taking may be part of any research as in any human relationships.

The second issue is for whom are ethical issues being considered. There can be no doubt that ethical understandings are necessary to avoid exploitation in which researchers profit commercially and professionally from their work; but it is not work which has been carried out in their brains but through the medium of obtaining information from informants who in their own asocial situations would see the provision of any information as requiring reciprocities. As those who are researched become more aware of the outside world, there are complications over the ownership of information. A French documentary film maker Nicolas Philibert spent six months recording the life of a small one-room rural school with its one teacher having got release forms from all those involved; The film *'Etre et Avoir'* became a national and almost by accident a commercial success and is now the subject of prolonged litigation over a share in the profits. So it would be sensible for researchers to remember that tape recordings and photographs and films are seen as different to the taking of notes.

There are ethical issues in every human relationship and in primary ones behaviour is subject to social and psychological acceptability. What is acceptable outside the social, religious and political sphere in which a researcher may be at home must always be largely unknown. If researchers do not behave in ways which would be unacceptable in their own societies, they may not go far wrong and of course in this the national has perhaps greater knowledge and sensitivity to what might or might not be acceptable. Professor Evans-Pritchard's previously mentioned injunction that researchers should behave like gentlemen is probably transferable as an ideal in any culture, but this requires much imagination wherever one is working. The permutations over what might not be acceptable are perhaps largely a necessary anxiety.

Research ethics may involve issues that in a cross cultural context are intrinsically wrong such as researchers conniving at incest within primary groups or ritual murder. It seems unlikely that any anthropologist knowingly rather than accidentally has eaten human flesh. Most ethical issues are negotiable since they occur in social contexts and it is difficult if not possible for researchers who are participating in communities to do this. Kumar must have showed her disapproval in witnessing the Banaras police beating up a suspect (Kumar,1992) just as Chakravarti in researching a community in Rajasthan, was appalled by the villagers wholly pragmatic amoral acceptance of the necessity for bribing officials for achieving specific objectives (Chakravarti,1979). Any researcher should be aware of ethical issues from start to finish and it should be an advantage for the project to have had these considerations treated as important (Skinner,2002). In the end what we have in social science fieldwork is a continuous series of situational dilemmas that involve every researcher in ethical issues. Whether they require decisions or not about how the researcher should behave may depend on their own culture based feelings as to the appropriateness or not of what they intend to do or to be involved in.

There is another ethical factor which is of greater importance than in the earlier years under colonial governments when communities did not think that they had any power over admitting researchers to their areas. The Pueblo Indians of the United States after generations of being tourist attraction and venue for researcher, now control all such access through their own tribal councils. In most countries applications to do research by non-nationals require national academic as well as local government support. The activities of non-national researchers can become the subject of diplomatic concern. However, national researchers working in their 'home' environments and publishing their results in national and international journals may well have outlines published in the local language and in local newspapers so the subjects of such research may well see the results; this is a new development. As mentioned earlier the key figure in Redfield's

'The village that chose progress' (Redfield,1950), had not seen the book or his photograph in it. On the other hand a Swedish researcher of Sukuma customary law over the use of water sent all his main informants copies of his book even though they could not read English (Drangert,1993). Now those who have been part of research feel that they can knowledgeably discuss and comment on such published results from their own points of view (Wolcott,1981) and personal understandings; they have all been through the same social processes such as going to school, taking part in local politics, being ill and having family crises. In this context there is an ethical need to consult the community more widely over what the researchers have or think they have discovered. The Sukuma who had been informants were well able to understand their own behaviour and to distinguish it from that of other cultural groups.

When it comes to the 'truth', the protection of key informants has to be of paramount importance to social scientists which is best done by dividing up the information so that it appears to come from several sources at which informed readers can only guess. Redfield's photograph of his principle informant and the key figure in the changes about which he wrote would have been disastrous for that person in any modern community in which connected persons could have read what had been said about their own roles and those of others. An informant at the centre of ordinary day to day affairs has to be disguised. Just as much as those involved in issues which verge on possible corruption and overt criminality. Informants can have their information split up as coming from unattributable commentators, as coming from other published research or from public records or by having more than one field of study so that information cannot be attributable to any single source; in this extent it is always sensible to keep informants names under cover of a pseudonym.

15. Confidentiality and social distance

Redfield (1950) may have thought that his respondents did not speak English and he did not anticipate a return there, and that there was not much point in sending them copies in which they could at least have taken pleasure in discussing the photographs and perhaps getting an educated visitor to translate some of the chapter headings and picture titles. Perhaps acceptable then in terms of the American dominance of social science research in Latin America but bad manners then and perhaps socially and politically unacceptable now. We can also speculate as to who thought of the other as friends or was this just the means to advancing careers? Some researchers (Fuller,2003) have had their research work translated, in this case into Tamil, and it was thus made readily available to those involved which led to several complications. It became difficult for him to think that he could continue research there as his abilities to do this depended on the priest controlling the Minakshi temple in Madura who did not like what had been written about their activities. There was also the factor that the 'us' and 'them' distinction in the studies by non-national social scientists were in the past as much a reflection of inequality as a methodological necessity. It may have become inappropriate to maintain any such distinctions when researching such a narrowly defined community and in this case the intellectual level of the Hindu priests. These situations raise key issues in the confidentiality between the viewpoints of national and non-national researchers and the national objects of their fieldwork as particular identifiable people. Non-nationals live with the results of their work in another social and professional environment and in practice they do not have to consider the feelings of those about whom they wrote. There is an enormous social and geographical gap between these two environments. While there may be a vague moral obligation to keep those researched in contact with what has been written about them, it is usually felt sufficient to thank them in the introduction to any publications. Most would feel that it is difficult enough to follow the criticisms and requirements of their professional colleagues, superiors and editors without adding the much more difficult issues of explaining to people who are not social scientists the reasons behind much of what had been written about them. The longer researchers stay in a community or in an institutional setting the larger becomes their circle of clients, people who have become tied into the researcher's and their own needs. These networks become valuably inclusive since they are mutually supporting, while excluding those not in these 'magic circles'. National researchers on the other hand because they are much more proximate both socially and geographically will have to be much more careful about what they write and more importantly what can be connected to actual people and places; they have to live with the consequences of what they have written in terms of their own national careers.

16. Research and the need for covertness

The word duplicity implies a conscious awareness that actions serve two contrary ends which are in conflict; that it is wrong to play one role while pretending to be in another. Perhaps this is to simplify a normal social situation. Everyone performs a number of roles and in order to avoid the plain fact that they will always conflict, duplicity is a normal social necessity. Any research process is composed of innumerable inter-relationships which may be grouped but are just as likely to be socially specific. Working for a government may involve the need for some degree of covertness and of course the use of results may involve some degree of deception over the extent to which information supplied by informants is considered by them to be private property.

What does arise in discussing the ethics of research are not so much the issues of duplicity but the need for it. If researchers actually did explain the objectives of their fieldwork that they were doing research for a higher degree to enable them to earn more maybe an approach that they would understand and of which they would be envious. However to suggest that they are trying to find out more about what is happening in a community in the hope that it might lead to their betterment and the injection of some development resource is to become involved in political processes from which they are professionally barred and in any case unqualified to participate. Also to tell them that they do not know what they are looking for is unlikely to create enthusiasm for participation. Over all this hovers the problem that it is likely to occur at the beginning of fieldwork which if there is no shared language will involve interpreters with a limited understanding of what they are supposed to be translating. There are no doubt ethical issues here of balancing personal gain against the use of others who are usually of a much lower standard of living to that of the researchers.

No social science researcher at work is without built-in role conflicts. There is no obligation to be straight forward if for no other reason than that a clear statement of what is being undertaken would bias the information which would be supplied. Then there are role conflicts within the individual researcher as to whether they are using what they find out for the betterment of mankind or for their own advancement, but even that is somewhat disingenuous when almost all researchers are paid for the work they do. It may be that role conflict is so much a part of the research situation that it is not so much unnoticed as constantly adjusted to as the day to day work progresses. Role conflicts may be internal, the workings of conscience or external relating to particular persons or groups from the obligations which come from inter-relationships with persons or families, a constant process of negotiating priorities.

The professional social science researcher must always be, to some extent two-faced; they cannot disclose what their trends of thought are without in fact causing their informants to react to this information. Even in applied research in which those involved know to some extent what is being proposed researchers cannot disclose too much without risking premature political and social reactions in which they do not wish to become involved. In abstract research there is a build up of understandings which those being researched may or may not understand. It is perhaps a professional obligation to disguise thoughts in order to avoid the necessity and difficulties of explaining what is not yet clear in the thinking of the fieldworkers themselves.

It is probably safer to disguise personal status; it is certainly safer for a woman researcher to be married or to state that she is married even if then she has to suffer the criticisms of why she has left her husband to work alone in a strange environment. The unmarried woman may sometimes be the object of undesirable attention although Kumar in all the years in which she researched alone in Banaras stated that she had never experienced unpleasantness of this sort except in one crowd situation. Temporary partners should state that they are married particularly as it gives them a more acceptable status in the eyes of local authorities. It would certainly be unwise in many cultures for researchers to make the community aware of gay or lesbian relationships. Colin Turnbull's companion was seen as just another researcher who conveniently happened to be an Afro-American. The unmarried man of a mature age who is not married has lower status than the married men of equivalent age in many cultures; it might be suspected that he is not a complete man and this might be passed on to his status in the research area; a suitable fiction explaining the absence of a woman might be a reasonable deception.

Much is known about the few researchers who have consciously taken on a role which deceives those who are being researched. The most extraordinary example of this must be that of Burton who in the nineteenth century successfully pretended to be a Muslim in order to get to Mecca; he was accused by his contemporaries of betraying his character as a European gentleman and a Christian (Burton,1893:xxi). Almost by definition this may only be possible for nationals who are better able to enter their own communities in disguise. Researchers have become involved with criminal gangs or get admitted to prisons or mental institutions with secret identities in which there is obvious danger as in pretending to be mentally ill. In other cases the pretence of being homeless for example can so take over the researchers that they have difficulty in maintaining any distinctions in their own minds as to who they actually are and this can lead to the destruction of their normal private lives. Any researchers even beginning to plan such total and prolonged deception need to take professional advice over what is involved irrespective of the importance of the research topic.

However, in many situations the researcher may become

involved in levels of deception which are both recognised and unavoidable even when those being researched are told from the start that the researchers are not what they pretend to be. In researching the working conditions of agricultural workers both national and non-national researchers can state that they are working on farming methods and information of their penury or the behaviour of landlords becomes a by-product of these enquiries.

Once the field of enquiry moves out of the seemingly pragmatic into the metaphysical there cannot be any such acceptance of thinking within a single framework. Asking about anything in the spirit world draws attention to its existence not its possible reality. No doubt this is an interesting philosophical issue about the reality of abstractions, but for non-national researchers to ask about witchcraft, the malevolence of ancestors and magical protective and aggressive rituals makes these outsiders acceptors of these ideas as physical realities. National researchers might not be placed in such dilemmas as they would clearly belong to families, groups and cultures in which such ideas are part of the ways in which they come to terms with inevitable misfortune. They would not be cognitively hostile to these forms of reality, while social science researchers who tend to be both pragmatists and agnostics have to disguise their feelings.

In examining the issues related to researching religious communities or sects there are the problems of these clashes of belief. Those being researched may not see the researcher as just an outsider interested in what they do but as just another person who should be proselytised as a possible convert to the True Faith; in functional terms a redeemable or possibly already a redeemed person seeking the light (Englund,1996). It might be that Hindus and Buddhists would find this situation less involved in deception as their faiths are tolerant of variations while the religions of the book have a sharply defined yes or no to issues of faith.

17. The limitations of trust

For someone to provide information to another who is not a close relative or perhaps not even then, involves trust and that is a scarce social commodity. It is extremely fluid in what it implies and to whom and moreover when; it is always situational. The finding out of information from someone who does not offer to provide it, is always a long and tortuous process. At its worst it is confrontational and at its best the information is provided without the informant knowing that it has been provided. Why should a stranger be trusted and indeed even a close relative who may well be a rival for affection and property. It is a commodity which is often used to obtain some social and psychological advantages. For someone to provide information out of the goodness of their hearts seems unlikely.

Since information is generally considered to be a commodity, what is wanted and from whom has to be assessed in social, psychological and even economic terms as trading; what is being asked for and what is being given should be assessed as profit and loss balances for those involved. To the informant this way of looking at relationships and information is always something into which they have been unconsciously or consciously socialised. The seeker of information in social science fieldwork is almost always an outsider even if temporarily socially acceptable for reasons which are likely not to have much connection with the aims of that research.

Asking direct questions in most social situations is bad manners and might be considered bad research practice and rare and usually associated with officials who usually want answers to their questions as quickly as possible so that they can get on with whatever matter initiated the questioning in the first place. Social scientists in their fieldwork are not by training patient because of the time limits imposed on most research and since they may not know what they are looking for and are as such similar to crime detectives searching for clues. Worse from the point of view of creating enduring social relationships they have time constraints whether they admit it or not; their fieldwork has to be paid for and their time there is costed in terms of academic leave, visas and their personal lives. It would be best therefore to collect information by observation and some type of participation made possible by similarities in ethnicity, age, sex and assumed equalities of education and intelligence. But this is a waiting game particularly when the information which is wanted is interior to the individual and invisible rather than socially exterior and observable. So a short cut is the creation of trust not in the concrete sense which can be developed between people who are tied together by marriage, blood or affection, but as a created commodity between people with different interests. It seems unlikely that this can be created across the board in a community, as there will always be gradations of possible closeness just as there will be gradations in reliability.

A possible way in is through those people who are attracted to the researcher as a source of social and economic enrichment or who see any association with an educated outsider as an intellectual stimulus. Initial questions might be asking them how the researchers should proceed in making them initiators as much as respondents, giving them ownership through making them co-producers of channels of information. But of course the way forward through some sort of reciprocal trust depends on securing self-interested informants in the first place. There is another aspect to the gaining of 'interior' information and that is getting to be known in the community not so much as inoffensive but as a socially incompetent who should be helped; this can sometimes be done by the use of inquisitive children. At all events trust cannot be assumed or created for detached and impersonal reasons. From this author's experience working with prisoners as informants involved them in securing information of value to them and cigarettes from the researcher which could then be marketed among other inmates.

Most people in researched communities with some understanding of the outside world would assume that the information which they have provided directly or indirectly through the ever-present notebook, is going to be sold in someway to the advantage of the researcher. It is rare indeed for researchers such as Verrier Elwin with his Indian wife to have a reputation of commitment to those researched in representing them to the outside world.

18. Ethics and when is a confidence to be respected

The basic issues must surely be that both covert and overt social science research is almost always intrusive on personal privacies as matters are found out and publicly exposed without due consideration for those about whom such exposures are made. The parallels to such exposures of private matters must be private confessions to priests, lawyers and doctors who are all bound by professional and moral obligations, in some cases protected by law, not to disclose further what has been told to them. A researcher must consider whether their informants are aware that what has been told to them in private could be used in writings available to the general public and even in some cases available to those about whom this has been written. It is not difficult to assume that much of the information retailed in this way would be damaging to the informants if they were known to have passed this on to a stranger; under what understandings were they passing on these facts or possibilities?

Secondly, are there aspects of confidentiality which researchers should observe and is it adequate to just change names when every community knows well enough who has been talking to these researchers and for how long. There are also legal aspects to the confidences about which the researchers become knowledgeable which give them obligations to inform the authorities or become liable to prosecution. Most researchers in the course of long stays in a community become privy to much of the darker side of social life about which they should remain silent.

Outsiders generally in neutral social environments such as bars, trains or waiting areas for flights often receive personal information from strangers as it is assumed that they will never meet again. Except for research which only involves seeing and listening such as behaviour at political meetings or public rituals involving no person to person contact, social scientists are professional extractors of information from people who are not known to them in any personal way. The difference is that what is reported to them is written down or recorded and becomes the property of these people over which the providers of this information have no longer any control at all. This is why some potential informants are suspicious of researchers' questioning them as they would no longer have any control over their property or privacy. 'It was too easy for European anthropologists to be accused of betraying their informants' (Werthein,1972:321-24), and it may well be that contemporary national researchers may become subject to the same accusations unless they stay clear of government officials as much as they can.

All fieldworkers are professional outsiders accumulating data collected from individuals while over-viewing the respondents' behaviour in relation to others which none of those involved would want to be openly disclosed much less discussed. This is rarely data which would be accepted as scientifically valid or legally acceptable as evidence and would probably be better classified as an accumulation of gossip, dangerous in itself in any small community. This raises the whole question of the validity of qualitative research which is not scientifically valid, it is only an indicator under heavy influence of the researcher in the way that the information was gathered. The anthropologist who learns a great deal about his neighbours in confidence must respect that confidence except for overwhelming reasons, though it is of course conceivable that there might be occasions when it should not. No hard and fast rule can be laid down and 'these are matters of conscience rather than of reason' (Beattie,1965:55). Broadening the public knowledge for academic reasons can scarcely be such a justification for disclosure even if the non-national researchers work in another country and never return to the areas of their original fieldwork. So researchers accumulate confidences and they have to decide for themselves how this data should be used. Notes and tapes can be and have been stolen by thieves and possibly the political police so it is prudent to remember that these materials can very easily get out of the researcher's control; research material can often be lost under suspicious circumstances and possibly used for political purposes. It is therefore often a reasonable precaution for any researcher to post fieldwork data home on a regular basis.

Then there is the problem of what goes into print and thus

becoming public knowledge for anyone who happens to come across the publication. Unless the published research is biographical and has the permission of those involved to publish as was the case with 'Baba of Kano' (Smith N,1954), it is prudent to use anonymity both for people and places without too much circumstantial identifying data. This is particularly important when the researched activities are illegal such as drug addiction (Tanner,1966a) or cattle theft (Tanner,1966b) or just deviant to the standards of any society's dominant elite. The further distinction is between publishing about people and places in the 'home' environment who are likely to read or hear about what has been written or using a medium in a distant and specialised location where it is unlikely to be read by anyone who does not have a specialised interest in those matters.

There can be no clear dividing line between national and non-national researchers, but the latter have less to worry about over the use of confidences. However, the national must always be aware that they and the researched, in some respects, share a common social environment and every community now is likely to have several people of high intelligence and education who are quite capable of following up what researchers have written about their community and then seeing these results as hostile to their sensibilities. An Indian researcher with a minority religion, intending to work on his own community received death threats against him and his family. This author's own published research on European behaviour in East Africa, as if they were part of yet another tribally identifiable community, resulted in social distancing (Tanner,1964, 1966c).

There are other issues to consider. Social scientists are not working in a social environment which is not their own just to get on with strangers, although 'just getting on with strangers' might be a valid research topic. These researchers are working for a particular aim which should have little to do with their personal feelings except that they have chosen that particular topic or agreed to do it on the suggestion of another academic. They have to ask themselves how far they would go to get the data which they consider to be essential.

When the nearly complete fossil skeleton of the proto-hominid Lucy was found in Ethiopia, a modern knee joint was needed for comparison so one was obtained from a modern Afar burial mound although warned that the locals would be outraged if they found out this desecration (Johanson and Edey,1981:158-59). It seems very unlikely that had an Ethiopian professional been present that such an act would have even been considered. It would seem likely that some non-national researchers see the 'other' as outside the boundaries of any necessary social and indeed ethical considerations. Non-national medical researchers in East Africa were interested in the El-Molo living round Lake Rudolph who had a low fluid intake with a high saline content; it was suggested in a Medical Research Committee meeting at Makerere University College, Kampala that some teeth should be extracted from the living in order to study the effects of such unusual and limited fluid intakes. Uganda medical students sitting in on this research meeting became restive that this should even be considered. Researchers may go outside the boundaries of acceptability even in their own societies in the interests of science as in the recent British case of hospital doctors retaining the organs of dead children, but this is more likely to happen in communities and under conditions of extreme social distancing.

19. The social networks around researchers

The non-nationals working in their own countries will be creating around themselves networks of people who will be useful to them in their research work, largely composed of fellow professionals but also some informants in key positions. These networks do not include relatives and in the Western world the number of relatives is much smaller than the extended families elsewhere which have multiple uses. It seems likely that in the developing world professional and familial networks work much closer together and that real and classificatory relatives are valuable links into many different fields of research. National researchers are going to think in terms of these relationships as part of the process of thinking about choosing and putting into action a field of research. Aspects of kinship which would help them gaining access to particular groups of people in their locations and perhaps to gain access to records. They would be foolish not to do so when this is a common feature in most institutions and communities. Whether informants who are relatives are more reliable then those who are unrelated is a matter for speculation because they will be working from their own points of view about how their joint family is working (Gupta,1979); perhaps it is prudent to use some social relationship however distant as ways into sources of information and not to keep the relationships closer than is polite and necessary. This is a bilateral process and just as the researchers use relationships to further their ends similarly the researched will see any connections with the researcher to further their opportunities in the outside world of which they are presumed to have influence (Dua,1979).

20. Ethics and the boundaries of acceptable behaviour

Every culture and community has boundaries defining what is acceptable and unacceptable behaviour. This accumulation of rules, practices and customs never has any unity except in the codifications of written legal systems and even then a whole profession has developed to argue in public that particular words and sentences coincide or do not coincide with particular laws and abstract definitions with which actual events are not often narrowly related. So researchers are often faced with what are in effect personal boundaries within which they themselves behave in their primary relationships; these become increasingly attenuated with social distance; it is totally wrong to steal from your parents, alright to borrow and not return from friends and not matter of any serious moral concern to cheat over your tax returns or steal the property of government or international companies and to take souvenirs from hotels when on holiday. This of course applies when people lead relatively static existences. A national from a Hindu background researching the priesthood in a southern Indian temple is working more or less in the same environment as existed in their own communities, but the non-national is in a totally alien environment (Fuller,2003).

However, researchers do not have any simple array of standards to which they can conform even if they are working in their home communities, but there are still the standards which are laid down by the profession to which they belong. A non-national researcher outside any broad understandings of their topics really has no idea as to how an outsider should behave among men and women pilgrims to Lourdes, or market women in West Africa. The success in getting research data depends on establishing a very special role in any community which legitimates a kind of information getting behaviour which is not previously part of local social expectations; this requires that these researchers show both personal rather than professional sincerity and diplomacy (Hotchkiss, 1967, Pelto,1970;218). Most social science fieldwork is not done within home environments except for the large amount of in-house research which is done in universities using students for tests which has led to a huge accumulation of data about their psychological attributes with much less known about those who live outside universities.

So most research is done in social environments about which the fieldworkers may know something in theory but in practice they are faced with an endless series of social and mental inter-relationships for which they have had no imaginative preparation at all; they have injected themselves into social environments in which they are in every sense aliens. Most researchers should be sensible to the fact that they are pushing their luck when they publicly behave in ways that almost all the locals would see as suspicious. Geography researchers working near the Lebanese border wearing khaki, carrying maps, binoculars and cameras looked like every child's conception of spies (Howell,1972); they should perhaps consider themselves fortunate not to have been shot at.

Then there is the issue of personal standards beyond which the researcher would not go. This author was working on the East African coast and was given the role of male 'midwife' to a key

Swahili informant for his marriage which appeared at first understanding to have parallels to that of 'best man' in an English wedding celebrated in a church. What he did not know was that the first act of sexual intercourse between husband and the wife was felt to be under the possible malevolence from outsiders who might make the husband impotent by witchcraft, that this is performed by the husband's midwife. When faced with this situation in the marital bedroom, he retreated as this was not a personal boundary which, as a married man, he was prepared to cross for research purposes. Even for a Hindu researcher wanting to get acceptance into a Tantric cult involving sexual excesses might well face similar personal reluctance.

Most social scientists develop a liking for the people they study and have an understanding of the conditions under which they live and cannot avoid. This humanism clashes with the formal and detached ways in which they write their reports. It is not only that they experience a double identity, an almost biographical element and an almost conscious sense of betraying their respondents because they are not able to present this in their professional writings (Stewart,1994). Colin Turnbull, for example, developed a dislike of the Ik through a combination of their social behaviour and possibly the fact that he had not chosen to study them in the first place (Turnbull,1973).

21. The national researcher and the ethics of research

Professional standards for social scientists must remain a matter for individual assessment if they are not accredited to any professional body which might support them in any conflicts of interests in which they might become involved. Only a few researchers work from private means or as an off-shoot of their usual occupations as with missionaries and doctors. Nevertheless 'having an ethical code is a necessity in the folk epistemology of professionalism, an epistemology widely shared by anthropologists' (Pels,1999:102). They have no corporate status in law nor can they be considered as comparable to medicine and professional bodies exercise considerable control over both doctors and lawyers as to what they can and cannot do. The professional standards of these two professions in most countries are based on not just the centuries of development of their status, their links with learning and the acceptance by the state and public opinion of the facts about their activities; since society accepts these professions for what they are, their standards are maintained as much by the society itself and not just by their members.

In many countries now specific permission is required for non-nationals to undertake research which would involve negotiations with a national University or similar body. Even if a code of ethics should be created it is 'inert knowledge' that lacks the concrete problems in which decisions have to be made. More importantly it is the University or grant donor who could not train potential fieldworkers in making the choices between conflicting principles (Appell,1978:35).

It is not a science within the accepted meaning of the word in which hard facts accepted by other scientists not connected with the original research work. Indeed it could be called a field of work in which opinions are formed rather than conclusions based on evidence collected with any personal influences. One can question whether sociology can ever produce hard facts, they are surely just indications. It would seem therefore that any discussion of the potential existence of an ethical code for social scientists is no more than an attempt to gain professional status because it comes down to what individual researchers decide that they should or should not do in the circumstances facing them. Outside the provisions of the law which apply to all people living in a country other than diplomats, there are few if any restraints which can apply to social scientists other than their own consciences. There is an immediacy of choice in many small scale ethical issues which perhaps only in retrospect can be interrelated to major issues of ethical importance. Most fieldworkers are involved in issues of duplicity between what they say they are doing and what they are in fact doing, a day to day factor in everyday life and unless it is blatant and tactless this probably remains an individual matter of conscience.

Any broadly applicable ethical issue must always be situational and capable of wide ranges of interpretation but in all nations there will be ethical codes maintained by institutions which are imposed on the individual researcher over which they have no choice. There is firstly the legal code of the nation, then the political system as it is practised and finally, the unwritten ethical system of whichever community is being researched as to what they do as well as what they say they do. Thus ethics will always involve the individual researcher in choices and in this the national ones are less likely to blunder into difficulties either because of ignorance or because their personal values have not been created in another cultural setting; they know the social realities of caste discriminations, mother-in-law avoidances and the restrictions and possibilities of private and public inter-relationships. Perhaps the most difficult aspect of social science fieldwork is when the researchers as individuals feel that they are participating in something which is morally wrong but find that they are unable to make any material gesture in opposition as when they are accessories after the fact in the planning of criminal acts as might be the case in the researching of deviance. National researchers cannot distance themselves from these situations in the ways that the non-nationals can by leaving the field.

There may be two levels of ethical principles which prevent researchers involvement or ethical difficulties in the first place such as pre-emptive ethical choices in which they decide not to get involved or are forced to keep silent. Then there are the ethical principles which may guide relationships between researchers and their informants and the community from which they come. The moralities of everyday relationships which are a mixture of those learnt when they were growing up and what they have learnt professionally. Again the national researchers have advantages here as they may not have the same extreme contrasts to cope with as must be the case with non-nationals who may well have fixed intolerances when it comes to what they see as the unkind treatment of women, children and animals, and then their own personal taboos. Intellectual and personal understandings of 'their complexes' fit uncomfortably together. It takes a very long time for a non-national adult to be socialised into the patterns of behaviour of a different culture. It is an interesting fact that if their children accompany the researchers into the field they may well be envious of the way in which they appear to fit into these supposedly alien cultures.

22. The internal expatriate

The smaller the nation the greater the possible coincidence between individual social science researchers and what might be called national averages of social understandings and behaviour; this would appear to be the case in Bhutan and Luxembourg, but this cannot be so in nations of many millions of people and enormous geographical size. People from northern India and northern China have little in common other than nationality with their southern compatriots in language, social behaviour and their equally important geographical environment. They share a political identity under a single flag and a sense of pride in their nation when its status is challenged by outsiders. Beyond this there may be little that is sufficiently shared to make a marked difference to what researchers may discover; political ideology may have little influence on behaviour at lower levels.

Sometimes these differences are so extreme as in the case of Kenya in which the two largest tribal peoples, the Luo and the Kikuyu, do not have languages within the same linguistic group as well as the pastoral Karamajong and Maasai who are outside this national binary political structure altogether. So we have here potential categories of researchers who are nationals but in terms of their areas of fieldwork may be in practical terms 'internal expatriates'. There is nothing in common between the researchers and the researched except perhaps physical appearance as well as possibly some measure of acceptance simply because they are all nationals and that people of status with whom the researched community have regular dealings will have confirmed the researchers' status and thus removed a potential source of difficulty.

Beyond some primary feelings for their acceptance these

researchers may have to go through the same prolonged processes of filtering themselves into the community of their choice as if they were non-nationals. There would be a tendency for nationals belonging to well defined segments of the national population to carry out research within that group for Punjabis and Baluchis to work with their own kind just as it would be for the northern Han to research north of the Yellow River in China. But these could be classified as soft options so that researchers with different intellectual ambitions and divergent topics of research will go into communities with which they have no natal affinities at all. An example of this internal expatriate research would be the work of someone from West Bengal researching the feeding habits of Tibetan women refugees in northern India (Bera,2004). Is not this reference to 'feeding' rather than eating an indication of a discriminatory viewpoint?

23. Legal issues

So we see that the whole field of social science research with its initial Western intellectual dominance often remains within this same framework so that a top-down structure of thinking and behaviour may still predominate. It is hard to think of a situation in which a person's thoughts and behaviour recorded surreptitiously or in what they thought to be more or less private interpersonal conversations, come available in print to the general public without objection being taken to such disclosures particularly when it is published in a commonly understood and readable language. Where this has occurred in Western environments over work done by Western social scientists, there have usually been such reactions. There was uproar in the United States over the publication of a study of a small town (Vidich & Benshan,1968) as many there felt that the use of pseudonyms did not effectively disguise particular individuals; privacy and trust had been betrayed and reputations put at risk. The 4th July parade had a float with an image of Vidich bending over a manure pile. Two studies of an Irish community (Messenger,1989; Scheper-Hughes,1982) attracted an even wider range of criticisms from the people in those communities. When Oscar Lewis's series on a Mexican family (Lewis,1951;1959) was translated and published in Mexico despite the actual family being unnamed, there were widespread hostile objections in the national press and a suit was brought against him and his publisher for obscenity and slander. In this the core factor must have been the fact that Lewis was an American, a non-national researcher who had denigrated the Mexican national image. Certainly, the criticisms are likely to be stronger and more widespread when the researchers are non-nationals and one can imagine the protests if the Irish research had been done by a Kenyan or Bengali. But it is prudent to remember that a national researcher may be equally foreign in terms of social distance. In fact an Indian researcher (Pandey,1979) has worked on American Indians from which there do not appear to have been any hostile reactions perhaps in part because of physiological similarities between the researcher and those researched.

Who is likely to read research findings and who is the audience for which these publications are intended? There is always the factor that self-criticism is acceptable but its exposure is not. It may be the dilemma of popularising research or at least publishing it in a widely read and readable language which causes the problem from its public nature. Crudely, few social scientists would pass up the opportunity to have their works published and read widely if there was an obvious financial return. Is the situation likely to be different in developing societies where the researchers are nationals and who may have to be much more careful in considering what may be the local reactions to popular publishing of their fieldwork. As a general framework of thinking this is clearly outdated as those who are researched become aware of their rights to be consulted over all stages of the fieldwork. So their consent to be researched is now an important factor in the preliminaries to fieldwork; the right to be researched or not becomes important and this involves some disclosure of its objectives. We have to consider the extent to which permission to be the subject of research by an outsider covers the publication of the results to the outsider world whether in the local or a major foreign language.

If the results make the location recognisable and the identification of specific individuals possible then their informed consent becomes increasingly necessary. It would clearly be difficult to require informants to sign consent forms to the use of the information which they supply (Fleuhr-Lobban,1994) although this has been attempted. This would entail all the difficulties of translation from another language or linguistic code over consent to what; a court case or indeed an appeal to a government official about such generalised permissions would lead to embarrassing rulings about getting the researched to sign blank permission. Even the implication that those researched have some say in what is published about them, gives them power in effect to restrict or to enforce restraint on what is published. This knowledge and its acceptance as an ethical requirement encourages or at least makes necessary a greater openness or disclosure by the researchers. This itself suggests that researchers will have two levels of information; one which they are able to discuss with those involved and in this context published and a covert one for which they will not be able to disclose their sources or their conclusions. A more collaborative relationship between the researched and the researchers whether they are national or more particularly non-national is certainly a contemporary necessity. This is a difficult issue since one can assume that in any fieldwork there is some sort of a collaborative relationship without which little indirect information would ever be obtained. It is not difficult to see the conflict between the extent of such obligation and the necessary professional secrecy to reduce the extent of bias.

24. National law and the publication of research

The research activities of social scientists are as liable to legal action as any other social activity and many universities in developed societies vet research proposals to avoid possible civil actions against their researchers and the results of their research. Research undertaken through a university requires the agreement of special committees and probably that of a safety officer as well as clearance from another committee that covers ethical issues. But this is only for research requiring funding and the use of the university's name for getting access to a particular field of research. National researchers using their own funds and time, may not require such permissions in furthering their own interests and careers as with doctors who are continually researching their own specialisations.

Often it is assumed that there are no civil law issues in researching downwards in the social scale because few of those researched would be likely to see much less read any publications coming from research on themselves. There are now several possibilities in which researchers might become involved in disputes about the 'truth' of what they have written and published. There are growing numbers of literates in all societies from the most tribal or urban slums upwards; they can seek out through relatives, friends and journalists some ideas about what has been written about them. The universities take in students from every section of their nation who are bound out of curiosity or professional interest to seek out what has been written about their own communities. Finally, national publishers in both developed and developing nations are often looking for new markets and there is an automatic attraction in publishing research on local social behaviour in a popular local language for an expanding middle class readership as well as the fact their universities pass out more students with professional interests in such local affairs.

Whether an issue becomes a legal dispute might depend on how it relates to the national culture and more particularly the nation's dignity internationally. In Indian culture there is a tendency to use the courts as a weapon of attrition whereas in China the legal system has not been designed to encourage litigation and in Japan there is if anything an avoidance of the courts since their use is seen as a failure to conform to consensual requirements. In this context non-national researchers would not be affected since they would not be in the country long enough for such dialogue or action to take place; they are certainly not likely to remain there or indeed to be allowed to stay there long enough to defend such civil cases, and they might well find it difficult to undertake such types of research again. It would be unwise politically and economically to get entangled in any such legal processes over copyright and accuracy of published data. Much the same would apply to non-nationals researching in the United States which is as notoriously litigious as India is reputed to be since a high proportion of elected politicians have graduated in law. So non-nationals if they became involved in any legal issues over their research would almost certainly be advised by their embassies to leave the country.

Two legal issues might result from fieldwork. Firstly, some verbal statement made to a third party about an individual who then feels that they have been slandered. In this we must remember that very little data on which social scientists rely would meet the required standards of correctness to be acceptable as evidence in court; except where the informant is reporting his own activities, as much data is second-hand. There is a distinction between whispering something beside a fire or across a counter and seeing it printed for the world to see and as the subjects of research said over an Irish project, 'why did you have to write it in such a way that we could read it and understand exactly what you were saying (Scheper-Hughes,1982). Secondly, that something written under the researcher's name is not the truth and then the person about whom that statement was made sues for libel. This is the most dangerous of civil suits because it does not rest on words which may or may not have been recorded, have a limited audience and therefore become easier to defend but on something in print which has potentially an unlimited audience and no time limit. It is thus over printed reports that future legal cases might arise, not so much from those who were the subject of the research, but from those of equivalent status particularly when the research has involved persons of high status or work in powerful institutions.

There are a number of these issues which have arisen in Ireland where published research has been read by those about whom the researchers have written. The reactions to a published study of mental illness in a particular rural community where the reactions of their most literate and self-reflective people was harsh, swift and unsettling (Scheper-Hughes,1982) when the researcher had contradicted a hitherto unchallenged native interpretation of their lives and created a public shame. As the village schoolmaster put it 'It is not your science I'm questioning, but this; don't we have the right to lead unexamined lives, the right not to be analysed? Don't we have the right to hold on to an image of ourselves as different to be sure but as innocent and unblemished all the same', The Irish might have been over-sensitive to what they saw as the criticisms of outsider researchers, but might not this apply to many situations in Africa and Asia where non-national researchers have touched on the dignity of those involved once data is put to print. The Irish reactions to the published research of an American social scientist ranged from accusing him of being ill-informed, ignorant, illiterate, malicious, anti-Catholic and pro-Anglo-Irish to warning about never trusting what an Irishman says and to more general statements about the impossibility of anyone, but especially an outsider, gaining an understanding of the Irish and concluding that everything he says is true but the reason he would be lynched if he returned to where he worked was that he had a right not to say it (Messenger,1989).The study by a national of Irish intellectual life was more complex as it touched on the connections between their university establishment and their party political connections over appointments so that Sheehan had to breakdown references to her sources of information by writing in an obscure manner in order was simply to protect herself from at worst legal threats and at best from a principal informant's undying enmity (Sheehan,1993).

The basic issue in any civil action is the truth of what has been put into print, and researchers collect information from a variety of sources of which only a small proportion will be documents obtained from an institutional source which may themselves be of questionable accuracy. Did the researcher obtain a signed release allowing the use of the material which is rarely done because it involves so much time consuming explanation? Researchers knit together information from a variety of sources of very varying reliability which is usually reported speech of little legal validity. Social scientists would always have great difficulty in proving the truth of what they have written as there is always an inherent dichotomy between what is told and what is recorded. This is one of the fundamental issues regarding the documentation of oral history. The social scientist would be wise to back away from any such possibilities by qualifying their conclusions with possible doubt as to their accuracy; in effect to avoid any assumption of incontestable accuracy in the scientific sense of the word. Is it not a fundamental that all research of this nature is both politically and socially debatable?

Chapter 3

METHODOLOGIES

1. The elements of fieldwork

The three basic components of fieldwork, the researcher, the project and the people researched are not covered in equal proportions in reports on completed fieldwork. Their introductions deal with the status of the fieldworker and a list of professional colleagues for whom they feel a combination of academic political and professional gratitude for what they have written. What else is known about these fieldworkers without whom the research would not have been planned, worked or reported on. Overall not much almost as if it was not important that they should be real people not more or less abstract entities. Their age can be guessed at, their sex can often only be known by their use of their first names in the author's title and possibly from the dedication whether they have partners, parents and children. Yet the whole report pivots round their personal abilities and circumstances; how on earth did Kumar manage her research on Banaras artisans when she was breast feeding her first child (Kumar,1992); this information only came from a later book on her Banaras fieldwork. Other than what is disclosed in their professional writings very little is available which would put these men and women into a humanized perspective made necessary by the simple fact that they did their research in entirely individual ways. There is no way in which fieldwork in the social sciences can be paralleled to research work in the hard sciences. Hypothetically the persons doing the hard research have no personal characteristics at all except for allowing personal bias, circumstance and inefficiency to disturb the process. In the field situation the researchers are faced with a great variety of personalized situations to which they have to react actively or passively. Whatever degree of success is achieved comes from their individual ability to cope with an ever changing variety of human activities in a strange social and physical environment.

It seems likely that results put into print have undergone some personal censorship of this version of events over the period(s) in the field. Where diaries exist, we have no knowledge as to the extent that they have been altered but there is certainly no guarantee that they are any more or less accurate than the fieldwork reported to have been done in the same period. The degree to which they coincide with actual events is certainly obscure. The diaries of Malinowski (1967) are regarded as important because of the author's pre-eminence as an anthropologist and because of their uniqueness but they have certainly been edited because of deference to his surviving family. His widow refers to the removal of a few extremely intimate observations but we have no knowledge of what they may have been. Haddon, a potential fieldworker, who went to him for advice in Cambridge in 1926 wrote to his fiancée that he had been advised how to get a native mistress (Gardiner,1984).

The fieldworkers have to be seen as people who have characteristics and abilities which antedate and coincide with their research abilities. In the hard sciences the apparatus and personnel through which the data is obtained are available for public scrutiny but in the social sciences this essential element tends not to be given the importance which their role as researchers necessitates. Everything depends on the extent to which they personally obtained and filtered the information which became available to them. It is not just methodology which is important but the ways in which individuals as personalities develop and use the methods which are available to them.

2. Researchers as objects of scrutiny

The dangers of researching off the tourist trail are of course realised but it is worth covering the ground again for the sake of completeness. Researchers are always and inevitably objects of scrutiny and speculation. It is curious that this should be so when they are entering environments which are new to them just as their presence there is new to those people; they are classified as strangers whether this situation is a village, occupational group or indeed a university. It would seem that researchers assume that because they have official and self-assumed status of social scientists that they can individually slip into a community in much the same way that a doctor or nurse might slip into a new posting to a hospital or clinic. They are often unwelcome in situations which communities or persons consider important and private (Laitin,1986:35); intrusion on Lutheran worship in London is no more acceptable than intrusion into Yoruba worship in Lagos. Unless they are researching under cover in the total sense rather than merely disguising the real purposes behind their presence, they will always enter the field of their research as strangers and remain so for most of the people they meet for most of the time, even if they have established equitable relationships with the community as a whole over a limited range of activities. They will always be objects of scrutiny more than of just passing interest because they do not fit into the available roles in any community; they have no job and curiosity is not usually seen as a benign characteristic of anyone. Of course, there may be communities which accept a resident researcher as just another example of migratory intellectuals in business or university living in as dormitory suburb.

Even if they may be accepted by the limited few who see the purpose of research and who get some personal advantages from being incorporated in such activities, for the majority they remain outsiders, and they will want to know as much as possible about how their personal behaviour fundamentally differs from their own. This curiosity about the behaviour of researchers will be much the same whether they are non-nationals or internal expatriate nationals. At this level the personal hygiene and body language of researchers will be of as much interest as their tentative and perhaps inadequate use of the local language. The greater the ethnic and cultural differences between researchers and those researched, the greater the extent of this scrutiny. In the background there is always an intellectual distancing between those involved which makes the scrutiny of the researchers more detailed and perhaps more hostile. No one likes to be visited for too long by people whom they assess to be or consider themselves to be intellectually superior.

Strangers who come and do not go like so many other visitors which the community may have experienced, will always be the subject to some adverse speculation. The lives of those researched whether subsistence farmers or urban casual workers who are dominated by the realities of day to day necessities mean that they have no reason to assume that researchers will be any benefit to them and so their scrutiny will usually be more than curiosity. In every research situation the researchers and the researched as individuals are different to most by reason of age, sex and ethnicity as well as by as many self-assumed roles as may be considered. The social scientists at work are always going to be objects of scrutiny on the grounds of suspicion or curiosity. It is probable that only within the parameters of university life or within institutions which have used such professionals before, can they expect to work unnoticed; it is a small circumscribed world in which their work and their ideas are as subject to as much jealousy and misunderstandings as would be the case with lawyers, actresses and doctors.

Whatever these strangers who classify themselves as social scientists are up to is interpreted by those they meet on own their terms. This probably means that they will not believe that strangers can be inconveniencing themselves to the extent of living away from their homes without substantial financial gain. After all in their own communities knowledge is power and information is rarely given away without something being taken in exchange. They would assume that researchers are being paid to do this work (Patwardhan,1979) as indeed they would be for work in their own socio-economic systems. They are very unlikely to believe that there is such a thing as the disinterested search for knowledge even if it could be translated; they would sense such an idea as hypocritical as relating to the researcher in their midst.

3. The spectrum of bias

Any assessment of bias is complicated enough without recognising that it is usually considered in the context of the researchers while it must be equally relevant to consider it from the point of view of the biases of the researched. There may be individuals but surely no communities or institutions which have not heard of or experienced enquiries about their affairs being conducted by outsiders as to what they are doing and moreover enquiries that rarely results in benefits to them. To the understandings of 'simpler' people this seems to be the sort of behaviour to be expected from outsiders; the agents of outside institutions, schools, churches, health authorities, police, agricultural inspectors, tax collectors and the agents of political parties. There is no reason to suppose that the receivers of these attentions from outsiders see the researchers as any different until they learn or indeed never learn to see them as otherwise. So there is usually a background varying from ambivalence to newcomers to covert and overt hostility. At another level there is some knowledge of what the researchers are trying to do or find out but there may be more specific reactions that they just dislike being the objects of investigation by a young outsider whether national or non-national. These predispositions might involve a variety of social distance factors; a dislike of the educated and newcomers based on previous experiences or just rumours. It is safe to assume that social scientists in their fieldwork are always biased and that the failure to recognise this inevitability comes in part from the refusal of the social scientists to free themselves from the incubus of thinking that they can ever emulate the approaches and procedures of the hard sciences in researching human behaviour.

It is not so much that North American anthropologists will produce white American anthropology (Hsu,1975:206) but that Chinese, Saudi Arabian and Indian ones will remain similarly structured within their own cultural paradigms. It would then be incorrect to refer to this as bias when it is the inevitable result of socialisation. Social scientists whatever their cultural origins will see the world outside themselves through these culturally provided spectacles, however much they may understand and appreciate the whys and wherefores of other types of cultural behaviour. It is theoretically possible to brainwash an individual according to the scenario of the novel 'The Manchurian Candidate' (Condon,1962), but whether this would be possible for a researcher remains no more than intriguingly implausible; who would agree to be being the subject of such an experiment and indeed this would result in just another form of structured bias created for a purely intellectual purpose.

It is not simply a question of getting a community's agreement to the research and the presence of the researcher in their area that creates an environment of rapport. The researchers themselves change with their experience and may feel more at home in the community in which they are living. Others may find that their lack of progress makes them show their likes and dislikes so that in the end they may long to get away to their own sort of people and a more self-accepting social environment. Those with whom they are working may also change in finding the intrusions in their lives much more unacceptable than they anticipated. Also the social and political climate may change so that they are seen as intrusive strangers allied to an alien or central government which has upset them. In socially divided societies such as India with its multiplicity of castes, religious and ethnic divisions, there are many opportunities to make objections to both internal and external researchers however well connected and previously acceptable.

The perspective of social scientists affects what is seen and in turn the results obtained, and indeed it could not be otherwise. This is simply because they use themselves as research instruments to record what goes on around them. Even if they were some type of permanently working machine recording what they see, hear, smell, touch and taste, they would still be recording in one direction at a time rather than another. The directed camera and microphone are accurate for the direction at which they are pointed by the fieldworker but have no relevance at all for what goes on behind their backs.

The most serious problem for the national occurs when both national and non-national researchers produce quantitative data which they state has come from their examination of their fieldwork data. The examination of sets of figures which have been allegedly gathered without any personal influence on the situation is debatable but at least initially acceptable. But once we have qualitative data in which informants produce their own data for the evaluation of the researchers, then the national researcher is faced with the dilemmas explaining this information through personal social filters which are biasing their conclusions more than they would possibly like to admit. All researchers cannot avoid being emotionally involved to some degree in their qualitative assessments of what they are observing and in this respect the national researcher may have particular difficulties (Jones,1994). What non-nationals might describe as no more than an evaluative analysis, might well be assessed as criticism by more directly emotionally involved nationals which is either not justified or should not be publicised anyway.

The language in which they report their findings may not be their own natal language and even if it was, this is often strikingly different to the researchers' own private use of the same language and the linguistic code used by their informants. At best there is a process of careful translation and at worst a built-in dissimilarity from the professional dialect in which researchers were trained and read in their textbooks. It would be rare indeed for a reader to find that what they read is what the informants actually said. It is rumoured that Mrs Seligman, the wife of the anthropologist, who specialised in Nilotic peoples, on seeing a naked Nuer man on the banks of the southern Nile, remarked that it was hard for her to believe that such a man could have the complex metaphysical thoughts written up by Professor Evans-Pritchard (1956). Leaving aside the possibly colonial bias in such a statement she has certainly touched on a valid point.

Of course, the non-national is biased by the enormous geographical and social distance between the researchers and the researched unless of course they are researching communities in which there may be a substantial amount of social sharing but they will still be some ones daughter or son and have the attributes of age and social status from education. If they undertake research in an institution such as a hospital they will still have the biases built into the roles they assume or are printed onto their identity cards; there can only be the questionable neutrality of such roles. The indigenous social scientist is not however necessarily aided by the absence of national and geographical distancing. They will always be part of the nation building process and are mentally involved. The fact that they may understand basic values is no more neutrality than the more academic pretensions of outsiders (Koentjaraningrat,1962). There is also the bias in which social scientists in their research treat one part of mankind as an object and in its less recent forms their endeavours come from an era of violence with the Western world's physical and ideological invasion of the non-European arena (Levi-Strauss,1966).

Then there is the bias which comes from the excitements of exposure to new situations which are almost a rite of passage for social scientists (Nash,1963) which cannot be confined to non-nationals. Only a few social scientists spend prolonged periods of study within a single community as family life and careers often clash with fieldwork away from home and work places, so there is usually the bias which comes from the brevity of their exposure to new situations. Clifford spent two years in his study of Bali, Morocco and Java (Clifford,1986), whereas Glazier spent sixteen years researching the Spiritual Baptists of Trinidad (Glazier,1993), so each will have had their personal forms of bias dictated by time spent in their fieldwork.

Any individual in their social environments will of course in the first instance see and experience their inter-relationships in personal terms; those involved do not have personal names, and the pronouns used are personal and not collective ones. At first it might be a Baluchi named Abdul who spoke to the researcher on a certain subject on a certain date and at a certain place and it is only very much latter that the researcher may refer to 'the Baluchi'. So it is only with expanding social distances that people come to be seen in terms of imputed and imposed group characteristics; Collective terms for the Bosnians, and the Sukuma have some of these stereotypical characteristics. All the time the social scientists know full well that the Baluchis individually may have some but not necessarily all of these characteristics. Stereotypes are maintained rather than created and used for the convenience of their users without much regard for the realities of what these people may do and think. How does this affect trained social scientists in their fieldwork? Of course, they will still experience their own particular interrelationships but whether they admit it to themselves or not, they will still tend to think in stereotypical ways about 'outsiders'.

Professional training has to proceed in ways that do not coincide with the social lives of either teachers or trainees; they learn about the generalities of human behaviour not only about how particular cultures go about their lives in general but that certain behaviour is to be expected that the mother's brother should behave in a certain way but they cannot be taught what happens when here is no mother's brother in existence. Professional training as a foundation never goes much beyond the 'typical' persons whether land-owners, AIDS sufferers, cattle thieves or priests. This training tends to overrate the influence of beliefs on behaviour, to overemphasize homogeneity, to overdo order and coherence and underplay exceptions and contractions and to emphasize structure at the cost of process (Hollander,1967). In the time allowed it is inevitable that the teaching of information should be grouped in this way. Any teaching that concentrated on a spectrum of behaviour in which the limits are totally dissimilar would undercut the requirements of teaching. Even though the social scientists know that there are five million individual Sukuma and that the same number of individuals live in Nagpur and that each household, housing block or hamlet will be dissimilar in personal characteristics and experience, collective behaviour will always emerge out of fieldwork projects unless the study itself is of a particular circumscribed group as with Lewis's work on a Mexican family (Lewis,1959). Then of course there comes the inevitable difficulty that the Sukuma and the people of Nagpur will say that they do not behave according to the stereotypes which the social scientists have created.

Any fieldwork which involves the forming of interpersonal relationships must of necessity mean the taking up of a viewpoint even if the researchers recognise that it is not so much that but a necessity of being participant observers. A Sukuma who happened to be a social scientist—would view hostility to the Maasai living to their east as part of the paradigm within which they are living. Just as Indian researchers would think within the caste paradigms of their own position. British children in the southern naval port of Portsmouth which was heavily bombed in World War II by the Germans, see their traditional enemies to be the French against whom a prolonged war was fought for the twenty years before the 1815 battle of Waterloo brought prolonged peace to Europe. The non-national researchers might accept that within their own country such biases exist, but that outside where they are now working, they would attempt to stand outside such paradigmatic biases without realising that to become participant observers, however nominally, they must become involved in the more prominent of these biases. A researcher living with Brahmins in a Brahmin dominated Tamil Nadu village had to accept his host's objections to Outcastes visiting him in their house. 'If I was to be given access to their homes and temples to which access was not given easily to outsiders, I could not violate their sentiments at will (Beteille,1965). Social scientists in their fieldwork may have to recognise that they cannot be both 'this' and 'that' at the same time and that their view of society may always be biased not only by these more sophisticated biases but by the simple fact that they are male or female, old or young and that some of those people with whom they are working appear benevolent and others malevolent. This is just the way that humans work out their lives in a collected network of preferences and avoidances from which even the most experienced social scientist is not exempt.

Bias is a hostile word and every researcher flinches from its application to themselves even though they have to admit that they see their research objectives through very individual perspectives;

they cannot avoid this without adopting almost clone-like assumptions about researchers in general. In fact we have to accept the realities of prejudice that we all have likes and dislikes in what we see and interpret and no amount of introspection is going to take this out of our perspectives. What we have to come to an understanding of is 'the prejudices of good people' (Esmail,2004. Sethi and Seligman,1993), which starts with the nature of research. There are first of all fashions in the subjects of research; what is chosen and what is not chosen. An examination of the Index Medicus shows that certain diseases are popular and attract research funds. In the social sciences great areas of social behaviour, such as excreting, have received very little attention because it is seen as a distasteful topic and not likely to enhance academic reputations. Much the same might apply to corruption in less developed and developing societies of which everyone is aware but which researchers and donors avoid working on because it is both dangerous and likely to lead to exclusions from countries. Researchers are often professionally blind to political matters which would adversely affect their own research interests.

An interesting comment on the nature of scientific interests was made by the Editor of the British Medical Journal (BMJ,2004), 'I am interested in the study showing a steady rise in the hospital admissions for acute pancreatitis because my brother had pancreatitis'; might that link have had something to do with the study making it into the journal, just as it has been suggested by that Editor that 'the British Medical Journal publishes on toe-nail fungus because so many of the editorial team suffer from it', and so on. Perhaps we should accept that the choice of research venues by non-nationals may well be influenced by travel brochures and articles in the National Geographical Journal or indeed just to get away from their own societies and partners.

There are also difficult issues in research environments where there is a dominant religion. In popular imagination however illogical it may be, individuals even if they are obviously strangers, are either believers or not; the status of being a non-practising member of their faith is understood well enough but having a detached view of any religion is not appreciated. We have seen the case of the researcher who was studying a Pentecostal community and was seen by them as investigating them with a view to their own conversion (Glazier,1993). There is always the problem when the researcher's faith can be inferred from their name as to whether they are Christians, Hindus or Muslims, and whether the researchers are still thinking unconsciously in terms of the faiths in which they no longer believe or practice. Professor Srinivas by his name was obviously a Hindu but it was assumed that he was an atheist because he did not behave as a Hindu in the village which he was researching and was castigated for not shaving after bathing, which was considered polluting. Since the grape-vine knew that he was a Brahmin even though he did not keep to the required caste diet, he was nevertheless given gifts at weddings in order to obtain benefits from his religious status (Srinivas,1979). The assumption for example, that a researcher is a Hindu lacks social specificity and there was a need in a Gujarat village for him to slot himself into a sub-caste so that he could be seen as normal (Shah,1979). Such a fieldworker's attempts to get to know all sections of the community they are studying, will come up against the restrictions of caste. There are always unforeseen consequences as when a researcher was seen to break his caste by accepting a glass of milk from an Untouchable (Chakavarti,1979), or after they have been seen talking to Untouchables they will be asked to wash their hands and feet before entering a caste household (Dua,1979). Once Hindu researchers are identified by their caste, then caste requirements channel the ways in which they can be communicated with and the permissible places for inter-caste contacts such as rice fields, semi-public bathing and temple bathing pools and the sites of bullock races (Unni,1979).

It seems likely that even the most liberal minded of male and female researchers will have a mind-set that comes from socialisation into a framework of thinking and this is likely to be even more so when the language used by men is in many respects different to that used by women. Their informants are usually predominantly male and even female researchers would tend to have much the same range of sexually biased informants except when they are working on matters from which men would be excluded.

This exclusion of women from the male dominated structures does not mean that even in the most sexually segregated of societies that they are without influence. The rise of education for women in Saudi Arabia is said to have been due to the influence of the only wife of a previous ruler for example. Male researchers have to learn that women have their own power networks which should not be classified as gossip so that their absence from public political life and their silence at public meetings makes their covert influence all the more significant. On occasions obvious blocks to the possibilities of research can be overcome by influence and persistence. A Bengali Hindu woman married to an American managed to do research in Pakistan on the few remaining members of the Red Shirt pre-Independence non-violent movement. She not only circumvented official objections but got herself accepted in the Muslim community where these men lived, because of an earlier contact in Delhi with the son of the movement's founder who got her a visa for Pakistan. However, even with this help it took two years to gain entry (Bannerjea,2000).

4. Restrictions from the expectations of others

Research in the social sciences is rarely if ever isolated from the political and popular needs of the surrounding societies. The only exceptions may be research that nationals are able to do in their own societies as part of their existing commitments or in their spare time and with their own funds. Certainly individually researchers are looking over their shoulders as to the effects of their research on others; nothing is done in a professional vacuum.

Are the results going to be acceptable within the boundaries of the field being studied and to what extent is attention paid to the sensitivities of those who have been studied; they should not be upset in ways that the researchers themselves would not like, but to investigate many cultural feelings and events can easily upset as we have seen in the Irish study (Scheper-Hughes,1982). What is being expected of the researchers by those researched as no one is providing information for nothing? Something is being taken away and what is replacing it? With particular informants these expectations are higher but also handicapping and in some cases socially and physically dangerous. They are inclined to see researchers as friends while the latter would not think in terms of such close relationships. The researchers have to think rationally of what returns they can give for information so that it is often an economic issue. But there can also be restrictions on access over which neither the researchers nor the researched can be said to have much control as in one Indian case in which the ominous behaviour of a cow prevented a woman researcher from getting access to the information which she wanted (Varadachar,1979).

Researchers have to consider their own careers and they are thus guided by the advice of their controllers who have their own specialised interests in mind and how their own professionalism can be advanced. There are upsetting topics which nationals are likely to avoid just as there are methodologies which are closed to non-nationals.

5. The biases of omission

The non-national researcher rather more than the national one are looking for aspects of social behaviour which they have been preconditioned to search for by training and their own cultural socialisations. Whether they like it or not they are inclined if not subconsciously forced to see the community which they are researching as bits and pieces of some cultural whole which is beyond assessment. Other than the fact that they can see people going about their daily routines, beyond this they see what is going on in terms of what their informants provide and what their own segmented observations lead them to follow up; their field notes reflect this and as trained social scientists they are not comfortable to consciously go beyond this limited data. There is an analogy here to the analysis of poetry whether it is in their natal language such as Japanese looking at their 'haiku' poetry with their condensed symbolism which is to Western thinking extremely brief, or the similar symbolism in the choruses of the Sukuma dance songs with their similar brevity. The words can be broken down into their dictionary meanings, their phonemes and grammatical connections but beyond that what they mean may be beyond the comprehension of national and non-nationals regardless of their intellectual capacities; understandings are in reality very personal.

Most researchers would agree at least in private that their own culture is at least emotionally superior to that of the people they are studying. It may well take years of close associations with these respondents to go beyond the closed world of information supplied by informants. It ends up as some understanding of what they may have missed. The readers of the reports of fieldworkers about the behaviour of these cultural strangers will rarely find descriptions of kindly people with some goodness in their behaviour. There will be details of quarrels and matters of dispute. It is perhaps hard to accept that the indexes of such studies whether they are written by nationals or non-nationals would be more likely to contain pages referring to sorcery, competition and bride-wealth than such concepts as love, kindness, consideration and altruism (Hau' Ofa,1975).

To a certain extent this comes from insensitivity since most people find it difficult to attribute to others the sensibilities which they would apply to themselves. It is more likely that this is a methodological omission; these attributes are not what are looked for and therefore they do not feature in professional writings. On the other hand there is ample work on the nature of altruism and pro-social behaviour from experimental work by psychologists working with small samples in developed societies. It is perhaps in that shared cultural milieu in which nationals are researching nationals that such benign attributes are searched for and found. In reverse there is a bias against researching 'evil', because it would be difficult to find a sample free from bias. Mao Tse Tung stated that without investigation of the lives of ordinary Chinese, no party member had any right to speak and that this knowledge was inseparable from their direct experiences among the people for whom they spoke (Mao Tse Tung,1969:11).

However national researchers in developing societies must often start with the assumption that those they are working with have positive virtues. The diminished cultural distance between upper and lower caste Hindus involves some sharing of ideas and a common humanism which would not be the case with a non-national who may miss much outside their intellectual understandings. It would be difficult to find a Kikuyu or Baluchi researching their own societies who would highlight the negative aspects of their cultures to the exclusion of more positive values.

6. The issue of representation

In the taking of photographs and the recording of activities on tape in which the operators have no role other than pointing the camera or microphone, there is the assumption that what is being recorded is as near as possible 'natural' to these situations. These machines are pointed and in this respect they do not pick and choose and are thus rightly considered to be neutral. It is not readily accepted that in the way they are pointed that they are taking up particular standpoints. And it is not just a question of how the recorded data is interpreted. An informant in prison research remarked that when there was a disturbance, it was probably a cover up for something which was happening in the opposite direction (Tanner,1970). These ubiquitous machines have an underlying bias in their use. The communication value of a photograph is also to a very significant extent controlled by the extent of the cropping of that image in the same way that the written word is controlled by the operator of the pen or the computer. A close cropped photograph of an African American in Alabama in 1950 will have a different value if the cropping is removed to reveal a white policeman. In such situations there are always some base lines of usable reality; this what the camera saw and this is what was said into the microphone. Two anthropologists, the photographer and a specialist in religious rituals. Victor Turner (Turner,1967) looking at a series of photographs which this author had taken of the initiation rites of the Sukuma form of the Buchwezi 'secret' society in Tanzania and saw different things; such 'facts' have to be interpreted. Even if we accept that photographs come from aimed cameras, they are still dated evidence of certain activities of some behaviour and as such can be interpreted sequentially by generations to come. The earlier part of the 20th century had the Western custom of exchanging large numbers of postcards in their hundreds of thousands and these are available in large numbers depicting insider and outsider aspects of life in India, Burma and the Sudan as well as for North Africa and Egypt and elsewhere (Tanner,2003). No similar data is available in such quantities.

Even if these researchers can pick and choose what is recorded mechanically they often choose what is unusual in terms of their own cultural understandings which are perhaps concentrated around their own high social status as intellectuals. What goes on in a corner tea-shop everyday is much less likely to be photographed than a religious procession.

There is another important aspect to representation which is that social life goes on round the clock, rain or shine but researchers see what is going on according to their own pattern of shift working. A researcher who attends the all night recitation of the Koran over children who are about to be circumcised, is not likely to be alert in mind and body when witnessing what goes on subsequently. No researchers have the ability to record data round the clock although they can use the data provided by security cameras in many public places and even if they had accumulated such data they would not have the means or the time to analyse such a volume of information. As it is most research institutions have mountains of data which have never been used and are unlikely to be used except by accident. No researchers have the physical means to be in more than one place at a time and none are likely to be actively researching round the clock even for a single week. We may have to conclude that all data is sectionalised by the ways in which it has been collected and collated.

Fieldwork is always a very personal process in which researchers collect evidence from those who are or who are felt to be connected with what they are researching and this is a filtering process which is more accidental than planned. The researchers try and direct their enquiries and even if they do not ask questions, where they walk and whom they encounter is largely accidental; the researcher may walk and walk and then by accident come across a rain-making ceremony, a funeral or an angry confrontation. Researchers' faces and their body language, are always going to show their relationship to the data collected. The recorders of information are never neutral sources of collected data; there is always the central 'I' and that 'I' is always choosing what it sees; fieldworkers are only in theory omnivorous and are permanently confronted with tbe dilemmas of choice and chance. These research outsiders whether nationals or non-nationals with their academic training and professional aims of being detached from what they are observing, are particularly likely to represent the data which they have collected in ways that are alien to the understandings of the people from whom it has been obtained. In addition there are always the contributing factors of age and sex as well as physiological and psychological restrictions and enablements.

The most valuable representations may be the 'insider' researchers who write and indeed think in the same categories as similar persons in the single researched society; there is no pretence of detachment and when the roles clash, they are often significant enough to be recorded 'My recording and photographing were not considered intrusive as long as they were perceived as an extension of kinship activity' (Seremetakis,1991:2). These dilemmas are well put in a discussion of the work of Marcel Griaule (Clifford,1983:125). It is simplistic to tax Griaule with projecting onto the Dogon a subjective vision while developing a research method for eliciting essentially what he was looking for, and the even more credible claim that Griaule overstressed certain parts of Dogon reality at the expense of other assumptions, the existence of a natural entity called Dogon culture apart from ethnographic inventions. Even if it was true that the key informant became 'Griaulised', that Griaule himself was 'Dogonized', that Ogotemmeli's wisdom (Griaule,1965) was that of an individual theologian and that the nature of the revealed knowledge was systematically exaggerate. But it does not follow that Griaule's version of the Dogon is false. His field notes for the crucial interviews with Ogotemmeli are clearly the product of at least some rewriting, eliminating specific linguistic problems, the presence of the translator Kogem etc. It must be abundantly clear that the social science fieldworker is not a simple recorder of information.

7. The sources of social science methodologies?

With the hard sciences there are tried and tested research methodologies which are always in the processes of refinement which may have some personal impetus behind them but are not ostensibly influenced by personal factors; the data is not going to change because of human social factors during the research programme. When we apply the same practical framework to human behaviour, there are few possibilities of such exactness being replicated. Of course, we can go through sampling from an assumed static universe and compare this to another group which has not been researched. This assumed static nature of humans' behaviour has to be accepted as an illusion. Although a village survey of the composition of households in Guyana among a community which was not classed as migratory showed that by the time the parish priest had finished at one end, the composition of the households at the beginning had changed significantly.

There are numerous factors which prevent any exactness, biological and psychological human change for example, on what were individual differences to start with, and vastly different personal reactions to environments which themselves were constantly changing. There are high cultures about which an enormous amount has been written and for which there is a mass of material available to the researchers. It is hard to accept that the same methods can be used in low cultures for which little is known; such as the managerial elite living in Mumbai or the Naga in Assam. In the former situation there is always the possibility of the researchers being partially included in that society but none at all with the latter. They are not much interested in changing their social understandings and methods for inclusion into another pattern of behaviour and thinking and the researchers are accepted as more or less their equals and yet there are more or less the same problems as in the management of companies in an Indian city of status, rivalry, exotic complications and, life-long obligations in the assumptions of friendship 'the anthropological insistence on equal respect for all cultures contains implicitly the possibility of pursuing it as deeply as possible which means...intellectual equality. In working in a high society he cannot rid himself of the implications of potential or actual membership in the society he is studying' (Mead,1972).

8. So what are the sources of methodologies used by social scientists?

It will usually not be what they wanted to do but what they are enabled to do in the circumstances in which they find themselves and there may be very little choice in this. It seems likely that the non-national may be less adaptable to these changed circumstances than national researchers who may have less urgent time constraints from money and visas on what they may be able to do. It is also likely that most national researchers will have had some social and political lead into what they are about to research; there will be a supervisor, relative or friend who will link them into someone who is connected with what they want to do. This will ease their introduction to what they are about to do but it cannot affect the constant changes in what they are observing. A distant relative or a school friend in a government department may make the availability of data easier so cause written data to be given priority over the chase to collect parallel data and a more sensible use of time. A willing informant pushes the researcher to work with what is thus made available. All researchers are guided by the opportunities confronting them and the relative ease with which it can be developed. The Roman Catholic Church for example, keeps its parish statistics in a standard form which makes comparisons possible while to find out what they mean in social terms is inevitably a lengthy process.

9. Parallels in literary research

We have to admit that the social sciences have singularly failed to produce any acceptable body of interconnected and validated laws, but then the definition of science in the Oxford Dictionary does not necessarily imply any universalism or falsifiability (Machlup,1956). Therefore empathy and inter-subjectivity are essential for understanding in any scientifically acceptable sense, the culture bound utterances and social behaviour of the Barotse for example whether peasants or aristocrats.

The social sciences aim at proof and to get correlations at various levels of acceptability within any field of analysis of which the literature of any culture must surely be a significant component. The literature of any culture usually combines biographical works from or about distinguished citizens which may well contain biased alignments. When these biographies cross cultural divides there is usually careful editing within the restrictions of quasi-friendship, careful linguistic analyses of what a single person has said as with that of the elderly Navaho's philosophy (Ladd,1957) or brief analyses of distinctive individuals such as Turner's description of the Ndembu diviner Muchona the Hornet (Turner,1967). All of these might be considered methodologically suspect.

On the other hand novels carry with them descriptions of imaginative realities which social science fieldwork may miss altogether. Most cultures have large quantities of literature which can provide data about behaviour and their background thinking. Prior to Independence there were over three hundred novelists who wrote about India in English and from any early date there have been even larger numbers of novelists writing in their vernaculars. In every country in which there is a literary tradition this is a virtually untapped source of data and it is accepted in both India and Europe that novelists have drawn attention to many social issues in their works of fiction which might have brought about social action just because they were popularly read.

While the information which is disclosed in their writings must be methodologically suspect, they are tied to particular people who lives may well disclose the backgrounds of which they write. We may know as much about them as we know about key informants on which much fieldwork will always depend; there is usually a link between their imaginative writings and their own lives; clues which are as valid as assumptions which are made about the grounds for the opinions of informants. It may be that the point here is that the novelist, not the material that they produce is the point of interest. We can presuppose that writers and critics even within a single culture will be socially and psychologically remarkably varied as between those who are immigrants, men and women, more or less educated and those who write about communities other than their own. Fieldworkers know that the greater the cultural differences between themselves and those they are studying, the more that they are likely to miss and this gap can be filled at least in part by novelists. There are some parallels between research based on literary sources and the social science aspects of behaviour which have some fixed foundations such as changes in the formalities of marriage and the social atmosphere surrounding relationships. Proust, Goethe and Tagore who wrote about the societies in which they lived are thus fixed in time. However marriages dated according to legal requirements carry no information about the circumstances surrounding them or the consequences; literature can fill this gap.

Nationals reading the literature of their own culture have an advantage over non-nationals who have to use translations which must always be to some extent suspect because of a wide range of grammatical restrictions which cannot be passed from one language to another, personal choices of words, linguistic codings and cross-cultural misunderstandings. These difficulties are likely to be less with languages in the same linguistic group than between Bengali and English which have no affinities at all. Of course with bilinguals there can be fluency in another language in which there is no conscious process of translation going on, but what is in their minds will be the remnants of their genetic heritage and their socialisation. The alternatives to learning a language in order to read its literature as a background to social science research is a long process, which is carried out away from fieldwork situations and carries with it the probabilities of missing the semantic nuances which the national might be aware of from socialisation. A learnt language never carries with it any automatic understandings of its behavioural correlates.

But perhaps there is another aspect to research using literary sources when the topics are metaphysical and provide the background to much contemporary behaviour which provides the imagery for greater potential understandings. The process of reading into data conclusions for which there are limited correlations must always be a part of any qualitative research which cannot be scientific by its very nature.

10. Implicit restrictions on the researcher's role

Indispensable to any social science research is the predefinition of what role or roles the researchers are consciously or unconsciously performing. There is first consideration of their official position as it is uncommon for researchers to be acting on their own for a personal quest into something which intrigues them for which they have private resources or family connections which allows them to act independently. Freedom in research is very relative as most researchers are either salaried employees or are dependent on grants. In the former case as university employees, whatever their citizenship, they would be confined within their university's regulations, teaching commitments and sabbatical requirements. If they are employed to do applied research the terms of their contracts would limit the range of their enquiries to what their employers consider relevant. In the latter case researchers would have to submit a profile of their intentions and in their work show some ostensible conformity to their stated objectives. In all cases it seems that within a theoretical paradigm of intellectual freedom, there are practical restrictions on what these researchers may be able to attempt.

All researchers realise that before they undertake any research or soon after they start fieldwork that there are implicit boundaries to what they may be able to do or not do. These boundaries are dictated by the demographic facts of age, ethnic appearance, sex, physical abilities and almost as specifically the cultural habits into which they have been socialised. They are likely to find themselves comfortable or not with whole ranges of behaviour which are not part of their own cultural repertoires. There are also biological issues of diet, choice of food and the regulation of their lives through eating and sleeping patterns.

What is r searched in any country which considers itself to be in the process of rapid social, economic and political development is always influenced by two factors. The country's ideological base and socio-political structure which narrows down to what can be researched in professional and physical safety, which can of course be shown in a straightforward refusal to have such research undertaken by non-nationals and the warning off of nationals. In this there is bound to be a sensible self-restraint as no non-national would think it worthwhile to even apply for a visa to study voting patterns in Zimbabwe, the policing of Calcutta or high level corruption in Indonesia. Since non-nationals need to be affiliated to a local university, these national academies will not want to be associated too closely with politically dangerous topics.

National researchers cannot afford to be too specialised since openings for this in applied work are not going to be frequent. Fahim for example, wanted to study the religious ceremonies and social organisation of the Nubians flooded out of their ancestral homes by the Nasser High Dam (Fahim,1982) which clashed with the urgent need for research on the problems of resettlement; 'he who pays the piper calls the tune' applies to social science research.

Decision making in human relations and researching human behaviour is always based on their exposure to the explanations of outsiders but the way that these explanations are reached are not similar to legal and scientific judgements in which the evidence is weighed up according to widely recognised rules. In the context in which these conclusions are reached, impartiality may be aimed at but in fact they are no more than the embodiment of a viewpoint from a different level of understandings to that of the people from whom the data was obtained. Opinions are layered and those living in one layer will not accept the validity of the views made from another. In some ways human societies are based on the constant dialectic of differences rather than similarities and the researchers are producing unacceptable and very partisan views. The researchers are assumed to be on one side or another since the societies from which they came are also binary by their very natures, flourishing on the oppositions within them. Outsiders will always be seen as taking an outsider's view because there can be none which are not aligned.

It is necessary to accept that where the researchers live in a community will both restrict and enable certain types of information to be accessed. The anthropologist Gulliver's tent was pitched among the densely populated slopes of Mount Arusha and disorganised local grazing, but Malinowski's tent was however pitched near to the house of the Trobriand headman who may have wanted to keep an eye on him. If the researcher lives with a family, it is always a particular family and when they occupy a rented house, it is not only a question of who owns it but where it is located; there is no such thing as a socially neutral locality such as working from a guest house (Baviskar,1979).

11. The absence of relevant quantitative data

Social science research, as with all research, is divided between quantitative and qualitative methods but this indeed is an entirely theoretical choice between two generalised forms of accuracy. However from a scientific point of view qualitative is never accurate. In practice the choice is between working in societies in which conditions allow such a choice to be made with some realism and one in which the choice may exist but lacks any such possibility of accuracy or availability.

Quantitative research has to be based on data which can be assumed to be at least partially accurate and conforms as far as possible to statistical accuracy. The Western and North American countries have had institutionalised bureaucracy as the basis of their administrations for perhaps a century. Every citizen is documented from birth to death on paper or within the memories of computers which can be mined for statistical data. These are rarely 'accurate' as they have been manipulated for political aims or as a result of the boredom of the archivist. This data shows the birth, marriage and death dates of almost all their populations and is tied into addresses fixed to accurate maps. In fact there is a web of increasingly accurate and complex information available for most citizens on which quantitative analysis can be based assuming that its falsification would be either personal or relatively rare. However it seems that in the United Kingdom it is estimated that the number of people who live in a particular area can vary by up to 10 per cant (Economist,27 Nov. 04), so there is no absolute certainty that social science researchers in these Westernised and bureaucratised countries necessarily have more basic accurate information than in developing societies. The fact that there may be varying accuracies in such data is usually taken to be a matter for evaluation with each piece of analysis rather than that all such data based on bureaucratic systems have built-in inaccuracies. Nevertheless in these countries there are enormous quantities of information which social scientists can use to fix their individual objects of research into social categories which are not only the basis for their own work but usable for the comparative use of their results.

This means that any Western or North American social scientist can start their projects with the same categories of basic data of what they assume to be acceptable accuracy. It provides a framework of thinking which makes them often think that similar bureaucratically created data is available elsewhere. It also creates a framework of thinking which fits in with Comte's theoretical approach to the understanding of human behaviour in that it should be able to chart this as accurately as in hard scientific work. It is also an approach which fits in with the fact that most non-national social science researchers are 'come and go' in the field and thus may not have the time for the more elaborate preconditions of qualitative methodologies. Thus they go for approaches in which data is an accumulation on paper providing them with personal satisfactions that some progress has been made in the acquisition of knowledge.

In this respect the non-national researchers start their work thinking that there must be already a considerable amount of data available in the country of their choice comparable to that available in their own countries which can be used as a basis from which to work. It is almost safe to say that in the Western academic environment there is something written on everything that the social scientists may wish to know more about that has come from Westernised research. In developing societies by comparison to their population size, there is very little data at all and whole areas with populations of millions may have little known data available which even if it exists is historically dated and almost certainly qualitative rather than quantitative in their authors' approach.

Census data in such countries and indeed probably everywhere, is always of questionable accuracy when local authorities may wish to increase or decrease their official population numbers and secondly, large numbers may not wish to be counted at all as they know that its nominal secrecy will not stop their governments from acting on the results. In populations which are based on migrant work many may be recorded as sleeping in two different locations and all totals will have a declining accuracy with everyday that passes from the official counting. This means that in developing societies social scientists have either to rely on hard facts of their own creating such as the traffic and accident patterns in the Indian city of Patna (Singh,2004) or accept that their research will be the first in that particular field and area for which qualitative methods may be more appropriate. The point of this assessment must be that Western or Westernised social science researchers are addicted to quantitative data and methods having an accuracy which must be aimed at and used in all their fieldwork; the first steps in the long journey between the start and finish of a project. For them it is the scientific element in their fieldwork which as far as possible should have dominance. Outside the Western and North American environments, social science researchers find themselves dependent and indeed grateful for any bureaucratically collected data which they can find. They assume a degree of accuracy in tax, hospital and police records which is probably no more than an act of faith (Tanner, 1970;1983). The usual alternative is for them to create their own data by combining observations with their own common sense abilities.

A basic fact which social scientists try to obtain for all their informants is their age although it is no more than a reflection of the more important fact of their social age in their community. Most of the world's population have no accurate idea of their ages because it has not been recorded and probably came be fixed by tying it to a popularly known date like political independence, the assassination of an important figure or some natural disaster such as the south-east Asian 2004 'tsunami'.

Children are of school age for example if they can reach their ear over their heads and a wrinkled stomach meant fifty years of age and tax exemption in Tanganyika. Migration can be recorded as movement from the south or wherever but distances are just guesses using an unknown terminology in developing societies. Death dates and reasons for death are even more of an unknown quality since social and biological deaths may be understood differently. Marriages may not be registered with any authority although it is possible to do so and as far as statistics are concerned the five million Sukuma do not marry at all. In these circumstances a qualitative approach is perhaps inevitable and that the approach of social scientists has to be one of finding out as much as they can and from that in the end developing hypotheses even pretending to themselves that this is what they were initially looking for anyway.

12. The difficulties of the scientific approach

Whatever the background of the researchers there is a sequence in how their fieldwork unfolds which is perhaps easier for non-nationals to follow when there is a clear distinction between 'us' and 'them'. Unless the researcher is without professional training they will have created for themselves some professional assumptions about what they want to do as well as its theoretical background and that when they get into the field that they will find this ready for them to develop. Next the researchers will put the information which they have collected from published information about the field in which they are interested into conceptual 'boxes'. Once they get into the field they will begin to realise that these theoretical and data -based preconceptions do not quite fit into the information which they are now collecting and then the process of reformulation begins to make their own data fit into new theoretical frameworks (Joshi,1979). This process is not too difficult for the non-national researchers who have no cultural overlap with the community which they are studying. They may regard them with some affection but they are rarely seen as refractions of their own cultural understandings. The national researchers will have more of a struggle in recording concepts which maybe part of their own understandings. Many research projects attempt a scientific methodology and these researchers find that what they have attempted to record, has to be modified by the realities of field-work; the researchers and the human objects of their enquiries may have different agendas and dissimilar objectives.

There is always horizontal behavioural variability in how individuals react to a single and seemingly uniform stimulus. How the congregation react to the repetitive Muslim prayers, the less similar but still largely repetitive Roman Catholic Mass and the totally dissimilar Sukuma propitiations of their ancestors, will always have those who are involved in a primary sense and those on the fringe whose only involvement may be just being there in the body but not in the mind. Behavioural similarities at best must be a gross simplification of individual complexities which is done largely for methodological reason. Then there is always the time sequences in which there is a behavioural progression and it is a simplification to state that what happens at the first waning of the moon at a pagoda in Myanmar is the same as what may happen at the next month's practices; there maybe similarities but not ones of scientific exactness. Also, whatever fieldwork method that is used to accumulate information there will be factors of choice and personal biases, sectionalised viewpoints and always the need to recognise that much has been consciously or unconsciously missed.

What people eat has always been important as a background to the possible improvement in their health and welfare, but except under prison conditions it may not be possible to assess the food intake of individuals. Under field conditions it would be impossible to find out how much of what foods are eaten except under quite impossible conditions of following people around, particularly as children have a 'grazing' pattern of eating and feeding from the breast. In a study of squatter children in Jaipur (Goyle *et al*,2004) mothers of children were asked to recall what their children had eaten on a day-to-day basis using standardised units such as a spoon or ladle and by assuming that there were standardised recipes for certain cooked foods. In this careful study we see the difficulties of collecting scientific data about the eating habits of people who may not have had any organised pattern of eating. It has to be assumed that to collect more scientifically accurate data would have been impossible and there must have been many variations in what the children actually ate and in the cooperation of those mothers. It seems likely that these researchers adopted a more qualitative approach.

There has to be an attempted balance between the two approaches satisfying the scientists in order for their studies to be found acceptable while conforming to qualitative understandings. In studying the seasonal food intakes of adolescent girls (Deepa *et al*,2004) there was scientific accuracy for what was biologically measured but other than referring to a daily recall of when they had eaten, there were few details of what they actually ate. This is particularly important for example in the extremes of festival eating which in Ethiopia caused so much stomach upset that it came to have a clinical title of 'injira' gastritis (Knuttson and Selinus,1970).

How far in fact can social science researchers fulfil scientific requirements? The difficulties come under three main headings. Firstly, the person and personality of the researchers in what is after all a social activity. Some men and women are unsuitable for getting information from people and are better adaptable to abstract work with inanimate factors such as statistics and paper records. These researchers are as subject to as many personal variations as any three people whom they study. There are of course no possibilities of correlating this to national and non-national researchers without going further into possibilities of cultural biases. Secondly, the social situations with which researchers are confronted and in this a detached view would tend towards thinking that the greater the social distance between researchers and researched the greater the difficulties of understanding qualitative rather than quantitative data. Thirdly, the fluctuations in the interrelationships of the first two categories on a day to day basis which could have been foreseen such as the ageing process in long term research, climatic changes, the agricultural cycle and the getting the 'getting to know you' processes. There are also the unforeseen factors of ill-health, mood swings in both researchers and researched, civil disturbances, local obstructions and personal dislikes. Overall scientific exactness and the categorisation of human behaviour do not ally well to the realities of fieldwork. As a rule of thumb it might be wise to assume that the decimal point suggests inaccuracies which should be assumed to exist and should be investigated by qualitative methods.

Can the weakness of much quantitative data on human behaviour be overcome by the use of modern mechanical appliances? Many researchers carry around cameras and recording machines of increasing sophistication during their fieldwork as well as apparatuses for measuring physiological changes; they are all materially finite as to what they can record and the possible correlations are limited. It is very satisfying to have amassed such material data as signs of progress in the research being undertaken, but it is all unidirectional and bounded by their own mechanical limitations. Then there is then the additional difficulty that they record a mass of data which parallel the realities of any social activity involving one or more persons. The researchers now have an overload of information in which there is much more to be understood than the simplicities of their primary factors; the façade covering the multiplicity of psychological and social involvements. Finally, there is the scientifically questionable methods used to analyse the massively complex data of a long interview, the recordings of a rite of passage or the photographs of a ceremony. Each viewer or listener will reach different conclusions and the greater the amount of data, the grater the differences. Masses of data are perhaps beyond the capabilities of single researchers to analyse and group analysis is time consuming and would not necessarily be any better related to the data.

Of course, in quantitative research there is a certain safety but at the same time a lack of emotion which must be part of any appreciation of the lives about whom social scientists are writing. Reducing human behaviour to statistics is in fact taking the human dimension out of what has been researched. The social scientists are under a professional obligation to be as accurate as possible and under a parallel pressure to express as much as they can of their data in numerical form. We have suggested that human behaviour is so varied and subject to so many factors under constantly changing conditions, that numerical accuracy can only be imposed on material by a number of methodological devices. The legacy of Comte has not been fulfilled to any great degree in comparison to the amount of effort which has been put into trying to work out the future of particular types of behaviour except in the very broad provision of trends to which scientific methodologies may not be well adapted.

Nevertheless positivism has a seductive appeal as in theory providing logical and invariable sequences which are immediately satisfying and in the process allowing the researcher to retain a sense of achievement. One of the principal difficulties is the amassing of much more data than they are ever likely to or indeed be able to use in an average work-life. Now onto the methodological scene has come the hand held mini computer into which vast amounts of data can be input turned into analyses in mille-seconds. It would seem that numerology will become more prominent in social science work because it is so easily available and receive professional recognition because it is hard in its visibility regardless of the weakness of its origins. It becomes hard data by the nature of this methodological process and not because the nature of the collecting methods has improved enough to make scientifically acceptable conclusions more easily provided. Computer generated data processing is producing sequential results whereas human behaviour is always contingent upon a wide variety of coincidental inputs. Researchers have an emotional need to pass from feeling and belief to accuracy but in the relationship between facts and conclusions it may be an augmented illusion.

It has always to be kept in mind that social science researchers seem to assume as part of their professionalism that they are impartial. In addition it is probably a quite conscious delusion of non-nationals that they have a superior ability to achieve this as part of the European Enlightenment tradition and the positivism of Comte. It is hard to assess the extent of a national researcher's impartiality since in some cases because his involvement is for status rather than professional reasons, or because his participation in the village has made him more than a researcher (Srinivas,1979) and in another situation he had become more than a participant in expressing a view about a land dispute (Chakravarti,1979). So it would seem that both nationals and non-nationals have the same problems of over-involvement with the social interests of those they research. We may conclude that impartiality is always an illusion.

13. Negotiating multiple viewpoints

It is reasonable to accept that social life is full of co-present dualism (Simmel,1971), public and private, rule and practice, freedom and constraints and form and meaning. By meaning (Rapport,1996) we can include purpose, interest, motive, drive, goal, desire and by form the mode of interaction through which and in the shape of which meaning attains social existence and these forms in language and systems of classification, categories and concepts, the synthesising processes by which individuals come to associate with one another. So in effect there are two aspects to consider in any analysis of social behaviour. The mega-structures which would appear to dominate much social science thinking—what they see as the collective realities of social life creating continuing social facts and the very diverse behaviour of individuals in small group situations which may or may not coincide with the social systems in which they are dominated (Wallace,1961:1962). The richness of social life in which the behaviour of individuals cannot exhibit much overriding consensus is not shown in the grand monotheitic theories of macro-structures. Those researchers who have a deep personal understanding of the people and communities in which they have been working find complex systems governed by circumstantial norms and complex systems of changing values and beliefs.

This was shown very clearly in the Sukuma approach to the writing down and publishing of their customary law (Cory,1953). In the prolonged discussions which went into the creation of this work, the thinking men who were consulted agreed the principles on which the solution to quarrels were based; they in fact agreed to the existence of a mega-structure of Sukuma customary practices. But when it came to dealing with cases which came up for appeal from the local courts controlled by the Sukuma themselves, similar elders who were used as assessors might well say that they agreed with the principles behind the judgment, but in the particular case being considered, they would interpret the principles in terms of the particular circumstances of the case. The outsider might well conclude that they are breaking the rules which they themselves agreed on, while those elders would have seen themselves as being entirely logical; they are micro-minded as indeed most people may well be in most of the situations in which they find themselves. This is an example of Simmel's co-present dualism. Even if we accept that there is a division in the social sciences between sociology focusing on groups and psychology on individuals, there is always the problem of individuals being in groups and that of individualism as a deviation from overriding group norms rather than just behaviour which is individual. Only Laing is prominent in suggesting that mental illness is the result of purely individual experience (Laing,1968).

If we can compare the groupings of social science as to their ability to recognise individualism rather than deviance from existing established group norms, there is the conditioning fact that most if not all social scientists are taught in terms of mega-theories in which there is little space for idiosyncrasy and individualism. For example, people consider themselves to be patri- or matrilineal whether they are nationals or non-nationals they are prepared to classify people in a similar vein. So whether they are nationals or non-nationals, they approach fieldwork prepared to group behaviour in the societies which they study and they indeed they find this easy to do.

To go further than this is to recognise that even in the smallest social groupings, there maybe certain common themes but that in most the behaviour of each inter-personal situation will rest on social environmental factors which can never be the same except in the crudest categorisations. In perception of these factors the national researchers will start with some understandings of local behaviour usually based on a shared lar guage into which they have been socialised and they will have a head start in seeing some behaviour as idiosyncratic. On the other hand the non-nationals even with the greatest willingness to learn and absorb the ins and outs of the cultures which they are researching, are never likely to know enough about what goes on, or to see some behaviour as socially acceptable in particular circumstances rather than as deviant and both deviant and unacceptable in other ones. Most research projects are small in scale and the researchers have roles ascribed to them by the communities in which they work. Once this has been done or negotiated, it is a static ascription which is only going to be changed to any greater degree of personalised acceptance because of frequent visits; a researcher may become our researcher. Achieved status turns into a form of ascription.

Whatever comes out of these covert adjustments the researchers will always remain of a certain age, sex and appearance as for instance on the East African coast a researcher's moustache was taken to correlate to a militaristic disposition. All or some of these characteristics may determine the roles they are able to establish for themselves when doing fieldwork which will also determine their access to information (Gronhaug, 1974:14); there may be some latitude in this but not much as they will always be of a certain age and sex. The researchers still have to negotiate and allow for the working out of their roles in different situations and this of course has always to be related to the basic fact that there is no single viewpoint in any assessment of a community's behaviour; it is a network of constantly changing ascribed and achieved roles.

A young male researcher will feel more at home with people

of the same age and sex although it seems unlikely that these demographically similar people will feel at home with him; relationships and viewpoints with some but not others has to be worked up. Why should such a person who is an outsider by any definition be acceptable to those who have no shared similarities in any of their more personal characteristics. Why should they expose themselves to those who are social strangers? Researchers always have the difficult job of negotiating and reporting on multiple viewpoints; for the latter they may have some professional training but for the former none at all outside the societies in which they have grown up and been socialised into its methods. Moreover most of this will be outside the range of any interpersonal negotiation.

Much of what researchers find out will come to them from those who are not directly involved in what they are passing on such as women, adolescents and children commenting on the society surrounding them. And, of course, this is the only way that they will find out anything about the more personal and secretive aspects of communal life. The claim by some social scientists that they have become members of specialised even secret groups with equal standing to the 'natives' is a polite fiction rather than a social reality. Much the same would apply in reverse to a young woman researcher who might have easier access to young women and children but we know from Margaret Mead's early work in Samoa that there are pitfalls in such working associations (Freeman,1983).

We have to 'acknowledge' that the process of translating the complexity of other people's lives and cultures into a seamless narrative takes enormous self- confidence, even intellectual conceit. 'At times it resembles a form of fiction writing when we suppress parts of what we think we know or force these insights to fit a hypothesis we have nursed too long to abandon (Sheehan.1993;78). How can these multiple viewpoints be handled between the poles of numerical listings and the reporting of specific dialogues, both of which leave the readers to reach their own conclusions as to the meanings attributable to numbers and the words of specific people. Numbers have always had an attraction to social science researchers battling with a fog of data in which there are few foundation facts; the columns of figures simplify the complex.

To record what particular individuals may have said involves editing. How else could this author cope with fourteen hours of reply to the single question asking a missionary priest in Tanzania 'what do you do'; however, this will be edited to illustrate the specialised lives of such people, most of what he said will remain archival material to be mined by others. Such informants who give lengthy explanations of what they do have overt or covert reasons for doing so and it is rarely likely to be in any way related to a disinterested wish to further social science understandings. They have views they want to put over and in the case of that missionary priest, he was using the researcher as a propaganda channel to the outside world. Informants can also just be talking and discussing for their own entertainment in lives that may otherwise be rather dull and without the diversions provided by gullible resident outsiders.

But in terms of understanding a community or institution in which everyone has multiple roles what the researchers' record must always be biased assessments; such interpersonal dialogues are not and cannot be from stratified samples. It is a hit or miss form of interpersonal opportunism and certainly hopefully a form of realism. There is always the ultimate factor that the researchers have to pick and choose what data they are going to use and the conclusions which they want to draw from the partial information which they have collected and more importantly which has been offered to them by biased role involved informants.

Also, social scientists may have lost faith in the core belief about the reliability, morality and usefulness of their own efforts (Shokeid,1997). There is the constant difficulty of biased sampling and apart from the methodological shortcomings of recording information from odd people, so that their publications can be valued as providing information of what may or may not have happened under very local dated circumstances. Should any social scientist visit a law court in a developed society they would have ample evidence of how information is abbreviated and changed to suit particular circumstances in which individuals are personally and professionally involved. Viewing those that give evidence there is no reason to suppose that if any universal values do exist, that they apply to informants.

There is another factor which may complicate the possibilities of reaching detached assessments of a community or institution. Fieldwork by social science researchers is no longer uncommon and indeed near to universities they might even be assessed as too common since communities are visited by students doing their exercises in practical fieldwork. It is necessary to know to what extent those researched have already had experience of being the subjects of research. Some American Indians have had regular visits from researchers and indeed one Zuni community even had a permanent room in one house reserved for the use of 'their' anthropologist. It is said that Coorg intellectuals even knew the anthropological terms describing their matrilineal customs. It is necessary to know the extent to which those questioned already know the 'correct' answers in the social science jargon to the questions asked about their behaviour (Varadachar,1979).

14. Social scientists representing themselves

Whatever their origins both national and non-national researchers do not come immediately from poverty stricken environments and it seems unlikely that they will have had any personal experience at all of living on the margins of survival on a day-to-day basis. Broadly any fieldworker will have some control over their own lives which are not directly dependent on the vagaries of nature or influenced by political violence as to what may or may not happen to them. Most subsistence farmers or unskilled itinerant workers are profoundly fatalistic and recognise that whatever efforts they make, may be futile as their work and prospects may be destroyed by locusts, plant diseases, floods, droughts and fluctuations in global market prices apart from the knock-on effects of broken communications; they are committed to the problems of day-to-day survival. All researchers particularly the majority who come from comfortable circumstances may have considerable difficulty in understanding the erosive effects on human well being of this persistent uncertainty and their malnourishment. It is prudent to accept that researchers may think unconsciously in terms of their own assured backgrounds in which there is only long-term uncertainties. There are of course exceptions such as Joshi who was haunted by the living conditions of the people amongst whom he had done fieldwork and he saw this as a moral commitment to their betterment rather than just research (Joshi,1979).

15. Micro and macro distinctions

If we can presuppose that the degree of accuracy in fieldwork is mainly due to the closeness of the professional social scientists to what they are studying, it would seem therefore that social relationships which they have achieved are likely to predominate. In these circumstances the researchers can be physically closer and more persistent in their enquiries but nearness involves the complexities of interpersonal relationships and the intricacies of bias so perhaps it is easier to negotiate for wider macro studies in which conclusions are more qualitative and as far as they go easier to validate and defend. In micro studies questions of the various forms of participation and personality factors in inter-relationships are paramount but the larger the scale of the study, the more inexact become the factors under examination and the national researcher would have less advantages over the non-national.

Macro studies can be accomplished with a combination of pre-fieldwork learning, the use of library resources and national archives as well as fieldwork in the project areas. The study of a community or tribe which in terms of its numbers and where they live can no longer be confined to a specific area nor indeed have any finite social or geographical boundaries at all, may be much easier to undertake or complete. Macro studies would be more impressive by definition since the numbers involved might well be in their millions.

The result of this approach might be that leading members of the communities involved will complain that not only were the researchers outsiders, but that their knowledge cannot have been thorough enough to support the conclusions which they have made. This was the case of research into Quebec nationalism (Handler,1993) in which the critics refused to accept the validity of the premises on which the study was made; in this respect it is a no-win situation for the researcher who was looking at the issues within the framework of what was internationally accepted rather than in terms of what these nationalist thinkers expected and wanted. In such circumstances it would have made little difference whether the researchers were insider nationals or outsider non-nationals.

The descriptive space of the social scientist in a micro fieldwork situation is not an appropriate context for working out the conceptual problems involving the larger surrounding frameworks, even when their research methods put a methodological line round the social group being studied; there are always profound influences moving backwards and forwards over such boundaries. But difficulties arise when there is no such possible theoretical boundary line as in almost every aspect of their lives, there are substantial links or influences which have not been initiated within that society. To follow a single theme or to treat the outside influences as a background (Marcus,1986) may not be an adequate solution and what may be necessary is a sustained tolerance of both incompleteness and indeterminateness of what lies beyond the greater accuracy of micro understandings.

A difficulty with macro studies is that their title suggests that the societies studied are self-sustaining and impervious to surrounding economic, political and social influences; they are not self-sustaining in the way that the Easter Island community may have been for several centuries. The influences of the state on micro communities has to be considered as the largest if not the most dominant element in local organisations and sub-cultures (Cohen,1982). Much of what occurs now in macro as much as micro communities is based on techniques, equipment and political values and motivations come from outside the boundaries of any research project (Vidich and Benshan,1968) and this applies more now than ever before.

16. Questionnaires and tests

The application of a high proportion of social science research involves the use of questionnaires and tests; it could be suggested that it is one of the fundamentals of social science fieldwork that it is possible to get reliable results from questionnaires and tests of both literates and non-literates. There is always the problem of the language and the format in which questions are asked. In a questionnaire which was administered in nine different languages to very large numbers of South African children aged 10-19 years (Anderson *et al*, 2004), it was noted that the phrase forced sex without consent as referring to rape did not exist in some languages. Firstly, if the word does not exist can we not conclude that the concept does not exist either; certainly in some communities to the south and east of Lake Victoria in East Africa violence is quite legitimate in initiating sexual relations. This is an example of the questions which inevitably arise in translations when what is wrong in one culture is transferred to another; there are always moral choices in what is used for a word across cultural boundaries (Tanner,1993).

It is prudent to remember that no known society has carried out formal questioning except as a response to the needs of bureaucracy or a public enquiry; it is imposed on communities which in their own social systems do not have direct questioning as an acceptable way of obtaining private information for a public purpose. Most people in all contemporary societies are suspicious of questioning of any sort and particularly so when it is done by an outsider with the results being recorded on paper. It is the taking away of information which is always seen as private property without any material return.

Most societies have tests as part of the move from one age grade to another and their aim is to aid the continuing incorporation of people into an existing social system. Perhaps even the word test is wrong for it is rare indeed for anyone to fail in what is required. The aim of tests is to find out what people might or might do in certain circumstances from which results individuals are put into discriminatory orders must always be seen by custom as a socially alienating process; of course societies put their members in informal sequences but not orders in which there are standards of success and some are publicly seen as failures. Thus tests even more than direct questioning are not something which fits at all into how people feel that they should conduct their lives. It is yet another imposition from outside.

Since most tests and questionnaires have come from Western social science methodologies and are seen by those to whom they are applied as verging on the ridiculous and done more as a favour for the researcher than because they see any sensible purpose in what they are asked to do, even if its purpose has been carefully explained to them. The following examples of socially inappropriate nature of Western methodologies resulted from this author's own fieldwork. The Rorschach ink blot tests were given to adult Sukuma who were asked to detail what they saw; the universal answer was that a multicoloured mess represented nothing at all. When pushed to answer whether they did not associate with red with blood, they replied that it was similarly ridiculous as blood almost at once became black. The Lowenfeld Mosaic test in which the Sukuma again were asked to make a design with different coloured and shaped pieces of wood cooperated, but saw no sensible reason for confining their design to the provided board of the size required for standardising the test and thus continued their designs onto the surrounding earth where their designs could not be traced onto paper. A questionnaire given to literate Swahili on the Tanzanian coast asking them to put each of twenty occupations on a status scale of one to ten, resulted in some thinking that there was a winning combination which would get them a prize. The Ishihara colour blindness test was given to the Hangaza, a Bantu community divided by the international boarder between Ruanda and the then Tanganyika, found high proportions who could not distinguish between red and green; a finding of great biological interest. When the same test was given over the border to people who were socially and biologically similar, the proportion was much lower. The answer being that in the curriculum of the schools instituted by

Belgian educationalists colours were taught but not on the British side. Finally, a test to find out what Tanzanians would do if faced with certain moral dilemmas, asked the question, 'which would you do?' often received the reply that 'I cannot choose as both are wrong'. A sensitive missionary asking their Christians what in their opinion is wrong with the Church today did not get replies that referred to the results of Vatican II, but the answer that the roof leaks or that there were not enough benches. The list of tests and questionnaires that have gone wrong because the ideas of one culture and that of social science do not coincide is probably endless. It is prudent for non-national and even their national colleagues to think very carefully as to what answers they are likely to get from questioning and testing and whether they are likely to coincide to any safe degree with their research objectives.

Rarely if ever is consideration given to the possible futility of attempting to extract exactness from information supplied in answer to direct questioning and questionnaires. Such behaviour if it exists at all under 'natural' conditions must always be related to the interpersonal and situational conditions under which it is supplied. An extreme example of this would be food intake which nutritionists attempt to weigh and detail all food taken by individuals for a fixed period which is impossible unless each individual is imprisoned. This means a great deal of effort is made to find out information about a very small sample and it is of course not known the extent which these people are average; this is time consuming and virtually useless except in the satisfactions which such work brings to the organisers; alternatives make get equally adequate results (Tanner,1956).

17. Fieldwork, documents and interrelationships

It is often assumed that bureaucracy and its products are conspicuously part of economically developed societies and it is only there that documents can be found that might aid research. This is not entirely correct and such records have to be searched for with imagination. Even in societies which are on the fringe of effective economic and social developments individuals and some groups see the advantage of keeping paper records of important events; these bits of paper enable and prevent even if their possessors cannot read what is written there, so that having these pieces of paper have a profound effect on their owners creating a talismanic effect of documents associated with the difficulties of modern living. Officials may ignore individual petitioners but paper documents have connections which are difficult to ignore; in social terms they are often the long-term tools in on-going interrelationships. Tribesmen in both Tanzania and north Burma will retain in bamboo tubes the judgements in personal and communal disputes given to previous generations which they cannot read but they know will be read by others; fragile reminders of past tensions.

All governments tax and keep records of receipts and what they hope to receive and these list names and locations which may disclose both religion and migration; addresses are listed and even if these cannot be located on maps, they indicate population densities and locations. Many people retain their tax receipts and licences for many years to prevent police harassment and as proof of identity and residence. What records there are in developing societies are kept until they disappear from the effects of climate and insects. Medical services keep lists of personal details for most of their functions which show age, sex, domicile and occupation and give the social needs of people for mass and degenerative diseases. Often doctors keep their own records of matters which interest them. This author's own fieldwork revealed a midwife in Mombasa, Kenya who kept records in her own handwriting in a single ledger of every birth in her clinic for seventeen years; an invaluable sequence of some ten thousand births. Religions are particularly likely to keep records. Many churches and shrines such as the Meiji Shinto Centre in Tokyo have boards on which petitions are place which disclose the nature of the distress and who is asking for help. Baptismal certificates are given and in Ngara, Tanzania Roman Catholic records showed that many twins were baptised in anticipation of their deaths so that expensive traditional rituals could then be avoided. Places of pilgrimage which involve pilgrims crossing international boarders keep records. The curators of temples often provide certificates of attendance there. Some independent churches in the Philippines keep close records of those who attend services on much the same lines as punching a time clock in a factory. National researchers with their superior knowledge of how the local bureaucracies work and of the sorts of records kept by individuals and families may be better able to find the paper trail which links individuals with the imposed bureaucratic structures which surround them.

18. Quantitative lineal research

The methodologies of the social sciences are ones that follow those of the hard sciences based on inorganic materials in which one defined event follows another and as far as possible the sequences are isolated from all other factors which are not considered to be relevant. Thus any piece of research is an exercise in logical accuracy and restriction. Human social activities are always immensely variable between individuals who have only coincidental similarities as persons and even more so in the environmental and social situations in which they find themselves. Social science involves the categorisation of human behaviour in as closely defined boxes as can be created which are usually based on the extreme dimensions which are stereotypical. This creates problems over the placement of data in the grey areas between these extremes and where the boundaries of behaviour should be placed in the first place. This author's fieldwork also uncovered a Tanzanian census enumerator who on being faced for the first time with the form on which sex was to be entered for each person counted, asked where he should put those people who were neither male nor female; he was worried over those whom he knew who were either or both biologically and socially inter-sexual.

Most social science research requires ages to be noted which for a high proportion of the world's population there is no documentary evidence and even when this is available the boundaries between one age and another as regards behaviour is entirely arbitrary. A distinction between 10 and 11 years of age may be clear in the middle of each year but the differences between children in a 10 year box who are 10 years and 51 weeks and those in the next box who are 11 years and 1 week is methodologically obscure. The list of such methodological problems is endless and is a feature of all quantitative lineal social science research.

This is the methodology taught in the Western educational system or in universities modelled on this same pattern in Asia or Africa which dominates social science research and thus most if not all social scientists trained in the tradition of the European Enlightenment and with the ideology of Comte lurking in their minds, will start fieldwork within the mental and institutional confines of this approach. It is a methodological system for examining human behaviour which has to exclude many more factors than it includes.

Initially data collection confines itself to primary factors and excludes the fainter relevancies on the fringes of more direct influences. It is often successful in what such researchers set out to achieve by providing as definite as possible answers to limited hypotheses based on carefully circumscribed fieldwork. Such fieldwork fits into the requirements of scientifically defined lineal research in which there are primary relationships between events and their instigators. As an oversimplification this is the railway wagon shunt approach in which one truck is hit by another both running on lines and with loads which have been provided. Thus this becomes an extraction of one reality from a situation which contains many more than this one form.

But there are always 'left out' factors either deliberately or because they complicate the research design; where in fact do you stop? The thorough and erudite study of Navajo thinking from interviews with a single elderly man (Ladd,1957) illustrates this problem of limiting boundaries in which the informant is isolated from the circumstances of his own long-life and the effects of relationships with his fellow tribesmen and the researcher himself. This is not a criticism of such research works for what they describe but what is not described. It is not just the fact that there are such boundaries but the impossibilities which grow from extending these boundaries because of their assumed relevancy. If a competent social science researcher can write hundreds of relevant pages about a single interview, then how much more would be required from a wider picture of this man's life and that of his community, let alone the 'white' dominated outside a social and political world. So it is not just a question of what they have left out for methodological reasons but what would be the consequences of including more data; the macro swamping of the seemingly efficient micro project.

But even in the small circumscribed carefully designed project there are always lurking unknowns; the 'left out' factors which make research results subject to doubts which are likely to occur in any research projects involving live animals or indeed humans. 'The equitable (air-conditioned) laboratory climate houses banks of computers capable of delivering an astonishing variety of sensations and intricate recording systems that measure, graph, correlate and analyse the eagerly awaited responses. But despite all the help that modern science can provide, even the reinforcement schedule of individual laboratory rates are not that easy to regulate, nor are their responses a cinch to record. A researcher discovered this when he realized after a semester's work that the 'interesting' results he was getting were highly correlated with systematic fluctuations in the city power supply into which his equipment 'was hooked (Bochner *et al*,1975). This is not just an interesting one-off discovery vitiating the findings of one hard working researcher but one which is symptomatic of most lineal research. The artificial nature of encapsulating any behaviour into a hypothetical or artificial environment created for the purposes of any particular research. Humans are far too varied to have conclusions on their behaviour artificially restricted in such ways.

Quantitative lineal research is thus not only by its nature circumscribed by its initial boundary-making definitions but is also tied to the time and place of its observations. Further, in the context of the developing and rapidly changing societies which predominate today, they are expensive to carry out in time and material and thus restrict the amount of data that can be made available. 'Scientific knowledge is perhaps the highest development of the linear mode, but the linear mode is only one mode possible to us. We find another mode of consciousness manifests itself culturally, personally and physiologically' (Ornstein,1972:96). We will now look at this in the context of qualitative research.

19. Qualitative assessment

Every researcher whether national or non-national forms opinions of the situations in which they are involved and moreover does so from the very first moment in which they are in contact; these opinions are constantly changed as the work progresses and they find out more about what goes on and they carry out evaluations leading to their conclusions. Since all research results are to some extent inaccurate and 'there are no whole truths; only half-truths. It is trying to treat them as whole truths that plays the devil' (Whitehead,1953), it is useful to consider qualitative assessments in the contexts of their possible value. If we take the example of a South African national study of the views of school pupils on sexual violence and the risks of HIV infection (Anderson *et al*,2004). This was a questionnaire based study in nine languages involving 269,705 pupils in 5162 classes in which every questionnaire was returned completed. Statistical responses were detailed to two decimal points for both boys and girls. What we have here is enormous effort and expense being put into evaluating the written replies to an anonymous questionnaire in which the answers could not be open-ended. The writers admit to no knowledge about the extent of exaggeration or peer influences. It is therefore no more than a study of the received written replies concerning a very personal topic in which no attempt was made to guess at how these replies may have related to inhibitions and the literacy abilities of the pupils. Apart from the difficulties of any questionnaire involving personal matters one can certainly question whether the results were worth the enormous effort and cost. A qualitative assessment from studying small and varied samples would have reached more detailed and perhaps useful conclusions.

20. Qualitative non-lineal research

Although most if not all people whether in quasi-traditional or urban-industrial social environments are logical in many aspects of their lives, the paradigm within which they see their own lives and the lives of others is not based on any such limited number of cause and effect situations which are related to immediate circumscribed events. In fact it seems likely that it has always been the case that individuals see their own lives and the lives of those with whom they are closely associated in terms of 'wholes', not itemised divided up elements.

In traditional thinking everything can affect everything else; there is rarely if ever a system of categorisation dividing up experience. It is not that these thinkers have rejected categorisation, but that they have never developed or rather never seen the need to develop professional categorisations with as sharp boundaries as they can create. They do not define the nature of the influence or nature of spirits or God(s) nor the benign or malevolent influence of their relatives or neighbours; these influences exist so that the definition is part of the concept. Categorisation is the tenet of a new approach to analysis resulting from a combination of literacy, independence from subsistence realities and the European Enlightenment of the 18th century. It is not part of what might be called 'normal' experience which includes rather than excludes all forms of experiential reality. Quasi-traditional thinkers in parts of societies which are only partially influenced or invaded by urban-industrialism, retain a perspective of wholeness in their understandings of existence. This has always been the understanding of Eastern philosophies and is also the approach of those involved in altered states of perception. Thus the approach as to how these people view their current lives and the difficulties in which they are involved is a permeable whole which can be edged into by any number of factors.

So research based on any lineal concept of cause and effect founders because those involved will always see an event not only in the framework of Western scientific lineal thinking, a fall from a tree is obviously caused by the branch breaking, but always their framework of 'why' will go far beyond any such primary cause reasoning. Their framework of reasoning takes in many and varied causes in combination in some respects a way of thinking and inevitability that predates DNA networks. In fact a thinker from a background outside the restrictions of lineal thought whether a quasi-traditional, semi-literate or sophisticated Eastern thinker in a centuries old school of philosophy, might well comment that he or she knew this long ago, so what's new.

Much of this thinking about the wholeness of existential being belongs to altered states of consciousness that many cults in Western societies have partially imported from the ideas and practices of Zen Buddhism (Suzuki, D.T. 1958:92), Sufi Islam (Le Fort,1966:105), American Indians (Castenada,1968:130), and other mystical systems which have existed outside the Western world for a millennia or more where they have been an accepted way of thinking within frameworks of primary realities. Similarly, within the ideas of spirit possession which are held in most societies outside the developed Western societies, individuals weave their lives in and out of contemporary realities and less substantial ones which for them have a pragmatic reality of equal importance to the narrow realities of pragmatism. In all these systems of thinking the reality of such beliefs is that the world of the mind and the world of what the Western world would consider to be reality, is a single unity within which people can chop and change either through learning or by invasions from one frame of reality into another. This is very alien to the framework of lineal thinking and should perhaps be assessed within the framework of 'straight' *versus* 'stoned' thinking (Weil,1972), in which its straight aspects predominate.

Spirit possession is not explicable as psychopathology since a high proportion of people believe that the other world is a reality, not as a parallel world but as a single unity. Certainly a recognition of a extra-humankind but part of a complex series of ideas and practices of a relatively integrated kind (Firth, R. 1969) which in many societies are made real in their constantly updated theologies. If the framework of metaphysical thinking is so porous, then the dependent approach to causation which must be considered is not only historically multifaceted but also, in general terms, far broader than the usual restrictions of lineal thought. Thus it would seem that the analysis of events would have to take in not just the possibility of multitudinous influences stretching backwards in a broadening pyramid of influences. But, more importantly also that these influences are not necessarily sequential but are acting all at once all the time with the instantaneous speed beyond that of the most sophisticated computer. It would seem therefore that a better approach to the analysis of human behaviour might be that the social science researcher should have a framework of thinking with those who hold a pattern of understanding which presumably antedates that of scientific restrictive methodology by many millennia. There can be no methodological confining of what humans think and do; the penumbra of influences should be accepted as a necessary reality for any understandings initiated by research.

21. Qualitative research and the novel

Once we are involved with qualitative research in which we record as far as we are able the thoughts as much as the behaviour of others, we are moving over the border line from a possible status of detachment into the arena in which the researchers' personal feelings must always be involved. Qualitative research can never be detached and its value lies in the extent to which it mirrors the rawness of reality. So it is perhaps the novel and autobiography which can express the realities of the lives of others rather better than any social science researcher.

If the power holders in any society, the politicians and senior bureaucrats are going to be influenced by reading anything about the problems facing their countries, it is going to be in competition with the realities of politics. On stage relationships have to be maintained in the spare time left-over from an overload of official reading, no matter how effectively these are précised by the support establishments of each politician and senior bureaucrat. This means that at most they will read one page summaries of complex and prolonged research work and only then if these senior bureaucrats thought it was related to the immediate political lives of their bosses. These reports make unattractive reading at the best of times and are hardly likely to be read because of their intrinsic attraction.

Certainly there have been research works of great importance and influence such as the detailed study of discrimination in the United States by the Swedish sociologist Gunnar Myrdal (1944). This illustrated the extent and consequences of colour prejudice and certainly influenced that country's intellectual and power-holding elite that it was no longer politically tenable in an internationalised world to maintain such practices by force of law. It started to move the white voter supremacy away from a political and social system which was no longer tenable. It is interesting to compare Charles Dickens who brought to the public eye a number of important social issues in 19th century Britain which might otherwise have gone unnoticed politically for much longer. For an understanding of political and social issues a power-holder is more likely to pick up and read something that is geared to a popular and up-market intellectual readership and which is thought to have an important and interesting coverage as with the Myrdal study on discrimination.

We can assume that popular novels with social themes must have some influence on their readers, but this may be an individual influence in most cases. Researchers are likely to read all that they can on the matters with which they are concerned and they are influential because their themes and style of writing is evocative. Beyond this novels may influence those in power but they are part of existing social movements to which they have contributed acceleration. Thus we have popular and intellectual writers producing novels which make interesting and evocative reading. A novel is defined as 'a fictitious prose narrative of considerable length in which characters and actions representative of real life are portrayed in a plot of more or less complexity' (Oxford English Dictionary,1956). Developing societies have produced a number of writers who have written important pictures of their own societies illustrating the supposed attitudes of those involved in matters of social importance. Some Indian novels written in English have touched on social issues well before Independence (Anand,1986;1994) writing about untouchables and coolies. Others in connection with the cinema have been influential (Diver and Patel,2002) in raising the awareness of the have-nots about what is depicted about the standards of living of well-to-do Indians (Desai,1980). The two Indian writers, Naipaul and Chaudhuri, have been most influential on Western intellectuals and could have been classified as expatriates who have either thought there was little in Indian society worth parsing (Naipaul,1964;1977;1990) or thought that the only part of contemporary society worth preserving had been introduced from the West (Chaudhuri,1967;1976). Without doubt Rushdie's work (Rushdie,1980;1988) has been the most influential since his writing provoked among conventional as much as fundamentalist Muslims an instant reaction against whatever comes from an author who is viewed as an apostate. Western intellectuals have found such reactions both ridiculous and dangerous to what they see as fundamental freedoms; such writings may well have contributed biased themes to the thinking of social scientists. The difficulty lies in whether what is portrayed is a reality attached to the characters who are in social terms otherwise silent or a reality created by the novelist. The vast majority of ordinary people in all societies whether literate or non-literate can be assumed to have complex thoughts and lives but do not have the ability to express this complexity in speech much less in writing. How far can the novelist go in attributing to this silent majority the complexities which their prose so cleverly highlights. The answer is hardly at all especially since the subject is non literate thus what is being portrayed is the fictional thoughts of the novelist. So is it a matter of intellectual trust with which we accept the prose which is supposed to be showing the thoughts and emotions about whom the novelist writes or are we reading what the novelist thinks he or she might do in similar circumstances and which they assume that their readers would be sensitive enough to accept as a form of reality.

The results of qualitative research are interpretations of information collected more on the basis of some mutually agreeable affinity between informants and researchers than on scientifically based needs and procedures. What is the difference between what social scientists produce through this type of research and that of novelists who actively research some aspects of human social activity and who consider this to be of sufficient importance for them to spend time working up their projects?

Some distinction had to be made between novelists who write to illustrate complex philosophical themes such as Iris Murdock whose works are likely to be appreciated for the complexity and lucidity of their prose but are not likely to influence the thinking of any influential political and social elite. On the other hand there are those who write to illustrate social issues of contemporary importance such as communal hostility, religious intolerance, corruption, social diseases and addict in their societies. If we take Jamila Hashmi's story of a woman abducted during the Partition of India (Hashmi,1998:102) about whose 'marriage' she wrote 'and besides every girl must one day leave her parental home to join her in-laws. Well maybe Bhai and Bhaiyya weren't present at my wedding—so what? Gurpal (her abductor) rolled out a carpet of corpses for me? Painted the roads red with blood? Provided an illumination by burning down city after city? Didn't people celebrate my wedding as they stampeded, screaming and crying? It was a wedding all right. Only the customs were new; celebration by fire, smoke and blood'.

This is highly coloured and imaginative prose. Clearly no woman except an exceptionally talented intellectual could or indeed would reflect in this way but is this the point on which we should evaluate this and similar writings. This authoress has pointed out and indeed clarified one aspect of an extraordinary episode of mass dislocation and violence. It is in such books that readers and we can presuppose that among them will be those who belong to the power elite, will find the depiction of important social issues in readable prose and it is in this that their importance lies. Novelists write about what cannot be known directly and yet these aspects are an important part of social life, these are the shadow lands of reality.

We can conclude that some novels and commentaries about African and Asian communities may have provided useful background information for both national and non-national researchers rather more perhaps than professional studies. In this sense they may well have been influential but only rarely can they be said to have had wider influences except in the case of Rushdie's works infuriating Muslims who were perhaps already in a state of contingent fury; a clear case of social and political acceleration.

22. The absence of fieldwork methodology

So we are faced with the realities of scientific methodologies which involve various degrees of exactness; the profession of science necessitates the finding of facts and the dialectic is always over the ways that facts have been obtained and the nature of the facts themselves. The aim is clear and these shortfalls are the subject of professional comment; they define and act within their own definitions as far as they are able. A demanding professional stance and one in which there can be shared methodologies; hard edged perceptions of the material being investigated. The standards involved are in some ways parallel to those required to keep machinery in operation; it works to conform to the objectives or it does not. There is a taut silence over the relevance of fashion in scientific endeavour (Kuhn,1962).

Most social scientists are very sensitive about their field experience and their methods; they have invested an enormous amount of personal effort in what they have tried to do; a professional investment about which they are almost aware of its shortcomings in the scientific sense and in the sense of what they wanted to achieve. These researchers want to maintain their professional image and want to be taken seriously both inside and outside their profession. There is however the built-in bias of personality; no two social scientists see and interpret the same event in the same way, if indeed there is such a thing as a similar event in the first place. Finally, there is the disadaptation syndrome; the conflict over the extent to which they as researchers are involved in the social events recorded, the conflict over their commitment to themselves, their employers or grant providers and to those with whom and on whom they are working (Wintrob,1969).

Can fieldwork claim to have methodologies which are scientific rather than social? The key to scientific accuracy is always impersonal and fieldwork is always related to the personality and abilities of the researchers themselves. Whatever realities exist in their approaches to the situations confronting them, these are social events without limited scientific exactness. The fieldworker can log events as participant observers but they have no means of knowing with any exactness, the social and psychological dimensions of what they have observed; they cannot record what they do not see or hear nor can they record the meanings of what they record except in the terms of those who choose to talk about it. Again this is social rather than scientific. Perhaps the only way forward for the social scientists is to accept that they participate as people; there is some social reality in such an acceptance.

There is also the problem of the distribution and frequencies of certain types of behaviour which come from interviewing informants. The variations in these which may have come from the variations due to the interpersonal reactions of those involved, sympathy or hostility to shared objectives and whether the information has come from direct questionings or draw from inferences. Finally, the even more socially difficult question of enumerating and sampling; attempts at scientific methodology within indefinable boundaries. Playing football in games in which the referee is changing the rules for each game and moving the goal posts as the game progresses.

The criteria for the goodness of any fieldworker's procedures cannot be scientific and can only be professional assessments of and personal confidence in the adequacy of the information recorded, the accuracy of meaning, precision in and completion of data combined with the efficiency and means cost of the information provided. The informant/fieldworker being seen as a surrogate census taker, as a representative respondent and the observer's observer (Zelditch,1970). How do national and non-national researchers fit into these criticisms and to what extent? The national working within their own particular social and cultural orbits have automatic advantages since they have some idea of the problems involved in any approach. A national woman would see her research opportunities in terms of the built-in restrictions of her sex. The non-national woman would have difficulties in adopting her methodology to what she might consider to be out of date practices which the Western world at least at the intellectual and legal level has largely circumvented. The most profound conflict for one American woman researching in Japan was the etiquette surrounding the serving of food (Kondo,1990:15).

Western non-national researchers who have an individualised conception of personal importance and a predominance of 'I' in their thinking may well see their methodologies in terms which clash with the methodologies appropriate to the behaviour and thinking of people who are more inclined to see their social systems in terms of 'we'. Surely the issue is not the scientific nature of what is achieved or attempted but the sensitivities of particular fieldworkers to the feelings of the people with whom they are working and this is a very personal thing. Indian graduate students can be very insensitive to their own fellow citizens, whereas the non-national may just not know the social dimensions involved in their methodologies; that questioning and talking to people with eye to eye contact is just bad behaviour and not just examples of out of date thinking.

23. Informants in developing societies

Fieldwork can be no more than pure observation, just sitting and watching even doing this with binoculars as the researcher in the Sudan who wanted to record details of how and when Sudanese excreted (Cheesmond and Fenwick,1981), but it almost always involves the use of informants who provide the links between what the social science researchers see, touch, taste and hear and their professional needs to explain and interpret. It requires exceptional diplomatic skills for any researcher to become independent of the people who have enabled them to start work. Srinivas writes of the headman whose permission was necessary for him to start work there at all, and not being pleased with him talking to those who belonged to different factions to his own (Srinivas,1979). This matter probably has to be faced head-on by explaining that in order to get a clear picture of the community as many members as possible have to be approached. But this still leaves the difficulty that researchers come into a community or institution under the wing of locally self-important people and these are their main informants from whom it would be difficult to shake free.

This is clear enough but the obverse is that these informants have little or any interest in the requirements of the social sciences. They are always interested parties to the relationship and they are providing information to the fieldworkers because it is advantageous to them to do so; what they get out of it is both impressively varied and without a doubt pays off for these providers of information. Even if it is only tea and biscuits, writing paper and ball point pens, these are commodities which may not be so readily available in their own lives. Research undertaken in a Kenyan prison used certain prisoners to record on sheets of lavatory paper any disturbances that occurred which would then be the basis for subsequent discussions (Tanner,1970a). The incentive there was their relief from the monotony of long sentences; at the best this was valuable information and at the worst total fabrications to keep the relationship going.

In other cases it is such a full time occupation that it is paid and it is in the informant's interests to keep the flow of information going; they will find out something about anything if they get clues as to what the researchers want to know. In conditions of high unemployment being an informant is valuable in much the same way as the police have paid informants for whom there are mutual advantages in keeping the relationship going and the more valuable the information supplied, the more money is paid to them. The parallels are too close for comfort in the contexts of high minded academic thinking. Informants will always have their own axes to grind by reason of their sex, age and their political, social and economic interests so that being an informant is felt by them to enhance something. There was a Zuni American Indian informant referred to earlier who had worked consistently with an anthropologist for decades so that there was even a room in his house allocated to visiting researchers to use. But the converse to this was this man's concentration on providing information about the past that he was considered one of the most conservative persons in the area (Blanchard,1977).

The classification of being an informant suggests a somewhat formal arrangement in which individuals provide information to those who want this. But the snag in this is that as the number of times the researchers meet their informants, the less likely they are to be treated as just informants; the relationships become more personal but not only in the sense of social closeness. They begin to know more and more about each other than is required for the passage of information which might well twist the data in certain ways. Of course being an informant, a recognised companion of an outsider may enhance the informants' status in their own eyes and possibly provide long term benefits for them or their families over what the researchers can do in return with their influence on the local bureaucracies. But of course there is the reverse that they provoke hostility locally by giving away information which in the thinking of the community is their private property. Headmen in African villages are often chosen by their communities as compromise figures who can negotiate a solution to local disturbances of social balance. The assumption by researchers that they have some institutional powers to help them is often incorrect and the taking on by any headman of any such role, may make his position more difficult and in some cases untenable.

Anyone who answers the questions of a researcher is an informant whether they are part of a stratified sample or are part of the casual collection of data under conditions in which any more exact methodology is difficult. Few surveys or fieldwork reports record the numbers or details of those who have refused to be informants or whether their participation is obviously deceiving the researcher. There are grades of informants ranging from those that casually answer questions to those who become key informants because of their intelligence or because they occupy key positions and are thus able to direct the researcher towards other more specialised and biased aspects of the required information; informants are not marginal people. So informants are not providing neutral offerings of information; it is an inter-relationship in which some will be attracted and others will be repelled by such opportunities. It is a personal choice which is so short-term that they may not want to disclose too much and perhaps lose control of some aspects of their own families and communities identity by making their familial information some part of foreign understandings (Hennigh,1981). It seems unlikely that informants who are attracted to helping researchers are discovered by them, and that the researchers will ever share their research objectives. Informants will always have their own agendas and the presence of curious strangers may provide opportunities for them which would not occur otherwise in their lives; perhaps they have the intellectual capacities of opportunists.

Any individual in any community will have personal views but they will also have more that as they are representative of groups to which they belong. Consciously or unconsciously they will put forward particular partisan views, not so much of what may be the community's ideological aims as to what is right or wrong but what goes on in particular situations; abstract thought is not likely to be in any informant's mind until considerable sophistication has developed. So it would seem that a particular viewpoint would mean that those who do not agree with it, would oppose it and speak against those who oppose it. Informants are not disinterested providers of information.

There is a time and motion aspect for informants and the provision of information. The elderly have more time on their hands and able bodied men have more time on their hands in the dry season as do the unemployed in towns. Married women probably have less time than anyone else unless they are able to talk to researchers while they are engaged in domestic work. There are always times in which it is more convenient to talk, bar girls in the morning and cattle herdsmen at night. No one has the equivalent to the 'spare time' of researchers who are not committed to any work schedule.

What rewards might an informant receive? The stimulus of talking to an outsider is possibly more attractive for women who have, with women researchers, an opportunity for socially submerged aspirations to be aired. There are others who see in the presence of a talkative outsider an opportunity to put the record straight on a matter about which they feel strongly. The informant who has a close relationship with the researcher, particularly if they are non-nationals, is a human storehouse of marketable information about the personal peculiarities of this stranger's behaviour and the inadequacies of his understandings of what is going on locally (Unni,1979). Certainly paying informants may always be unwise and can create the provision of dishonest replies in order to keep the relationship going. The courtesies of shared teas, biscuits and cigarettes may suggest to informants that there is there a measure of social equality. The result of paying some informants however is that this may lead to all providers of information wanting payments (Buechler,1969).

Informants may well be difficult to fit into a particular sampling framework as in most fieldwork situations they will tend to be age and sex related and they may be unable or unwilling to provide information outside these demographic and social boundaries as was the case with Margaret Mead's informants in Samoa (Freeman,1983). Of course, under ideal conditions informants should be chosen for both their knowledge and their representative positions in a society which has a range of behavioural varieties. This may be easier for national researchers to do as they will enter fieldwork with some idea of the social dimensions of the communities which they hope to study. They can choose informants with some understanding of who might be best able to provide what they want. Non-nationals might not be able to do this as they are usually in a hurry as they have only a predetermined time in the field and they might not know much in advance about the community which they are entering. They need informants at once in order to learn the language and to start finding out what they think they need to know; the volunteer of information should always be suspect. By the time that the particular biases of an informant may be known the relationship has developed into a personal one from which it would be difficult to break free.

One of the most difficult aspects of research is separating informants from their roles as providers of information and their characteristics as individual human beings with quite distinctive personalities. Damyanti, an important informant narrating the horrors of the Partition of India and Pakistan 'at the time she was eighty years old, a diminutive energetic woman with mischief and humour linking constantly in her eyes. She said that she would come to my home 'so that you can give me some coffee and lunch and I can have some fun' (Butalia,1998:109). How alive the following discussion becomes since she has not just been described as an 80 year old unmarried Hindu woman, but her information had also to be tied to the fact that her work for many years had been with the recovery, rescue and rehabilitation of abducted raped women as they returned to India and the anguish to which this unusual woman had been exposed. It is no reflection on the moralities of non-national researchers but it is surely easier for national ones to see informants as particular people whose characters are understandable in terms of what is at least a partially shared culture.

24. Being involved in the personal lives of informants

If we accept that emersion in the social lives of the community being researched is only possible to a limited degree for either national or non-national researcher, then the relationship between the two is unequal; whatever the ostensible poverty of the researcher in their own communities since they are on a limited and limiting grant or salary they are bound to be better off than their informants, when they are in developing societies, and both sides know this with the former perhaps embarrassed by this and the latter well aware of the opportunities that this gives them in having among them comparatively well off strangers.

So informants provide information and access to what the researchers want to find out and indeed where and when interesting events may take place. They may well feel that they have relationships with their researchers which should go beyond just this passing on and receiving of information; most come from cultures in which the possession of knowledge is a commodity and is only divulged for social and economic profit. They expect a reward perhaps beyond the pleasantries of everyday association which most non-nationals might consider was necessary as it would have been in their own cultures. From the viewpoint of the researcher they may be quite prepared to go along with a state of pseudo-friendship but their informants are more likely to see the relationship in terms of their own social understandings with its basis as an ongoing network of reciprocity. There are two aspects to this: In the short-term the researchers are a resource for medicine and help for trips to the doctor if they have transport and in fact it is often a drawback to have any transport at all other than a bicycle if that is how the people themselves get around; if they walk you walk and to have a car is to lay themselves open to constant requests for emergency help which they cannot evaluate and for which certainly the non-national does not have the semantic talents to talk round without giving offence. The researchers also assess these requests in terms of their research needs and their own budgets while the informants see the situation in terms of their own cultures in which it is accepted that in any real or classificatory relationship the richer help the poorer and if they do not do so they suffer social distancing. The researchers have to evaluate to what extent these demands detract from their research objectives; driving someone to a hospital or a family to a funeral has to be seen in light of what research data can be gained by attending these events or are they just precious days lost from their project.

So to become a resource to the large family of important informants needs to be one of evaluating reciprocal values. To hand out money as loans is no more than the passing out of gifts and is rarely sensible. For national researchers who do not live too far away from where they are researching informants can be a mixed blessing; researchers can keep in touch with them but informants can find them much more easily In researching behaviour in a high security prison certain convicts were invaluable informants and on release the found their way to the researcher's home asking for additional help (Tanner,1970). Being an informant is far from being a one-sided process as they can turn their relationships with researchers into a valuable personal resource; they can pass on to their friends and associates all sorts of interesting information about the behaviour of outsiders; what they see, hear, touch and smell while working for these outsiders is as much a commodity to them as their information is to the researchers.

25. Research assistants

Important non-national researchers may bring their research assistants with them or use their partners in this respect but for most researchers whether they are nationals or non-nationals if they want consistent help in their work they have to use nationals. Any fieldworker who has stayed in one community or social group or has worked on one project for some time usually acquire some key informants who are usually cut down to one or two persons with whom researchers form long-term relationships. Although these people may have become basically important to the researchers for reasons that have nothing to do with their research aims, they have developed a talent for collecting information about which their employers are interested, and a mutual personal liking for the outsiders for whom they work. But it is more than this as after a time they are then working with them rather than for them.

Since researchers are always outsiders except in very exceptional circumstances no matter how psychologically and socially integrated they may feel subjectively. In line with this inevitable outsider status researchers are usually questioning within a framework of thought which is unlikely to be shared with those who are being researched unless they are researching their own community or professional colleagues. An answer from within one cultural setting is unlikely to get an answer compatible with another. The wider the social distance between the researchers and the researched the less likely are answers to coincide with what the question intended to elicit.

The intentional use of ambiguities in the communications of Ethiopian Jews living in Israel (Weil,1995) may make the researchers' collection of even 'basic' data difficult. In fact in many cultures there is a built-in resistance to providing outsiders with information of which the Ethiopian Jews way of behaving may not be all that much of an extreme example. In most situations research assistants with ascribed understanding of how their own cultures work and an achieved one of the researchers needs, are important and indeed valuable assets in attempts to interpret data provided in one cultural setting so that it can be understood in terms of another; they say this but I think that they mean that.

So these essential relationships develop which the researchers may assess to be only professional but which these assistants may begin to assess in terms of something approaching a required equality in a contemporary world in which the ascribed status of the non-national may not be shared with these national research assistants who may expect more than mere employment from someone who is here today and gone tomorrow. Additional requests if not demands may be made on researchers for help in their private lives and more importantly since the researchers would have been to some extent socially blind without their prolonged help, that they should receive some acknowledgement of their status in publications and not just a passing reference to their help in a foreword and indeed who has the copyright on information which they have collected and then passed on to outsiders (Fisher,1946:351).

Wider consideration should perhaps be given to these key personnel however self-trained they may be in projects which have been subject to research for some time under several fieldworkers; they are assets which may be important for some institutions to keep on their books for subsequent researchers to use as lead-ins to their work. In this situation the nature of pseudo-friendship becomes clearer. The researchers are getting more than what they initially may have wanted so they should pay and provide more; in cost-benefit terms there has to be some move towards inter-personal equality and this is even more so if the researchers anticipates returning to the same area or community; it would be short-term foolishness to neglect such assets.

On the other hand it is perhaps an illusion that people in small communities lead lives of idyllic balance and self-restraint, but in fact are leading lives of great boredom relieved by periodic rites of passage and religious festivals. Their main preoccupation may be rivalry with a muted hostility to their neighbours, the enlivening of their existences with factors over which they have some control. This means that in small communities informants who are chosen or make themselves available are committed to many minor differences of opinion over land, marriage, children, built up from jealousy and inherited animosities. Factionalism should be seen as the normal social condition in which small community life has evolved and this has certainly made any unified approach to development more difficult (Shah,1979), while creating a dialectical situation favouring change. Fieldworkers who start by trying to get as broad a picture of their researched community, may well find themselves, through their informants, becoming part of these ongoing feuds. If only the researchers could be well informed, it might be a case of choosing a particular viewpoint rather than attempting impartiality, or at least making certain that the whole community knows that having talked to one side, they have talked to other side as well.

If the aim of applied social science is some betterment of the human condition, then the information solicited from informants is vitally important. Are they unusual people in their own right or just ordinary people who have become involved with the researcher for a variety of covert and overt reasons? Should these informants be seen as positive deviants who have patterns of thinking and behaviour which could be valuable to their own societies, then the information they pass on could be beneficial. The definition of positive deviance is the observation that in most societies there are always a few individuals who follow uncommon lines of thought and behaviour and who may experience better outcomes in their lives than their neighbours who share similar risks and backgrounds (Marsh *et al*,2004). Although this refers to medical practices, it could apply equally well to any members of a researched community who appears to have coped adequately and even better than their neighbours experiencing the same problems. It is these people who should be sought-out as informants because they provide interesting clues as to the possibilities of social change in general and innovation in particular circumstances. These people are as likely to be found in every undeveloped societies as in universities.

26. The inaccuracy of informant information

There seems little doubt that information provided by informants in response to the enquiries of social science researchers, is likely to be fundamentally flawed in terms of its scientific value and as evidence that something has or has not happened unless it is corroborated from another independent source. In an unfortunate parallel the police would never accepted uncorroborated information and the social scientists have just as much reason to be on the guard against ingratiating informants.

The accuracy or the extent of this inaccuracy has been shown repeatedly in Western social environments and there is no reason to suppose that the accuracy of informants is likely to be any more or less elsewhere in less economically developed societies, indeed even more so when there are ranges of reasons for informants to please the seekers of social information. Our question must be whether inaccuracy is more or less likely with a non-national or with a national researcher. The constraints on accuracy must to some extent depend on social distance; how much does the questioner know about the circumstances surrounding the question, an additional form of bias, and what might the informant gain from such answers, yet another form of bias. The question and answer always has to be seen in its social and continually varying environment. There is little reason to hope that informants will be sharing the fieldworkers disinterested search for local truths.

The researchers can themselves detect suspect data if any by assuming that any data is suspect if it does not fit into what the majority of other informants have provided. Then they have to use their own professional understandings to assess its value; after all there are likely to be very few devastatingly new ideas that are likely to appear in any fieldwork. Suspect information could be referred to other informants but this is surely unwise when informants are likely to know each other and such enquiries might well initiate local difficulties over favouritism. Researchers usually find some informants who are more willing than others and thus want to believe them, but informants will always have their own agendas. It seems likely that the national researcher might well be able to detect misinformation more readily than a non-national and more so because they might not be working to such a tight timetable. Non-nationals will have their suspicions trained in another cultural scenario with its own duplicities.

The basic difficulty must always be that asking questions is a rare and indeed an impolite and unpopular way of behaving. People do not like being questioned and there is in this a quasi-instinctive reaction of evasion. No question is seen as simple that can be detached from a social environment. Most will have had experience of questions which were very much part of an ongoing and disadvantageous part of a wider picture when it came from government inspectors of various types and probably missionaries as well. In most people's minds there is no such thing as an innocuous question. It may be that the national researcher becomes the subject of more suspicion and evasion than the non-national just because they are connected with the immediate outside world which already contains so many interferers in their lives. They may well be assumed to be potentially malign until after some lengthy association when they have proved themselves to be less so, but not necessarily benign. The non-national may be seen as too socially distant to be a serious social worry and thus less inaccuracy may be put into their answers.

Questioning will always be seen as an intrusion and unless the researchers go to the trouble of finding out the degree of inaccuracy related to certain topics and certain categories of researchers, it might be better to accept that serial questioning in search for quantitative data may not be evidentially worth the time and trouble taken. What is surprising is that these constant and substantial inaccuracies which come from serial questioning have not resulted in any reluctance by social scientists to spend much time and effort in asking specific questions for which they anticipate specific answers; 'did you or do you' is going to get a proportion of unreliable answers while an open ended question may be less so. These same doubts apply to asking questions of key informants or of so-called sample populations which in their turn must be the subject of substantial doubts as to their reliability.

So why do social scientists continue to have a questionnaire approach to their enquiries when the results are questionably accurate. If we can accept that their work is always tied to a limited time span which is almost certainly inadequate for learning about the topic of their research, then questioning provides at least some evidence of progress. They have to ask questions because they cannot wait for answers which might come from participant observation.

27. The limitations of interviewing

In the course of social life in any society it is rare for anyone to be interviewed in which one person asks another for information in a face to face situation in which both are seated and then it is a formal occasion required by banks, government officials, doctors and applicants for employment. This formal process is a consequence of bureaucracy in which people are employed to find out information outside the range of their usual associations and with whom they share to some extent some common social characteristics. The social science conception of the interview is of a one-to-one situation conducted in private on the lines of a consultation with a lawyer or doctor. This cannot apply in the field in which there is little or no privacy even less so in developing societies where compounds, houses and rooms are not enclosed spaces -protected from outside listeners. The community will all know who is being interviewed and for how long and in many cases there will always be hangers about who to all intents and purposes sit in on the interview. Those who are being interviewed are very conscious of the fact that their neighbours are aware that this has happened as well as their soon knowing what has been said and the probable criticisms what they have said. So it seems that interviews will produce information but not necessarily private or indeed personal information. Even if they are conducted in ways which the interviewer considers to be friendly with the provision of a cup of coffee and a cigarette and this must be an entirely one sided assumption, it is a confrontational process. It is a one to one relationship in societies in which such an isolated interrelationship, much less a formal one, is uncommon. To ask questions about anything from a person who is to all extents and purposes a stranger for whom one has no social obligations or connections is not normal in any society; it fails to recognise that the answers are socially based and are not information which is necessarily correct in another situation.

People who live ostensibly social lives and indeed think of themselves in terms of their on-going interrelationships may almost instinctively dislike being isolated with a stranger however socially acceptable he or she may be for other reasons. We have already pointed out that direct questioning is an intrusive activity at best and is usually considered to have hostile parallels, and it is a new practice initiated by colonialism associated with contemporary civil servants whether police, health inspectors or tax collectors. A set piece confrontation of any type does not often occur outside institutional requirements and is usually structured within the needs and functions of literacy.

Interviews usually follow a schedule of the information which it is hoped to obtain and is an imposed way of making the recording of social facts conform to the orderly nature in which such information is recorded on paper. It seems unlikely that people who are interviewed will give straightforward answers on matters which are not more or less obvious in the first place. Interviews cannot provide uniform information except in so far as this is created by the structure of the interview itself.

Any direct questioning in an interview is bound to get answers tailored to what the interviewee considers should be the 'right' answer either for them or in terms of what the interviewer wants to find out. Why should the interviewer assume rather than just hope that the answers to questions such as 'are you married, how many children do you have or what fields do you cultivate or own' should be correct in the understood meaning of the terms which the interviewers have in their minds; men and women, old and young will always answer the same questions differently and within the boundaries of their own social honesty.

Answers may be given according to the strategic thinking of those replying. An elderly Sukuma woman claimed that it was not difficult to fetch water from a dam a kilometre away during the dry season while a younger woman said it was a heavy task to fetch water from thirty metres away. The reasons for these different replies might be governed by their thinking over the use of water, a simple need for the older woman to show that she was still able to fulfil her woman's role as a water fetcher or it could have merely reflected the fact that the younger woman used more water (Drangert,1993:57). It may well be that answers in terms of what the interviewers had in their minds are few and far between since those answering always have a clear idea of where their interests lie. Answering a question is not the same as having a chat. Those answering the questions of a researching stranger know full well that what they say is being recorded even if it is not being done mechanically. The informants will be considering what their fellow community members are likely to think and how they will react and of course, they are thinking about the uses to which their answers are likely to be put; it is never a recitation of normal behaviour (Vansina,1985:111).

A more profound experience of interviewing and observing can help to improve the quality of information from informants. A skilled interviewer is also a skilled observer able to read non-verbal signs, sensitive to how the interview setting can affect what is said and to be ever aware of the strength of the interactive relationships between interviewee and interviewer (Patton,1990:32). If the interview takes the form of an informal conversation that again is a social definition made by the interviewer since the situation is bound to be formal or at least formalised by the taking of notes or the use of mechanised recording. Even if there is a flow of unsystematised information in reply to such open ended questions as 'what do you do' or 'what did you do yesterday', the replies are likely to vary in both quantity and quality. In the interviewee mind, there is always the process of thinking about what he or she should disclose in this confrontation with a stranger.

This author in working on the lives of national and non-national Roman Catholic priests, nuns and brothers in East and Central Africa asked only one question 'what do you do'? He received answers which varied from a twenty minute stand-off from a non-national priest who had decided that it was none of the researcher's business to either ask or receive answers to a lengthy reply from a national nun who wanted the nature of local vocations to be publicised. At the other extreme a missionary priest in Tanzania who had at one time served as curate to a Ugandan priest talked on and off for fourteen hours.

Then the interviewers have their own reasons for interviewing which may have little to do any particular methodological approach chosen for abstract social science reasoning, but because a particular researcher has only a limited time in the field and needs to collect information, any information as quickly as possible. It seems necessary for the interviewers to ask themselves why they are getting any answers at all and if they are getting cooperation what may be the interviewees thinking behind their presence in the interview however informally it is carried out. Taking the argument in reverse, how can the interviewer know that the interview has been a failure other than obvious non-cooperation; she may expect minimal replies but may not be able to distinguish them from realities? It seems certain that interviewing is an alien and alienating way of obtaining uniform information which must be of doubtful statistical value. It is best avoided except for quasi-formal discussions with valued informants who come some way towards meeting the requirements of the researcher seeking for insider knowledge.

Social science research in the field is seemingly tied to one interrelationship and the viewing of group activities which are opposed to each other in practice. The individuals with whom they relate spend almost all of their time in group association activities and certainly very little of this time is ever spent in one to one relationships with strangers. One to one relationships in interviewing whether the person being interviewed is a stranger or someone already personally know to the researchers has another more human handicap. Even with strangers, since they are not interviewing someone about crime, there is a personal distaste in going further in their questioning than would be considered polite in their own social environment. To be intrusive with people of the same sex and age as the researchers might be acceptable but not with older people and of the opposite sex. Yet methodology as based on Western social science is skewed to this source of information obtained from individuals who are not used to expressing themselves in such social isolation. An alternative is a group approach in which informal groups discuss with the social scientist all the issues pertinent to their enquiries (Gwaltney,1981). But this has a built-in reluctance to speak in front of neighbours, relatives although this itself would reflect the reality of communal life and be useful data.

Information might then be obtained in the same way as public opinion is formed in small scale communities which is by public discussion over what is the public reality of what the researchers are interested in. The fantasy that the informant can be isolated from the groups in which they are always located and that such persons can produce information paralleling that used in hard science, needs set on one side. Then the researchers are freer to discover a different and yet more sensible realism detached from the individualism on which they may base their own lives. A group of informants would be harder to align with unrealistic theories and forms of bias because it would either be contested or align itself with some part of existing social thinking through social activities already taking place.

At the back of every interview is the aspiration of the social sciences that the source of information should be individualised, and characterised as coming from a particular person usually with some abbreviated characteristics. This is not possible simply because someone is invited into a room and the door closed which is a very particularised research environment; there are always listeners. In practice it is rare to have such one to one isolated discussions as even if someone else is not present there are always those from whom the informant has just been separated and with whom they will be united again after the interview. In fieldwork an informant often comes along with a friend or a classificatory relative who contributed influence if not words as well. The result is 'often as with many interviews, most of which is collective rather than individual; the whole then would turn into a conversation with everyone pitching in' (Butalia,1998:109-110).

28. Informants and their memories

Social scientists in their field research want to uncover information about which they already have some ideas as to what they are likely to find out; their minds are not blank slates confronting a piece of inorganic matter and they already have formal and informal thoughts and theories. In the same vein can we assume that their informants are blank slates or do they already have some inkling of what the researchers are after? It seems very likely that even after a week of a researcher's presence in their community or institution, gossip will have spread far and wide about what they are doing and what they want to find out.

No community would ever expect that they are there for months at a time just to enlarge knowledge about what they do as just one community among thousands in that society. This author's research in a Kenyan prison was considered by the prisoner gossip to be related to unsolved crimes and that he was working undercover for the police. All the information which he received from prisoner informants for a year at least was geared to this assumption; the prisoner informants created relevant memories of what they and everyone else involved may have done in the past.

An individual's memory is not some store house of everything that has happened to them in their lives similar to computer memory, but it is a collection of information which they find of value in their current lives. In fact we can asvume that memory for something outside the necessities of their daily lives is likely to be very inaccurate. A project to evaluate the length of exposure of 43 Guatemalan women and their children to cooking fire smoke became significantly inaccurate after a day had elapsed from the first observations (Engle *et al*,1997) and a study of 40 Egyptians recall of their domestic routines found 56 per cant inaccuracy in what they recalled (Rici *et al*,1995). This was an easy domestic activity to remember and one can imagine the inaccuracies over what may or may not have happened in a previous year or even last week. On average about half of what informants report is incorrect in some way (Bernard *et al*,1984).

It is incorrect that non-literates have better memories than literates because they have not let their memory capacity weaken because they can write down all that they have to remember. The English carpenter, the Sukuma farmer and an American Indian Navaho sheep herder have detailed vocabularies to deal with wood working and its tools, plants which have material uses and the colours of rocks and soils. The error free recall of details of a series of cattle transactions by a Swazi cow-owner was almost error free and in related areas there was evidence that he had superior numerical skills (Bartlett,1995:267), not that he had a good memory for everything. Everyone remembers and keeps in current use in their minds what they find useful. We forget what we do not need to remember and this certainly applies to most of what we learnt so industriously in school.

When confronted with the researchers' direct questions or responding to what they know are their areas of interest, informants are going to dredge up from their memories relevant data for just that. After all it is not difficult to speak about matters of which virtually everyone has some experience. There will always be informants who can meet the researchers' half-way in matters of general interest and invent examples of their folklore and how people in the past have reacted to these. This will be no more 'truth' than the testimony of community members in a dispute who know all the background and have reached their own conclusions long ago. We and our informants notice what our cultures guide us to notice, retain and construct. The informant and researcher have a shared interest in the commonalities of their lives, marriage or matters in the abstract. But when there is a move from 'a' marriage to 'the' marriage of particular people, the memories of informants will be slanted towards their views of what that particular marriage involved. Memory is inevitably based on the contemporary value of that particular memory. Above all we cannot assume that these memories on which the researchers draw is something exact. Memory is not a fact producing ability but a faculty for presenting the past in ways that show its contemporary relevancy. Bartlett's conclusion about the nature of memory have great contemporary value which social science researchers should remember. 'When a subject is being asked to remember, very often the first thing that emerges is something in the nature of an attitude. The recall is then a construction made largely on the basis of this attitude and its general effect is justification of the attitude' (Bartlett,1995:207).

It is not that informants are in any formal sense deliberately or indeed unconsciously deceitful but memories of particular events are mental processes which align with current needs which may or may not have happened in the ways that are currently remembered. In this accurate inaccuracy informants are telling the researcher a modernised 'truth' not an old one produced in its historical form for the researcher's use. Informants memories should always be seen as suspect until proven otherwise. The difference between memory in the past and memory in the present is usually obscure. As one young woman informant in Jordan stated 'I really wait for your visits. I can tell you all sorts of things and I know you understand and that it will not cause me trouble the way it does if

I talk to someone else' (Shami,1988). But what reality is being passed to the researcher here and in what particular framework will it be interpreted? The 'truth' is not something that can be expected to emerge in any situation as it depends on what contemporary reality is being filtered into the researcher's field notes.

29. The role of observer; I know because I was there

Any fieldworker has the very basic human conceit that as observers they are reliable. To what extent is such a consideration acceptable in the contemporary world with the implication that the social scientists whether they are nationals or non-nationals have some specific professional ability which makes their observations acceptable as facts; they have some access to the true nature of social situations. There may have been some implications of this in the past with the solo anthropologist researching a 'primitive' community for whom this is the only information available, but this can have little contemporary relevance. We have to consider the natural egoism of those who are unexpectedly confronted with what is unusual from their perspectives. Whether we are national or non-national social scientists we see what we are socialised to see or more particularly what we want to see. There is no adequate reason for social scientists to be accepted automatically as reliable witnesses.

In contemporary situations researchers collect their information from many sources other than their own coordinating observations either from informants who provide second-hand data or from witnesses who have seen or experienced the events themselves. In the latter case they are acceptable witnesses in law and this is acceptable as an accurate record unless challenged under legal procedure. This is rarely possible or indeed advisable and the truth of eye witnesses' accounts in fieldwork situations has often to be taken on trust according to the researchers evaluation of the informant's character for which they are not likely to have had any specific training; they may have suspicious natures or they may just be trusting people. When researchers state that they know because they were told so by people whom they regard as reliable, the truth of the data is on shaky ground. Few social scientists record the grounds on which they regard their data as reliable or at least on some scale of truthfulness. All we can know reliably is that the social scientists did have contact with certain informants who made certain statements.

What informants may have said to the researchers who wrote it down or recorded it, is not admissible evidence as it is hearsay. We do not know the relationship between what was actually said and what was written down or taken out of the recording as would have been the case for the informant who talked for fourteen hours in reply to the question as to what he did as a missionary priest in Tanzania. We cannot assume that there is some necessary coincidence. A recording is evidence if it is taken as a whole but once it is cut into pieces to assist in the accumulation of evidence on particular points, then it is no longer direct evidence This knitting together of data might even be called fiction in the sense of something made or fashioned (Clifford,1986:6).

Whatever the social scientist does in the field, they are observers; they are outside what they are researching but by the fact that they are there, they are part of what they are researching. Theoretically they are not part of what they are observing since they are practising detachment so right from the start they are caught in an obvious dilemma; denying their own situational importance as the centre of what they are observing. Observers are only different to the machine recording data in that they can change direction quicker than a machine but nevertheless they are still basically unidirectional; they cannot look in different ways at the same time nor be in two places at once as when they are present at a religious ceremony when who does what and when as well as where would involve a mass of conflicting and not necessarily unified activities much less thoughts, they both give divided information about singular events.

In the early days of social science fieldwork researchers were usually quite separate from the researched; it was an 'us' and 'them' situation in which social distance factors kept the research role quite distinct. They felt themselves consciously as much as unconsciously intellectually and socially superior and benignly motivated. As global thinking has developed whatever private feelings they may have about social distance, it has become less socially and politically acceptable to have researchers abrogating to themselves any separatist way of behaving and thinking.

What may be often forgotten in considering observation is that the unusual particularly as regards the difficulties of getting information is often given priority over more commonplace activities. There has perhaps been a concentration on the unusual by researchers with particular talents and commitments. The study of homosexual activity in a narrow public environment (Humphreys,1970) or being sane in an institution for the insane which of course involved deception and its ethical balance with the need for such research.

Although in some senses all human activity in its individuality is unusual and can be made so by research concentrating on this dimension, this probably is a form of bias. Most behaviour in the context in which it occurs, is an aspect of what may go on in that community within the boundaries of their understandings and is accepted as some sort of normality. Of course there are some who are beyond the boundaries of this acceptability. Societies have always had murderers, homosexuals, and the mad and so on which they have coped with in their own ways that satisfy the requirements of their current lives. So perhaps the concentration on the unusual or the creation of the unusual in terms of methodological isolation is what the observer finds interesting in terms of their own socialisation, is unfortunate. It avoids the fact that by the standards of those involved such behaviour is normal or at least explicable in terms of what they consider to be normal. This observation involves all types of personal dilemmas; eating people is not always wrong (Conklin,1995). Observers should start from this assumption and that by such standards, their role does not involve any great need for more than the development of what strangers, for that is what the intrusive social scientist is and must remain, can and cannot do outside the confines of their own natal understandings. It is however more complicated than this seeming simplicity and much depends on the role that has been imposed on researchers as national or non-national people (Gold, R.L. 1969).

Except in the rarer situations in which the researchers are undercover, most fieldwork is undertaken by those who are known to be strangers and doing something which is not quite clear to those who are being observed; unlike medical research which is usually quite clear in its objectives at least to the researchers themselves; it is by no means clear in social science fieldwork who or what is being researched, much less why. So the awareness of those being researched that there is a stranger in their midst must lead to some alignment or bias of some of those involved with the researchers; there are possibly bound to be both negative and positive reactions with the majority in the grey areas of indifference apart from the entertainment value of having an outsider in their community. Thus fieldwork divides between the various forms of participant observation dictated more than anything by the variations in what can be observed and in what situations the researchers can observe with what degree of interference with the scene.

30. Participant observation

All social science fieldwork starts with some form of participant observation as they enter the community in which they are going to work. Whatever their backgrounds this is going to be a new experience which is going to have as primary influence on their thinking. Even the "half and half" researcher is going to be immediately struck by the changes which have occurred since they were last in their natal environments.

It is one thing to accept the need for the fullest possible participation in the researched community or institution, but it is another thing to actually do so. Most researchers as intellectuals are not necessarily athletic and full participation involves physical effort for which researchers have a little or no training. Briggs working and living in the Artic with Eskimos found that they adjusted what they gave her to do to what they saw as her physical ability (Briggs,1970). Joshi when he wanted to work with Indian farmers found the villagers amused at his lack of physical stamina (Joshi. 1979). It is clear that for every degree of involvement which opens up a new channel for information gathering, the researcher takes up a specific role which on the one hand opens up some new and previously unknown area of understanding but at the same time imposes restrictions. It may be both interesting and exciting to be more or less accepted into one of the Sukuma so called dancing societies but by belonging to the Bugisu one, the researcher is excluding himself from the Bugaru one. An unmarried German researcher in India lived with a Hindu family in which all the women observed '*purdah*'. This made it impossible for her to move around freely until she made friends with a child welfare government official who acted both as her language teacher and chaperon,(Bellwinkel,1979).

In a strange way most social science researchers want to be socially invisible for professional reasons and yet be able to collect information but whatever they do and wherever they are as nationals or non-nationals, they are always someone. In some cases they have no choice in what they are by age, sex and ethnicity but in other matters they can choose their roles by selecting what they want to do and how they go about doing it. There are serious methodological issues in using the data from the point of view of some form of participant observation. It is always the observations of one person at one place and at one time and however long that participation may be in that community, it is participation in only one part of that community sequentially experienced. So the basic methodological problem in observation is that it is restrictive and no researcher can observe what he cannot see. This leaves reliability to those who live in one part of the community as a round-the-clock resident and perhaps not living in a separate house or tent. Observations received second hand from informants are even more suspect as they have already been filtered through their personal conscious or unconscious biases.

It has to be accepted that any form of participant observation is ego-centred and it cannot be otherwise and thus carries with it all the problems of such centralisation. As in any cultural setting one sees what one is culturally trained to see and what one is physiologically capable of seeing. From this author's experience, a literate Swahili Muslim in Mombasa, Kenya on being asked to write down the genealogies of all the people known to him took months of persuasion before he started to record the matrilineal rather than only the patrilineal connections.

Being an observer does not make even the most talented of social science fieldworkers into something approaching the camera or recording machine. What they observe may well be a reflection of the roles which they are performing by just being there even unobtrusively; they cannot be all things to all men. Their age, physical appearance and sex predetermine much of what they can do and probably what they can see and hear. It may be more methodologically sound to be a sole observer with a known role in a highly structured environment like a hospital ward or a school class-room or one in which there is no restricting structure such as a railway station, market or airport in which the control is one of the area through which there is a flow (Baker,1998:chap 7) in which the observer is undifferentiated among the many who are present.

The range of acceptability of a researcher being in a community and more specifically in a family in which they seek to become a full-time round the clock member is likely to change in overall ways as well as in relationship to particular day to day issues; one may be differently accepted in different seasons by different people since the family members are not one but have different roles. This odd person who wants to live with them for reasons which they will treat with some scepticism will go through a scale of partial acceptance from stranger, guest, potentially trainable childlike adult, simpleton unable to do the simplest of tasks known to their children at an early age to being a straightforward nuisance, an ever-present nuisance and a handicap on their usual way of carrying on who has to be endured (Briggs.1970). If the researcher expects to become one of the community which they are researching then immediately their self-sustained and understood status of an educated intellectual disappears. They become just what they are in their host's terms by age, sex and property. They may be given classificatory kinship roles which must inevitably be daughter or son and this of course is immediately qualified by the material benefits which they should providel for their fathers and mothers. It is not a situation in which researchers can sit on property and say that it is mine and mine alone; their limited stocks of tobacco, personally preferred food and petrol in their transport will soon be exhausted. In such situations the researchers are not renting social space in which they can expect to be left largely alone to carry on with their mysterious research activities. They cannot dictate the conditions of their residence in this way unless there is already there a well understood role of lodger who then is clearly not classificatory kin. Living in a family means sharing their work since they are sharing the cooking fire and for this they may not have either the ability or the stamina. It is no doubt upsetting to be treated as a daughter or son and then to find that one cannot live up to the obligations that go with such roles; being kin does not allow time for writing up notes when there is work to do in which the family expects participation as a matter of course. The bilateral adjustments which would have to occur are probably virtually impossible to solve except at a peripheral level; the social science researcher cannot be that and a full-time participating native at the same time if either role is to be performed satisfactorily.

If they become truly absorbed into these cultures as classificatory daughters or sons then they will be addressed in the peremptory tones appropriate to their roles. They will not have been addressed in this way since they were in primary school and they will have long grown out of any form of kin related subordination in their own cultures. Most households do not have provision to accommodate a lodger who is a researcher; the very fact that they are in the house or compound even if they are given classificatory kin status, means a substantial amount of rearrangement will have to take place. Subsistence farmers in Eastern Africa usually have a saying that a visitor is a guest for a day and on the second day they are given a hoe to work on their farm as 'payment' for the food they are consuming and their share of the firewood and water used. Overall for a variety of personal and social reasons it is perhaps prudent not to attempt full time participant observation.

31. How far can a researcher go in participating?

It is virtually impossible for a researcher to enter fully into a different role just for the purposes of research without grave and indeed irreparable costs to those who attempt to do this. To cast off one role and take on another is what actors and actresses do all the time; on stage and off stage behaviour is rarely confused and if it were we are in the area of disturbed minds and psychiatry. It would certainly mean the abandonment of existing relationships as when a researcher became a homeless person, the member of a criminal gang, drug addict or paedophile. How long could this schizoid duplicity be maintained and at what cost. The American anthropologist Cushing certainly went as far as he could to become a Zuni, but whatever ceremonies of supposed inclusion that he undertook such as rat eating, did not make him a Zuni. He only became a special sort of American in that he seems to have kept the secrets which he obtained by initiations and after all he still kept his university connections (Gronewold,1972) and he did not appear to have created a Zuni family by marrying and having children or indeed living permanently in a Zuni pueblo.

Certainly researchers can go along with many of the social activities in the communities which they are researching, but this is socially probably not much further inside than any social stranger or long-term visitor would be allowed by their usual acceptances of outsiders; the parameters of good manners, just low-level commonplace behaviour. To what extent was Srinivas, the distinguished Indian anthropologist but a Brahmin able to avoid psychological and social clashes within himself and within the community when his contacts were Dalits? However going beyond the commonplace behaviour which the community may permit any stranger, including the research outsider, would involve high-level behaviour of a different order altogether. This would involve crossing a behavioural line which few researchers might not want to go and even consider going in the first place because it would involve unknown activities and reciprocities.

All national and non-national researchers are socialised from birth into particular patterns of behaviour and thinking which sets out what should be done in certain situations and those situations in which behaviour is negotiable. More importantly not so much what may not be done if they have the time to reflect on it, but the behaviour which is 'instinctively' crossed off as outside behavioural possibilities. The greater the social distance between the researcher's natal understandings and that of the researched community, the greater the likelihood of these total barriers existing. Of course, there may also be differences as to how researchers' interpret these high and low level behavioural boundaries. Most researcher's can overcome food taboos which are often known in advance because for most non-nationals these are matters of choice and not always linked to religious requirements.

Few researcher's would deliberately break statute law or widely accepted moral requirements such as being involved in using human body parts in protective or aggressive medicines or killing while participating in cattle raids. Similarly few researchers would choose to go towards the extreme edges of acceptable social behaviour. We can assume that almost all research projects are designed to stay within the boundaries of cross-cultural acceptability. Difficulties will always arise which may not involve male researchers because they stay within the male dominated spheres of activity, but female ones who move from a university environment may find themselves excluded from much that goes on.

32. Participating as a part-time observer

It might be methodologically more accurate to admit that all social science fieldworkers are part-time and that except when they are living in a single house as a full-time member as a lodger or even when studying their own community from a house in which they are accepted as kin, all their work is part-time. They are researching when they are not eating, sleeping ,washing, reading or writing up their notes so that in fact they have working hours. Moreover these working hours involving research in comparison to someone who goes to work and comes back home in the evening, may not be very long as so much time is taken up in self-maintenance since in most cases they are living alone. They come on stage for the purposes of research and then go off stage back into the personal environment in which they will certainly feel at home. They do not want to be a cultivator or factory worker or the wife of one, but they do want to be regularly in a personal environment which they have created for themselves surrounded by their own personal possessions. Fieldworkers are lonely people and there are limits to the amount of time that they are capable of spending in the research environment. Overall part-time roles are easier to sustain since it involves going and doing something within time restrictions. Raymond Firth's wife in starting fieldwork in Malaysia with no particular research theme in mind, used to sit beside the path near their house until curiosity got the better of passers-by and she was then in business as a field researcher into the social behaviour of Malay villagers (Firth, S. 1972).

The image of homogeneity which often becomes part of the participant observers thinking as some justification for their conclusions, is grossly overused because it is so tempting. The long standing coastal village of Mkwaja in the Pangani district of Tanzania with some hundred houses is divided into two by a shallow wet season stream; adults in one half were sometimes unaware of important life cycle events and their ramifications in the other half. Ancestor propitiation ceremonies among the five million Sukuma of Tanzania were all dissimilar except for the collective title imposed on them by the observing anthropologist. In fact it is an arbitrary sample based on the movements and use of time by one intruding person. Even if the researcher is assured that what is being observed is typical, such an assurance is probably valueless. Informants who are supporting this participation would see little advantage in extending this into areas in which they do not have the same social pull; it is to their advantage to make statements about typicality.

The acceptance that particular activities are not typical means more effort all round for these researchers to find a balance or variations from the originally observed activity. Researchers usually do not have the time or the money to do this except for those such as Margaret Mead who had a long term commitment to one New Guinea community which in the end may have become one of affection and obligation rather than of continuing research.

Admittedly many fieldworkers realise the narrowness of their own observations and for this reason narrow the scope of their enquiries and conclusions. Although this may reduce the margin of error in assuming typicality, there is always the methodological unacceptability of giving weight to a sample of one. Participant observation of one family in a community of perhaps a hundred families, will be a representative sample by chance.

What the social scientists may see and record is probably almost accidental. They may come across much of what they observe or are directed to observe simply because they happen to be walking that way at that particular time. Just seeing people walking towards a valley could lead to seeing a rainmaking ceremony to which otherwise they might not have been informed about because such rituals are disliked by their government with its developmental framework of thinking. Few fieldworkers would walk in different directions to a set plan of sampling. In a systematic attempt to record variations in activities which would be environmentally variable anyway.

Even then the context is likely to be abbreviated. Little can be assumed so the context has to be collected through another linguistic code with all the attendant restrictions of time, place and association. This raises the question of how much data can be obtained and from whom in the time available without jeopardising obtaining other data elsewhere. The obtaining of a context from informants is a long and tedious process in which researchers swing between the patience needed to be accurate about very little and their impatience to get some sort of overview. Even if we can assume that the fieldworkers have had some sort of general training and have some specialised understanding of that particular culture or area, there is inevitably less surface knowledge in comparison to the knowledge that a national researcher may have amassed since childhood about their own culture.

33. Participating fully as an observer

The social scientist comes to live in a particular community which they have chosen for their research or is already resident there. Whether the people there may have accepted him or her as a transient stranger with or without knowing what research is or what they intend to do, there is always some partial covertness since they may not know what they intend to do so they cannot explain to others what they do not yet know themselves. Of course, when they are researching something in which they are already a part such as the hospital in which they are working or the police force of which they are already officials, the research has to be covert. When they are just inquisitive strangers and have just appeared on a particular scene, the receiving community will give them a role and the researchers do what they can to get information by sitting, listening and looking. They are obtrusive but not disturbingly so and some of the community will see for themselves that there are advantages in having such a stranger around. This is probably the ideal form of participation for both nationals and non-nationals working outside their 'home' environments. 'Though I am willing to participate as circumstances allow and am fully committed to the anthropological tenet of experiencing another culture first hand in order to learn about it, I also realize that personally I feel most comfortable and most productive in the role of an obvious present but relatively uninvolved observer. I feel particularly fortunate when I am treated as a privileged observer, allowed to witness more of the total scene and entrusted with more personal opinions than are shared with the casual visitor because my purposes for enquiry are regarded as legitimate and sincere (Wolcott,1981).

In this way researchers take on a round-the-clock role in the community or institution which they have as the objective of their fieldwork. For example, a nurse researching the hospital in which she works, the teacher the school in which he works and the farm labourer in an agricultural community. If they already have these roles and are functioning in these occupations before the research started, then the problems may only be moral ones and whether they can subsequently remain in the community or institution on which they have reached public conclusions is debatable; the role of whistle blower even in the cause of research is not appreciated by employers or neighbours.

If in order to do research the outsider has to become or tries to become an insider there are built-in complications. To what extent can the outsider go 'native', crossing over wide cultural barriers which may be ethnic? In the case of Cushing, he did not become a Zuni but went so far in identifying with them, that his role as an anthropologist became confused (Gronewold,1972).

It is not possible to hold two divergent roles at the same time and in the holding of the role of a homeless tramp, or member of a violent gang is indeed possible but at the expense of other public and personal roles that they may want to retain. It is something like living in a witness protection programme, expensive and a necessity but not a matter of choice. It certainly has repercussions for mental health as there is no relief from the necessity of pretension and in some cases involvement in which the researcher is expected to participate in acts which he or she would see as morally wrong. Among the Swahili of Pangani both bride and groom have 'midwives' of which the groom's one has first intercourse with the bride to avoid hostile witchcraft on the groom to make him impotent. The moral and political implications of this for the researcher requested to be the midwife have been discussed earlier. Many social scientists may well have been faced with such moral dilemmas in which to become one of 'them' has built barriers to absorption; the involvements in criminality, physical violence or personal moral problems. This becoming one of 'them' in terms of almost all research projects is unattainable; for the researcher to become one of the researched is an exercise in autobiography. They will give themselves away by their body language, inappropriate semantics and personal idiosyncrasies. To take over a role so completely that others are convinced of that person's occupational and social identity must mean that they have had a considerable period of lead-in training. In this process they will have been extensively socialised into what they should expect more particularly into what they might be expected to dislike. It is learning to behave in a way that is different from their own socialisation. Actors however can stage manage their appearances so that for short periods they can be totally convincing.

Living in a village with that community as the object of research, requires a very long lead-in before such a researcher is accepted, but accepted as what? An American hired to make a long range assessment of a rural school in a small township found that he and his wife had no roles that they could perform other than that of normal American short-term citizen like so many others there and as such they had to perform social roles related to that type of life which most of the community shared even though quite soon everyone knew that he was a hired researcher (Hennigh,1981). This married couple were required by their contract to live in the community to be researched for five years had two serious methodological problems. Firstly, they became their own principal informants of what was going on in the community and secondly, the community used them as a source of information. The community had an acceptable place for active newcomers and they were treated as any other community resource because doing so was 'natural' for couples like themselves. Their active participation in community affairs as a requirement for local citizenship was a highly efficient way of gaining information. It was a defence against the isolation of long-term fieldwork but in this case their absorption into the community was enabled by the fact that it was a community in which there were large numbers of newcomers in a community with a rapid turnover of population. An accidental but significant short-cut into a local culture which was just a part of their overall national one also cuts out the traumas of withdrawal. This form of active participation by a researcher is only possible when their characteristics are broadly similar to that of their field of study. Once there are social differences as when an educated Bengali researches Bengali farmers, the research problems are broadly similar to that of research in a foreign country.

In any situation in which the researchers become full-time observers there is the foundation of deception. Of course, they may disclose what they are doing but with this will come some forms of 'on' and 'off' stage behaviour and thus cutting down on the detachment need for adequate assessments. It has to be recognised that for anyone to maintain an acquired role to which they would not have in the normal course of their lives is always extremely difficult but also involves considerable social and psychological strain as they would have to watch every word and gesture, paralleling undercover police work and resistance involvement in an occupied country. It would involve separation from all other relationships which might give the game away. The personal cost of isolation must be high. Except for situations of deception within a national's home social environment, it is almost certainly futile to think that a stranger can become a native. No one can be 'with it' beyond the boundaries of their own cultural understandings and beyond the limitations of their own physical appearance.

Whatever the researcher attempts to be and become, it is always limited by demographic realities that he or she will become slotted into the community in the roles appropriated to their sex, age, ethnic appearance and even height. So we can see that total immersion is in fact to take on a role that handicaps any total detached appreciation of that community which such researchers might have retained by other forms of participatory observation. Whatever the real research benefits coming from such total participation, it also depends on the capacity to memorise; notes and photographs cannot be taken and observations can only be put into writing when the researcher is alone and often in secret. Thus there are many serious methodological objections to full-time observation beyond accepting that it is the main means of qualitative assessments.

34. The fieldwork technique of immersion

The ideal methodology for any form of research is for the researcher to become totally immersed in their research. We have established that this is not a possibility without severe personal sacrifices and is so much a one off scenario that the results may become so much of a personal story that it become more a tale of a social scientist's resolution dominating that of the community which is being studied. It is just not possible to be on stage the whole time round the clock.

The average researcher, of whatever social or ethnic origin, is incapable of paying the personal costs of such total immersion in a research project. When they are partially successful and take on some of the understandings of 'them' rather than seeing them as being professional 'others' then their partial participant roles has paid-off. Whatever the degree of immersion there is bound to be some loss of a detached perspective from a growing sense of sympathy with those who are being researched. Even if this does not occur there is the feeling that they are 'my' people, informants, and families or at the very least the objects of 'my' research.

35. Active or passive participant observation

Except in the rare cases in which the fieldworker is silent as to what is their hidden purpose and not known to be researching, they are always social additions to the community; they would not normally be there and even the 'half and half', the researcher who has taken on a new nationality but retains links with their country of origin, is seen as a visitor living in a social world of which they are no longer a functioning part. The receiving community has to decide the ways in which they are going to accept these intruders which is always short of full acceptance; they will be attached to the socially qualified role of a previously unknown form of stranger.

In some cases communities have been the subject of so much research over the years as to have developed just such a separate role of researcher different from itinerant health workers and other entertainers. Some Zuni pueblos decide on what researchers they will accept and on what terms and it is said that some Todas of the Nilgiri hills in India know the anthropological terms with which their social structure has been described. It seems likely that some American Indian communities accept the attention of some new generation of researchers with bored resignation and hopes of personal gain.

So most fieldwork is a variation on themes of intrusion. Some actively participate as strangers and to some extent seek out what they want to observe whether it is religious ceremonies or agricultural fieldwork or house and water channel construction. They are actively intrusive since they are additions to those who would 'normally' be there and we have no means of knowing how much of their observed behaviour has been put on for the benefit of the visitors. Do the Ikoma men of north-western Tanzania in their dances usually wear dark glasses and decorate their faces with blue dye or was that just because they knew that 'their' anthropologist would be there with a camera. At the other extreme there are passive participant observers who just sit and watch in environments in which everyone else is just sitting and watching because there is not much else they can do. There are variations in this as in sitting on any available benches and watching in the Beijing International Airport as the police move on the cars of the well-to-do Chinese who attempt to wait in a no-parking zone or being in a Madras church and noting the proportions of men and women who take communion. In the latter according to the researcher's sex they will be sitting on the left or right of the church and always facing the activities of the priest at the altar so that they cannot see the faces of those taking communion. There is no intrusion in these two examples as airports are full of strangers of all shapes and sizes and strangers often appear in 'foreign' churches as passing ethnically similar or different visitors without having any particular attention paid to them except that it is likely that they put more money in the collection plate than locals.

On the other hand going to live in a Tanzanian village in which the researcher and her family take up living-space is intrusive, and even if she just sits and watches it is still intruding, as it is not normal behaviour for any of these people to just sit and watch;

they should be doing something and perhaps only those with connections with the spirit world would be sitting and thinking and even then there would be speculation as to what they were thinking about when looking at particular people. Gulliver when he wanted to work on the densely populated slopes of Mount Meru in northern Tanganyika, pitched his tent with local permission on a small patch of grass and upset the grazing pattern for his host's goats.

So the basic issue is the degree to which these fieldworkers are visibly intrusive by just being there and whether they can become less so by renting a house in the community where others do the same thing on long term visits. In this the national and indeed the 'half and half' researchers have distinct advantages but not exclusive ones. They are more likely to look like those they are researching as the Afro-American researcher among the Yoruba who felt more at home there than a white woman researcher and was seen by the locals as in some way coming home. Kondo describes the difficulties which she experienced in moving out of a family home in which she was treated as a classificatory daughter and into a separate rented house where she had to learn to practice neighbourhood obligations in that particular street (Kondo,1990).

If non-national researchers are accepted into a community, it will be on their conscious or unconscious terms not on that of the strangers which will range from their entertainment value to the economic and political advantages from getting themselves better known to the outside world. The community's attachment to Margaret Mead certainly paid off in the material help that she arranged for the Tokelau people with whom she had worked on several return visits. Perhaps the issue should be framed differently to explore when and under what circumstances strangers become acceptable, and in what social situations. The social scientist researching tourism can be just another tourist and many types of strangers attach themselves to Indian *ashrams* in search of spiritual truths. An educated Indian seeking to research an urban slum in Mumbai or an Egyptian researching Nubian migrants after the flooding on Lake Nasser will never become insiders however well they get on as pleasant individuals.

36. How far can the researcher be a participant

This is far more complicated than the simple aspirations of the researcher to be 'with it' in the lives of the community which they have chosen as the site of their research; few are cynical enough not to have this hope. Researchers just as everyone else want to be liked. How far will the researcher go into being both a participant and observer of the community being researched. Many researchers do not face dilemmas such as the one previously described in which the anthropologist was expected to have intercourse with a strange woman as part of his role which he had accepted in ignorance of what was to be involved. Would the fieldworker accepted as a classificatory brother in a clan in a New Guinea or the Amazon community which eats parts of the bodies of their dead relatives do just that; it seems very unlikely.

Most do not get involved in the experiental realities of the communities in which they have chosen to live or at least become acquainted with in the long term, they can always leave. Turnbull describes with limited compassion and in detached detail the marginal lives of the Ik in northern Uganda (Turnbull,1973) in which their short term interest in their own survival as individuals made them live lives of what to Western eyes would have been amoral anarchy. There is little evidence that he used his knowledge of their condition to get them help or to save particular individuals who had been helpful to him in his research which advanced his academic career. In fact, he came to dislike them and recommended that they should be 'rounded up' and moved to as distant area so that they could not return (Grinker,2000:160-169); a British administrator would not have reached any different conclusions without considering that it was necessary to take into account the wishes of the Ik themselves. It would seem that Turnbull's fieldwork among the Ik is an object lesson in understanding that the limits on the cultural sympathies of even the most dedicated fieldworker are very near to the limits of their own socialisations.

In the first instance many fieldworkers may not be placed in the position of having moral dilemmas as they may define themselves as outside such issues. Of course, they see the physical cruelty of a husband to his wife, local thieving and illness as just parts of how the local community functions, but turn informants into friends, and these dilemmas crop up almost at once. The woman researcher sees that an important informant whom she has come to think of as more than that is seriously ill, and yet her relatives will not take her to hospital so at what stage if ever does she break into this iron ring of locals practices (Bowen,1964)? The Indian researchers who found that boys were taken for medical treatment before girls, did they never feel that they should interfere in some cases rather than see the whole exercise as one which required detachment (Malhotra and Prasad,1966)? On the other hand they may research market practices in which they themselves are cheated but that is part of what they are researching.

Commonest of all is the plain fact that most researchers are on the fringes of the societies which they are researching and in practice they do not want to be involved further in; it would be moving into areas of social darkness. They are not involved in moral dilemmas as their work does not go as far as to involve them morally. The division is most clearly seen in the realities of the national rather than non-national researcher. The latter is never going to be fully one of the researched community however much they may state that they have been initiated into one of its sub-sections. But there is always the dilemma of knowing too much and thus putting themselves in a difficult legal position as accessories. Thus in researching criminal activities or the activities of criminals, the researchers will always have to decide for themselves when and if they should inform the authorities of potential dangers and if violence does occur why they did not relay that information.

In researching religious groups in which the fieldworker is not a believer, they may feel that they have to pretend to believe which raises moral difficulties which are not quite so clear as deception in secular research (Neitz,1987:15). In this case the researcher stated that she was sympathetic with what members of the group were experiencing but that she felt could not go through the conversion process in all honesty.

It is one of the most hopeful truisms of social science research that researchers have the ability to empathize with those they are observing without actually agreeing with them (Douglas,1972: 28).This must remain an unproven hope. Most researchers would certainly assert that this was so but it seems unlikely when one considers the range of biases to which all researchers are perhaps unconsciously subject as well as the unpleasant aspects of their host community's behaviour in terms of their own socialisation.

37. Looking for the major significances of their research

In practice almost all social science research is micro-cultural; researchers work with a small number of people in circumscribed surroundings since as single individuals their viewpoints are limited. It is difficult for these highly motivated individuals to recognise the smallness of the field from which they are drawing conclusions and the limited range of these conclusions. Nevertheless the concern of most social science researchers is to have the relevance of their work seen as an important part of macro-cosmic understandings. This is due in part to seeing less developed societies as less complex than the Western societies from which many researchers have come or been trained in.

If we take such seminal Indian research work as Srinivas's study of the Coorgs (Srinivas,1952) it seems unlikely that this detailed fieldwork can have had any application or implications for a wider Indian society except at a very general level and as a stimulus to other Indian researchers to follow in his footsteps. The Coorgs have a social system of great complexity which is certainly not reduplicated elsewhere; this work and this knowledge stands alone. A competent study of an Indian or Chinese village or community is always just that and its particular composition and interrelationships are unlikely to be duplicated in any other village within the enormous land areas of those countries The permutations of behaviour in a single village of a thousand people will be virtually endless.

There is perhaps an emotional inability of most fieldworkers to appreciate that their work will be of limited significance even if they have emotionally appropriated a particular field of studies as their own 'property'. Researchers in developing societies unless they have been hired to advise on a particular project and thus have little latitude over what they conclude, should feel that they are creating the building blocks for subsequent researchers to develop and represent achievements in their careers.

38. The use of informants and communities

No community is going to accept the researcher's initial assurances that they are there only to find out how they carry out their daily lives; they do not believe that there is any such thing as the disinterested search for knowledge. It is rare indeed and whatever stand the researchers take, a successful and complete piece of research is a credit to those who worked at it and remains a milestone in their careers at the very least within the academic community from which they came.

The presence of researchers in most fieldwork situations has to be overt since they are known to be a strangers and this has both advantages and disadvantages. The fact that they are obvious means that right from the start they are the subjects of local gossip. They are intrusive but their intrusions as strangers allows their use by individuals to talk about matters which they could not do in their own interrelationships; we have seen that in the relationships of Turner with the Ndembu healer Muchona the Hornet (Turner,1967:131-151).

Of course the researched at whatever level have initial reservations over being the subject of research; no one likes being the object of attention from strangers about whom one knows very little and are suspicious about what they are doing. They have to be won over by interest in what these strangers are saying and start to thinking about how they can turn this relationship to their advantage. These strangers are better off than they are in most situations and they have connections with the outside world which might be helpful to them and their families, because they would not be there in their communities unless they had had such connection; there are material benefits for helping these strangers as they are likely to be paid or at least rewarded for their assistance. It is a delusion to think that they are likely to help strangers out of disinterested wish to broaden the understandings of other unknown strangers about how their community conducts their affairs. Without some material incentives they are probably no more likely to provide masses of information than police informants; they hope to benefit in any number of ways from simple civilities of interaction with these strangers of higher status than themselves to direct material inducements.

39. The publicity dividend for informants

Of course the researched at whatever social level have initial reservations about talking to strangers regarding their communities and indeed even more about themselves; having a chat is one thing but seein what is said written down is another as even non-literates know that what is written down on paper is more powerful than the spoken word and can last for years and years in the files of those in authority. Their natural suspicions are augmented by the notebook and of course the very word research needs a lot of difficult explaining in languages which may not have such a word. But after this initial distaste and suspicion whether from the elders community or from senior business managers, the more alert of all the people approached by researchers come to realise that there is a personal and indeed social-cum-communal dividends from talking to these outsiders.

All informants have their own social positions in their minds and they can supply information which will affect their own status as associates of these strangers but also in bringing that particular business or community to the knowledge of the wider world. They and their community or business are publicized in research reports and other publications in which details of what has been found out through the cooperation of informants presumably reach the socially and politically powerful. Whoever reads what is written will be outside their normal range of contacts and it might well be to their advantage. The introductory references to their help and their community in all reports testify to their personal importance. Some researchers make a particular point in sending copies of their publications back to their principal informants even if they do not know much English or whatever language in which it is written. The personal prestige of informants who are documented in reports and even pictured in subsequent books such as the leader initiating the impressive social changes in Chan Kom (Redfield,1950) and more particularly those shown in Margaret Mead's work on the Manus islanders.

The socially disadvantaged in terms of national political and social objectives have in researchers an excellent opportunity to argue the case for their special needs and to put on record the results of perhaps their seeming neglect by local government officials, even if their situation is no different to every other community in their area. There is also a publicity dividend for some who are associated with a publicly acclaimed researcher. The fact that Margaret Mead returned several times to the same Manus community meant that she was able to get funds for their social welfare which they would not otherwise have received; it was no wonder that some who had been particularly associated with her, wept at hearing of the death of such a wealthy visitor whom in their terms they saw as a friend.

There are always potential political correlations to any social science research. If the research is done on a small and distinct social group, there is the possibility of that community seeing this work as an opportunity to build up their own political power. If the study is anything to do with the surrounding macro culture more or less coinciding with national borders, it can be seen as supporting a form of nationalism in opposition to neighbouring states (Cernea,1982).

If we then argue that any group of whatever type or persuasion which has been researched, are able to understand their own reality as prototype social scientists and thus able to influence and indeed provide solutions for their own communities, then we have difficult situations to untangle (Mott,1982); certainly some people in communities which have been over-researched by a succession of social scientists know the terminology used and how it is applied to them. But then there are cases in which a person there is as well qualified as the researchers to make observations about their own communities as the English speaking Trobriand chief who had a degree in economics from an Australian university. The researchers would be becoming public opinion leaders and in a way the ideological employees of those they have researched. It would be politic to inform those in the community who have shown a longer interest in the research project the details around which the project is going to be written up, not in the sense of getting their clearance but to ensure that any marked differences of opinion are given a hearing.

40. The case study integrating quantity and quality

Any fieldwork project, however defined, has weak boundaries over which people and ideas come and go in ways which cannot be controlled by even the most active and conscientious researcher. Designed experimental research situations have their own built-in disadvantages despite their convenience and the points they may illustrate under such artificial conditions.

It seems likely than the narrower a research project can be, the greater the control that the researcher can exercise thus studies often concentrate on individual events such as a Balinese cockfight, a Sukuma Buchwezi initiation or an English village separated from its neighbours by houseless fields or individuals who are sufficiently interesting and willing to be the objects of research. There are of course leakages and biases in such approaches but they may be more easily identified and allowed for.

The study of a single elderly Navaho's philosophical thinking on a one to one basis with all the biases that this may have included, leaves a residue of carefully thought out hypotheses. Others such as Baba of Kano are based on the friendship between two individuals but the problem there is not just that it is an interesting and sole biography of a Nigerian but we do not know what has been left out. There is also the fact that this is the work of a non-national and we do not know what points the authoress is making in writing up the 'story' of a single middle aged Hausa woman. This cannot be considered a case history as much must have been modified by friendship and the fact that the recorder/editor was a non-national (Smith, N,1954).

There are many situations isolatable, opportunist and finite enough to be of value to social scientists even though they may be partially handicapped by shortage of funds, time and other factors such as language differences. In a sense these opportunities choose themselves because they are self-defined such as the coronation of Elizabeth II or the killing of the policemen at Chauri Chaura (Amin,1995). A strike in a factory or the epidemiology of an easily identifiable disease which will both allow the collection of data and the follow up of specific leads.

There is always the problem of single events which are receding in time as in gossip related to the Uganda coup in 1966 collected within a few days of that violence (Tanner,1978). The wide awake social scientist in the field must surely take advantage of sudden impressive events even if they are outside their primary field of enquiry, even a sudden outbreak of student unrest in their own university or cheating in a particular examination. The non-national is handicapped because they cannot follow up a case study as easily as a national who can develop longer term networks of contact with related information and informants well beyond the boundaries of the immediate event. The single event case study is both more manageable and allows its integration into the larger picture of local events as well as the developing of details concerning key individuals. Macro studies usually end up as trying to detail trends for which it may be difficult to collect enough data which can be processed and verified. The assumption that the case study has greater accuracy may allow its conclusion to be more readily accepted as having a useful place in social science research than more grandiose fieldwork projects.

Chapter 4

The Foreign Researcher

1. The non-national outsider

The distinction between a citizen and a non-citizen in connection with social science research and indeed any other matter is an obvious one; it is a legal distinction based on national law which may both restrict and enable. This is only one aspect to consider of a complex set of correlations between the behaviour of those carrying out research and those on whom and with whom these foreign men and women are working. So from the start the non-national researching within national boundaries is a political issue bounded by time restraints of the researcher's visa, restraints over the areas in which research can be done, the subject to be researched and the researcher's previous record when they apply for a visa and research permission.

This distinction firstly involves the governments concerned; those receiving the researchers and the countries allowing them to go there; the importers and exporters of a special kind of foreigner and these are political issues. There is no such thing as the unsupervised passage of citizens from one country to another. The donor countries may not want their nationals to be at large in countries with which they are not in reciprocal nominally benign relationships since they could easily be arrested as spies and they would be outside diplomatic protection. Also many countries are considered to be from time to time so dangerous and disturbed that it is considered unwise for their nationals to be there even if they are members of well reputed universities or internationally recognised non-governmental organisations involved in aid. Some countries have been so disturbed for so long that there would not have been any social science research done there for several decades by either nationals or non-nationals. The governments may have legal status for international recognition but there are countries where their authority does not control much beyond the main regional cities. Other countries do not want the citizens from countries which are not overtly friendly 'wandering' about potentially providing or seeking information which in their eyes would be adverse to their regimes and their policies over corruption, inefficiency and maladministration. Perhaps a majority of countries have been on occasion closed to non-national social science researchers from the so-called free world such as, China, Russia, Myanmar, Sudan, Eritrea, Maputo and Mali. Similarly such countries may not wish to have their citizens in uncontrolled relationships with men and women from a different political environment even if money and sponsorship should be available.

So there has been from time to time a bilateral restriction on non-national social science researchers which has blocked off large populations and geographical areas as well as topics from any independent investigation. It may also be likely that the non-nationals may prefer to study people who are clearly 'others' by the fact of their different citizenship rather than study 'others' who are part of their own societies which involve complicated and less obvious but more difficult issues of bias. It is less attractive socially as well as intellectually than going into a strange environment which, at least theoretically, can be seen as exotic. This means that they start with the biasing assumptions of difference.

On the other hand nationals researching some aspect of their own societies have the major but more useful task of being forced to work out in their own minds what are the differences between 'us' and 'them' when both are reflections of the same culture however refracted by environmental factors (Lewis,D,1973). Non-nationals in their own countries tend to take a somewhat superior status of non-involvement in national affairs because politicians tend to be classified in pejorative terms. This is particularly valuable as national researchers cannot dupe themselves into thinking that they can have a detached view of what is going on around them; they are politically involved as voters, government policies influence their economic and social status and the health services affect their well-being (Stavenhagen,1971). Unni who was once an assistant to Haimendorff in his work in a tribal part of Travancore (Unni,1979) stated that they both found themselves equally as strangers in this environment and that he himself knew little of caste behaviour other than his own.

We have noted that Malinowski in his Trobriand village found his tent sited next door to the house of the headman and this was possibly no accident of hospitality as he wanted to keep an eye on this stranger and to get what he could from the relationship. It is easy enough to suggest that researchers should cut themselves free from such biasing restrictions and indeed to have this in mind as a professional obligation. This can rarely be done without giving offence and polarising researchers into being part of the implied local opposition. Perhaps it is best to retain such initial positions and sooner or later to explain to those with whom they are working that their position is too obviously on one side of social divides as in associating with a group of self-segregating socially superior students in working on the social system of a university (Sivakuma,1979).

2. The researcher as entertainment

Although social scientists find matters of interest to outsiders in almost any community, for the people themselves particularly those who have to work for their own subsistence and those who work under factory conditions, life can be extremely dull and continuous drudgery. A Javanese social scientist researching a Dutch fishing community was clearly an outsider but what these people saw as his peculiarities made this man an interesting person to meet and who found the experience pleasant. They enjoyed talking to someone who was a very distinct foreigner to whom the disclosure of confidences had no local handicaps particularly as he had learnt to speak Dutch in an Indonesian school prior to their independence.

So it was for him a similar situation to that of a non-national researching some part of Javanese society without the handicap of not speaking the local language. This researcher was not showing any new approach other than the one which he had been taught in a Western oriented university as far as fieldwork was concerned. He was received by these Dutch people not so much as intrusive which he certainly must have been in attempting to find out as much as possible of their covert and overt social lives, but as a diversion from the monotony of their everyday lives. And indeed one that was more exciting than the outside world which they saw on television; he was real and in their houses drinking their coffee.

In many fieldwork situations the non-national researcher is socially distant from almost every aspect of the lives of the people they are researching; it is not so much the questions they ask which become the subject of gossip but almost every aspect of their lives is of interest. Most communities have strangers who come and go and they are usually known by their functions as government servants, traders or religious functionaries of some sort, who inevitably want something from the community and its members which they are not particularly keen on providing; interesting enough but generally they are keen enough to see them go. Foreign researchers are different however as they stay and are thus the subject of endless gossip and observation because everything that they do; how they wash and clean their teeth, how they excrete, what they eat and how they keep their possessions clean. It is lived in entertainment. These non-national researchers are socially unknown and thus they are intriguing to all who come in contact with them or hear about them from their friends.

3. Lack of privacy and the foreigner

Privacy in any culture is socially marked out and varies between that understood to be acceptable at different levels of society. While most social scientists as part of a university social system have been in and out of the social life of a group, they have always been in social situations in which their privacies have not been invaded by outsiders. Most non-nationals have been used to a closed door pattern of living in which entry is subject to substantial restrictions; a knock on the door and a request for permission to enter is a particularly Western pattern of behaviour.

Non-nationals injected into communities of which they are not members and have no practical knowledge of how they behave or how outsiders such as themselves should behave are at a disadvantage. They are the objects of curiosity if not entertainment to people who have had little long-term association with outsiders beyond seeing them on television, in markets and as passing tourists or short-term visiting experts. Children creep closer and closer because the non-national is nervous about getting rid of them in a way that may not be locally acceptable. Many non-national field-workers find the absence of privacy stressful when they regard themselves as off-duty as they have had no previous experience of being encircled by the curious who have in the understanding of the outsider no socially acceptable reason for being there. Nationals will have had their own experiences of this type of sociability as they may have been brought up in families which carry on various forms of joint living and who are used to living in each others pockets. It would seem that the national social science researcher in this respect has at least two clear advantages over the non-national Firstly, they are not likely to be the objects of so much close curiosity and secondly, they are more likely to have personal knowledge of local socialising practices and expectations and not to find it intrusive.

The non-nationals may be followed about when they want privacy to wash and perform other personal bodily functions. They may also not even realise that it is much easier to be overheard in communities in which walls are unsubstantial and doors if they exist at all are usually open. The lack of privacy is something to which the non-national often find adjustment very difficult.

4. The foreigner and the Malinowski syndrome

For national researchers the difference between on stage and off stage behaviour (Goffman,1969) may be a matter of degree rather than of sharp contrasts which involve ethical as well social differences. For non-national researchers there are abrupt differences between what they do on stage and what they may do off stage in the privacy of their temporary homes and in their minds. These contrasts are often so extreme that for those who experience them, it raises for difficult ethical issues.

Occasionally the curtain of privacy is lifted and we find social science researchers expressing opinions which cannot be part of any detached view of the people whom they are researching. There must be many researchers who might think that those researched have brought their misfortunes on themselves. Diamond suggested that if the people of Ruanda over breed and overstock their small environment with people and livestock, then Malthusian solutions come into play and genocide reduces environmental pressures for perhaps a generation but then he wrote from the safety of his own country and established academic status (Diamond,2005). Most non-national field workers are conscious of the need to be politically correct according to the fashions of their time and they keep to themselves their often qualified dislike of the people and their practices with whom they are committed to work.

Diaries which are written for the private relief of the individual rather than as an extension of their field notes and thus may disclose the psychological and social problems of their writers, and a different understanding of their fieldwork experiences. Malinowski over several years researched Melanesian society from which he wrote his classic and rightly venerated studies of their life so much so that he is still seen as the father of systematic fieldwork by non-nationals or nationals. Then onto the scene came his diaries which presented a completely different facet to the life of this distinguished anthropologist; the 'on-stage' professional and the 'off-stage' person isolated in the Trobriand islands for several years.

It is not that his diaries opened up a can of worms to public view turning him into a man with racist views, reducing the value of his professional work but that the perhaps inevitable division between professional and personal life is certain to be always an important factor in the fieldwork experience of every non-national researcher. He was interested in these people but he did not see them in any sense of human equality and if anything he looked forward to breaks from this close association with people whom he saw as inferior to himself and the culture from which he came, by visits to European administrators and planters because he wanted to have a bath and a beer without being in the public eye of the curious.

The reading of many introductions to the published reports of fieldworkers do not often print the fact that there were persistent tensions between what they were required to do professionally and their wishes to be back in the psychological and social 'safety' of their own cultural environments. From the reading of these introductions it would appear to have been a situation of all sweetness and light from start to finish in which individuals and communities are thanked for making the fieldwork period so successful. On the other hand it has long been recognised that fieldworkers were inevitably subject to prolonged psycho-social tension (Gans,1982) and that the survivors who continued to be successful had rare characteristics.

The point must be that the Malinowski' diaries (Malinowski, 1967:247-49) disclose an important methodological point that personal experiential views are usually at variance with their professional ones: 'I turn in too late. I get up at irregular hours. Too little time is devoted to observation, contact with the natives, too much to barren collection of information...I must never let myself become aware that other women have bodies, that they copulate'. The reactions to the publication of this diary seems to have been that he was letting the side down by showing that he had certain psychological and social attributes which detracted from his professional status. Few gave public recognition that these disclosed differences were an important and ever-present issue in fieldwork by non-nationals.

5. The impossibilities of 'going native'

Going 'native' in its many and varying forms is always an attractive possibility for researchers wanting to be fully accepted in their fieldwork areas. Of course, this is an emotional response not far from the attractions of Italian and Spanish society to a British tourist on a short-term summer package tour which involves little of the requirements for mentally and physically fitting into a totally alien environment. Hypothetically it means keeping to their rules rather than the one into which the researchers have been socialised and that is a very tall order indeed. While many have thought of themselves as becoming a 'native' it is perhaps only the American anthropologist Cushing who has at least partially succeeded in becoming a 'native' (Gronewold,1972).This researcher in working for his acceptance as a Zuni kept whatever secrets he had access to despite his position as an academic to succeed as in this one area of acceptance meant taking on obligations in another which handicapped his own researcher activities which would have involved their disclosure in teaching and publication. But as one Zuni remarked the fact that he ate rat meat in an initiation ceremony did not make him a Zuni only an unusual white American. It was said that Cushing learnt the Zuni concepts of sacredness so well as to never have divulged what he had learnt even to fellow social scientists. This raises the question as to what can have been the point of such a transformation if he became committed to their moral controls; he ''half and halfed" himself into a special role of personal exclusivity in which he was neither a Zuni since he did not live with them and as an anthropologist kept their secrets, nor an anthropologist. This is not the same as those we have mentioned earlier who for a time did become one of 'them' in a totally encompassing sense. Kimball in trying to get himself accepted into an Irish community stated that he was not accepted until he said that he had seen a local Irish ghost (Kimball,1972) and he also stated that his co-author believed in fairies which may have helped his assimilation (Arensburg, C.M. & Kimball, S.T. 1940).

One has to accept that whatever the effort and cognitive honesty with which such a researcher would be involved in becoming a 'native' in a cultural as well as a native sense, for an ethnic Indian researcher to turn into an ethnic German or for an American researcher to become a Samoan is an impossibility in the short time span of research. Further even if it was possible it would undercut their research possibilities on ethical grounds; it is no more possible than to expect a marriage to work while the partners retain the practices and aims of the unmarried. In the long run there may be a legal change but it may take several generations but even then their different appearance or distinctive behaviour may make them continuing members of a recognised minority.

But there are of course small acts which appear to make the researchers more acceptable such as Kimball's seeing of the Irish ghost by which he was seen as having had a legitimate Irish experience and from that time on his research went ahead more easily. It may well be that a researcher who experienced the 2004 South Asian 'tsunami' in the community in which they were working will similarly have become more acceptable.

Most if not all non-national researchers want to be accepted by the communities in which they are working as fellow human beings rather than as more or less disruptive or intrusive foreigners; they want to be treated as if they were some sort of fellow 'native' and there is no pejorative sense given to the use of this word as it would apply as much to researching middle class professionals and their families in Mumbai or New York as it could to working with Igorot rice farmers in the Philippines. An Indian researching Indians, can to some extent, slip into the role of being one of 'them' without being so obviously different if only on ethnic grounds. This issue is one that dates from the earliest days of anthropological fieldk-work in which certain outsiders became so enamoured of another culture, that they endeavoured to become as far as possible one of its people and indeed at least for the time of their research abandon their natal status. Any check list of acceptability includes appearance and social criteria; how people speak and interact or are part of the researched social environ. An African American researcher in Boston would probably not get very far, but the same person in New York would have less difficulty. Conversely a white American in Louisville Kentucky would might have less problems researching in a black community than his opposite number in a white community in Boston. Much the same might apply to a Tamil in Amristar and a Sikh in Cochin.

The initial qualifying factor depends on how long the researcher expects to be able to devote to this absorption. There are certainly examples of individuals who have become totally absorbed into another culture such as the many thousands of women who were abducted by one side or another during the Partition of India and who could not be reabsorbed into their own communities for reasons of shame and pollution. There are others who have made the pilgrimage to Mecca by working their way stage by stage and having run out of money stayed on in Saudi Arabia or got stuck in the Sudan. In fact throughout history there have been individuals predominantly women who have become 'natives' after first entering these communities as 'foreigners'; women marrying in rather than men marrying out.

We are dealing here with people who for academic and emotional reasons have become or at least have attempted to become one within a community totally different to the one in which they had been socialised for the first decades of their lives. But we are discussing here people who assume a totally different dominant role for academic reasons and who latter attempt to assume yet another dominant role for emotional rather than rational reasons just to gain further insights into how their host community thinks and behaves. Those who change citizenship for reasons which have nothing to do with social science research are not in this category as it will rarely involve any change in social and psychological patterns; a transfer of political allegiance is a simpler process. There are of course extremes in this as with social scientists who become involved in guerrilla movements. The point under discussion is whether it is possible to become a 'native' in a totally new social environment with which such individuals have no previous connection for a limited period and then to return to their previous one. Within Western societies for research reasons there have been social scientists who have become a member of a violent gang, got themselves locked up as mentally disturbed, taken on the life of an itinerant vagabond and even become a stripper. There are examples of people who have taken on 'other' roles for personal reasons such as Burton's visit to Mecca and his literary exploitation of this achievement; apart from the difficulties and moral issues in such deliberate duplicity, this was not a social science exercise. The issue here is not the ability to take on another role which must be common enough in the personal lives of most people as they change their demographic characteristics, it is one of professional social scientists taking on a particular role for research reasons and then returning to academic life to harvest the information gathered during this period of detachment from their usual ones. Can this come and go role-change be achieved and at what personal cost? Indeed the issue is how far can any individual go in such a role-change without damage to their long-term identities socially, psychologically and physiologically.

Fieldwork requires a detailed involvement with the researched and their interests and at the same time an emotional distancing is necessary but it is of course difficult to detach personal emotional necessities from current work and from persons and matters that are not part of current professional needs. This is perhaps easier when the researched are not obviously distinct in some ways socially distasteful; there is not much likelihood of any emotional closeness in researching those who are extremely deviant from the standards into which the researchers have been socialised. Sympathy perhaps, but there is a limited tendency to draw together the concepts and feelings of 'us' and 'them'.

However, the researcher, possibly young and detached from their own society for the first time with which for a variety of reasons they may have developed some dissatisfactions, so that the culture shock of meeting new societies is often tinged with ideological optimism, and they enter another society of which they have had no previous experience, which they felt to be emotionally welcoming. Most societies have a cultural requirement of being welcoming to guests and accept them in a ready manner which is not all that common in Western societies outside the status groupings of those involved. There are examples of people who are quite overwhelmed by the ways they are accepted by strangers in a strange society as in Malcolm X's reactions to Arab hospitality during his pilgrimage to Mecca, which he saw as an almost extreme contrast to what he would have experienced in many situations in the United States and not as what might well have been a reaction to a fellow pilgrim. It could have been assumed that because he was a prominent figure he was accorded special treatment, but this seems not to have been the case. Most researchers do not stay long enough in their area of fieldwork to become emotionally involved or to attempt any transition into another culture, but it always there in the backs of their minds that they might become accepted as persons rather than as researchers; the partially possible is never the same as a total commitment in p actice into in an environment for which these people have nothing in common.

6. The permanency of the outsider status

Non-national researchers are always outsiders unless they are within the 'half and half' status of emigrants who are able to reinsert themselves into social roles from which they have previously excluded themselves by changing their nationality and becoming semi-nationals. These researchers must always be some type of outsider; how could it be otherwise when they look different both physiologically and in how they dress, talk and have received a higher education than is likely to be provided locally. This is apart from the fact that a social scientist researching some aspect of their own lives has inserted himself into an outsider role. The non-national is always going to have a personal sense of feeling 'out of place' since they are coming into a social and geographical environment of which they will usually have no more than limited knowledge from what they have read or been exposed to in the media. This sense of being an outsider will always be compounded by the comprehensive behavioural mistakes which they will make either in the normal course of social intercourse or in the course of trying to get information inappropriately for a person of their age and status.

Non-nationals are always out of place because there is no ostensible reason for their being there in this community or social setting, they are not cultivators, fishermen, married women, nurses or teachers. Social distance between the researchers and the researched is both a fact and a feeling. It is not so much a question of mistakes being made ,which could be interpreted as impoliteness or slowness to learn what every child of eight years of age knows perfectly well. It comes down to the degree to which they are acceptable by insiders and more particularly by insiders of local influence; in this a local headman is more important than being sponsored by a political leader in a wider area than the village. Of course should they continue to associate with one particular community as did Cushing among the Zuni, they can gain special status as honorary natives, a form of classificatory kinship in which they are seen as acceptable and indeed useful foreigners who may have provided help for some individuals and for the community as a whole and who may be expected to do more of the same in their futures.

However to think that they have achieved some special status parallel to that of people in the community of equivalent age is a delusion even if the people themselves go along with putting such delusions into play in the first place. It was written about Cushing (Gronewold,1972) that "although he became to an impressive degree a Zuni, he did not cease to be an anthropologist. He did not go native. He grew into being an anthropologist in a way not unlike the process by which he became a Zuni without the kind of formal education that is commonly followed in becoming either. The poignant truth is that having gained admittance to deeper Zuni secrets he learned so well the lesson of sacredness as never to divulge what at last he came to know, not even to fellow scientists". It seems reasonable to question whether he did in fact learn anything all that special and that his Zuni associates rather than companions went along with someone who showed such persistence and dedication to Zuni interests. It may have once been a particular facet of Western dominance that some fieldworkers have thought that they could translate themselves into 'natives'.

Except in the rare situations in which fieldworkers research their own communities and that cannot be the case with non-nationals, there can be no uniformity of interrelationships which would enable an adult non-national to become a national; even the most acculturated non-national would find that the fit between the two conscious and subconscious cultural understandings would at best be extremely jagged. But those who have put themselves into close association with small groups such as the woman who lived with a single Eskimo family through an Arctic winter (Briggs,1970) may have achieved a more concrete form of acceptance as real persons. Even then there are emotional and social barriers which block total social and psychological acceptance. How indeed could any Eskimo family accept as one of their own a young unmarried Canadian woman even if they could comprehend the total illogicality of any such person wanting to live with them under such conditions of propinquity and isolation; they must have thought her at best somewhat deranged. Perhaps all this adds up to the continuing delusion that the fieldworker is a pragmatic scientist so that the absence of personal details from their reports would seem to be a drawback to any realistic understanding, not necessarily detracting from their achievements. Personal details of how they coped with being socially and culturally isolated and having to use a strange language are an essential part of the what they are able to achieve, but is rarely given and usually absent in any detail.

What researchers are doing in practice, is going to live in another community composed of total strangers for so long that either he or she become members of a society in which they have no natal or affinal connection. For an incoming English man or woman to gain acceptance by the traditional core of born-there inhabitants when they have no connections there, is going to take decades, but in terms of research an Indian social scientist or any other non-national might have easier acceptance in that role.

Most social science fieldwork in developing societies is a continuing reality of come and go social encounters even if this is done quite regularly but the regularity is the researcher's not that of the researched. They for instance come in the dry and cool season so that their experiences would not involve hot weather or the rainy season and more particularly where they can get their grant and personal arrangements together for a long absence.

Non-national researchers can hope to be accepted in so far as it will enable their fieldwork to be potentially profitable but to go beyond that the power holders in the researched community would have to have their own reasons for more detailed acceptance and agreement on the terms of such acceptance, since in most societies there is no researcher role, he or she would have to be slotted into their social understandings this would have to be in their thinking some reason for the researcher's as classificatory kin, business associate and so on. As discussed earlier, the category of 'friend' which occurs so often in fieldwork literature as an abstract acceptance without any social overtones of reciprocity, probably does not exist. There have to be evaluatable reasons for assuming and maintaining any relationship; what is in it for me if some form relationships with a non-national?

However well-intentioned a non-national researcher may be and however careful their approach may be within their research environment, they are always accident prone. They will make mistakes in their behaviour according to the cultural environment in which they are cognitively embedded. They may be able to avoid serious ones because of their sensitivities to the major aspects of this new culture in which they are working, but they will show inevitable errors in understanding and behaviour in what they do and how they interpret it. There are always major aspects of a foreign culture which the non-national will miss, particularly in speech which will disclose that they do not quite comprehend the social dimensions of the language which they have only recently learnt, perhaps at a rather rudimentary level (Chilungu,1976).

But the outsider status is not necessarily one which is confined to non-nationals as a national researching some topic which is not automatically part of their existing roles as employer, nurse, teacher or farmer. In India a Brahmin researching village life which must by definition include castes and outcastes outside his or her status, is almost as much an outsider as a visiting Frenchman undertaking a similar fieldwork project. This quasi-outsider status of the national is not the same thing as that of the non-national. The former bring with them the almost predetermined biases of their own culture and it may be an advantage to have such identifiable differences available in advance. It may also be that this 'outsider' status can reduce the amount of 'working in' time which the social scientist has to be involved in before they can start work. They are what they are and thus have some neutrality to each group or person which they approach. As such they can be talked to without the built-in inhibitions of such talk within their own communities and this may help the gathering of data (Trice,1970).

This is not to suggest that non-nationals cannot make significant contributions to the understanding of cultures other than their own. There is always the suspicion that their writings, however sympathetic and dictated by prolonged contact with the people about whom they write, that they must have lived under special and altered conditions; both sides of any such experience will always have said to themselves that the researcher is not one of us. A French woman (Reymond,1958) who lived for a year with a Brahmin family, must have been either post-menopausal, or the Brahmin family concerned must have made substantial changes in their life style in order to cope with her ritual pollution on this one account alone or a degree of discussion which would be unusual in any family setting. In many societies the post-menopausal woman is no longer seen as a social or ritual threat and can enter some male areas as a neutral figure. The question of menstruation and indeed excreting is rarely mentioned in discussions of participation. It seems very unlikely that this family which she describes as strictly orthodox did not have to make substantial changes in their lifestyle to have had a non-Hindu foreigner living in their midst for seven months. The point must surely be that even such a well described and sensitive study of this experience, it is that of an outsider for whom this family made a place. We do not know what their reasons may have been for such a prolonged intrusion even of an obviously sensitive foreign woman as shown in her book; who got what out of this relationship and was the studying and recording one-sided.

So the non-nationals are clearly outsiders even if they did not look different and both sides of the research situation recognises this while probably covering this with some conscious and unconscious duplicity. These researchers have to get information which they may well be unable to get on their own and moreover they have to get it within the length of time dictated by the size of their grant and the conditions under which they are working laid down by the national government. For most non-national researchers there are limited possibilities of a return to the subject environment; local conditions change, further grants cannot be obtained because they were not as successful as they had hoped, and families have been started and become a demanding reality. It is wise to remember that if asked whom they would save if a boat overturned and they could save either a wife or a mother, those from developing societies tend to save the mother and those from developed ones the wife. This author has found this often when the question has been asked and this is perhaps shown in the answer from convicts and seminarians in Africa and Britain as to whom they will save. Cultural differences remain and perhaps understood, but the researcher's thinking still remains that of their natal culture; they have experienced another culture to which they remain alien. It is easier and more desirable to get employment within their own national environments as the attractions and excitements of fieldwork diminish with age.

Research reports usually contain a mention of thanks to the community on whom the research is based in general terms and in some cases to specific people who have moved from being informants to being research assistants. The thanks to those in their own profession and who may well be useful to their future careers almost always outnumber local thanks, often tediously so. It seems that the professional needs of the researcher dominate the more human aspects of their fieldwork in which they may have thought in terms of acquaintances while those researched thought in terms of useful friendship according to local ideas of practical reciprocities.

In practice relationships with local informants are based on expediency and seldom survive the enormous social and geographical distances which separate them once the fieldwork period is over; occasional letters are not an easy method of communication when there are differences in intellectualised literacy and the researcher gains nothing from such continuations except the embarrassment of having to refuse requests for financial and other help.

Since the researchers are always going to be social outsiders with some degree of social distance separating them from the people they are observing and even that word separates 'I' from 'them', there may not be any automatic advantage of being a national as there is always out-of-placeness. Research is undertaken because of a lack of knowledge, and this lack of knowledge can also apply to appropriate social behaviour. Personal adaptability is a variable ability since it is both situational and interpersonal. There is no reason in itself why either a national or a non-national should be better at this; who seeks out who and for what personal or professional purpose may not always be directly related to the research project itself. Convicts talked to this author because it was a relief from the boredom of long imprisonment.

But what happens when an Indian researcher goes to work in another country such as Japan; in most respects they are non-nationals in much the same way as a German researcher in Malaysia. It seems doubtful whether the theme that this Indian comes from another Asian country would have much relevance. There are situations in which a subordinate group like an American-Indian community who initially thought that the Asian researcher was one of themselves, then became unhelpful and only tolerated the researcher as a paying guest when he was carrying out his research (Pandey,1979).

Many if not all Asian social scientists have been educated professionally within the Western educational system and thus Indian researchers in Malaysia and Malays in Sri Lanka might use the same paradigms in approaching their research work rather than much older, and perhaps less relevant to contemporary needs, Eastern systems of philosophy. But moving into Japan or China is entering massively well organised and long established cultural systems which may have only been peripherally affected by western processes of thinking outside hard science and mathematics and in which the major Western languages have no part of the general life of these peoples. In Japan however there has been extensive westernisation of the language in relation to both social habits and technology, but not necessarily in patterns of thinking.

The Western non-national researchers have to combat their long-established feelings of intellectual and cultural superiority but an Indian researcher working in Japan has to cope with feelings of their own inferiority to that culture rather than just in comparing the mechanics of modernity. In addition it is perhaps culturally inevitable that an Indian researcher might look for something paralleling caste or deferment to age just as a British one might also look for class parallels, just as Marxists or Christians look almost instinctively for cultural parallels to the patterns in their own system of thought.

7. The handicaps of language

The outsider in social science fieldwork whether in their own societies or particularly in one in which they are culturally foreign, has one almost inseparable and certainly virtually endless handicap of language. Of course, non-nationals do their best to learn new languages and new linguistic codes but it is doubtful whether they get beyond a moderate competence because language use is always closely integrated into almost every aspect of social life. To learn and use a foreign language for social research and to see it as a way into comprehensive understandings of a culture is to put too much stress on the possibilities of learning a language up to the standards of a natal speaker. This author used kiSwahili everyday and all day for ten years as an administrator in Tanganyika and reached bilingual status as a translator for the International Labour Office (Tanner,1960) and of Machiavelli as a political classic (Tanner and Kamoga,1968) but he could never be mistaken for a natal speaker even on the phone. Trant writing as a tricultural French/English/American wrote 'It may be the ethnographer or the anthropologist again having his ears wider open to what he considers the exotic as opposed to the familiar, but I still feel I'm discovering something new in the use of language here almost everyday. I'm getting new expressions almost everyday as if the language was growing from every conceivable shoot' (Trant,1975).

Conversely while nationals may have problems over linguistic codes, they will have no difficulties over the major languages within which they and these linguistic codes developed. Thus non-nationals are faced with learning a language with which they have no previous socializing contacts. There must always be a long period before they can interrelate with the objects of their research in the simpler aspects of their mutual lives at a reasonable level of competence. They are also handicapped by the factor of time relentlessly running out before their grants or visas expire and all the time there might be some feelings of inadequacy and incompetence as a result of these communication issues. Evans-Pritchard was unable to learn to speak Nuer and since this was combined with their refusal to get themselves linked to his research requirements 'As I could not use the easier and shorter method of working through regular informants, I had to fall back in direct observation of and participation in the everyday life of the people. From the door of my tent I could see what was happening in camp or village and every moment was spent in Nuer company' (Evans-Pritchard,1940:15). Not only was this distinguished anthropologist unable to communicate directly with the people about whom he wrote but the more sophisticated aspects of his thinking on their religious ideas must be assessed as a personal and qualitative interpretation presumably by non-Nuer interpreters. So non-national researchers are likely to rely extensively on what they can see since this is easily done and free from the handicaps of limited language understandings. It thus tends to be the dominant mode of research throughout their fieldwork in different degrees of participant observation.

Seeing is of course a somewhat circumscribed method of recording reality since it depends on personal eyesight and obviously one can only look at one activity at a time, attention is limited in its efficiency. Seeing tends to concentrate on what the observer may evaluate as needing their attention; informants in this author's prison research on their behaviour of convicts stated categorically that if there was a disturbance which turned the eyes in that direction, then it was more important to look in the opposite direction as that was where the more important activity was going on; this is difficult to do in practice.

As a substitute for personal interrelationships based on speech in a common and mutually understandable language, it is just as reliable or unreliable as any other method of recording information. But combined with the adage that 'seeing is believing', there is no substitute for semantic understandings as an antidote to the various forms of deliberate or accidental deceit which occurs in many social relationships. Subjectivity gives a high value to what is seen and so does not receive the same critical evaluation as other forms of data collection. The insider national without linguistic handicaps can utilise their other senses in support of the understandings which come from sharing a language.

8. The foreigner and utopian frameworks of thought

There can be little doubt that any individual whether or not involved in research, who moves from their own cultural environment into another one finds it an exciting and stimulating experience. Whether from East to West or from developing societies into the developed societies of the West, such stimuli are inevitable. The writings of Nirad Chaudhuri, an Anglophile Bengali, which detail his cross-cultural experiences and reactions are remarkable examples of benign cultural shock (Chaudhuri,1951;1959). The non-national researcher is no exception to this and for them the sudden and overwhelming influx of visual stimuli in which virtually everything is new even though it may have been seen in the media and read about for pleasure or as part of their preparation for fieldwork. Most do not record their excitements from these first encounters with totally new environments perhaps because it does not fit into their required professional detachment from emotional involvements during research work. It is almost certainly a break from what researchers have currently been experiencing in their home environments, a form of mental freeing. Even in situations which are markedly poorer and even distressing in comparison to the researchers' own lives, there is often a feeling that these people have virtues and social practices which are better than their own and have been lost to them. These researchers have a utopian framework of thinking.

Since they are all from the urban environments of universities, meeting subsistence farmers and pastoralists for the first time is likely to see them in terms of a freedom which their own societies have lost, even some mentally held relics of Rousseau's thinking about the 'noble savage'. Their immediate reactions are not likely to be in terms of ecological and social interconnected instabilities, their low standards of living and high rates of infant mortality. The essential optimism of much thinking about 'strange' societies long before the ideological concept of development became commonplace has usually meant that this idea of utopia is always latent in the thinking of foreign researchers and brought to the forefront of their minds by the visual impact of previously unknown societies.

9. Involvement in politics

Whether they are involved in politics in their own countries or indeed have a political framework in their approach to their own countries and that of their host society, there can be little doubt that there is a file on them in the local police headquarters on the basis of information supplied by their superiors. They are not necessarily kept under observation but the social mistakes that they make and any rivalries that they create will be noted. After all they look different and socially behave differently and that can be disturbing to all involved with them if they have reason to object to something. The American teacher of English won a long distance marathon against Chinese competitors may have caused dislike and the choice of the literature which he used for class-work will certainly have been noted. Despite his knowledge of Mandarin his whole stay was on a knife edge of acceptability not only from the police but from the local political hierarchy over what he taught and the teaching staff over his popularity (Hessler,2001). This is not always to be seen as part of the hostile environment of an only partially acceptable political regime according to Western evaluations, but one of the potential embarrassments to the regime should they be injured as a result of a conflict.

We have already referred to a non-Muslim Arabic speaking American academic researching sermons given in mosques was edged out by the local police with great courtesy and much coffee drinking by being told that if Cairo would agree to his continuing then they would have no further objections; of course, on getting to Cairo, they played the reverse that they would have no objections if the local police had none. Eventually he realised that he would never get permission to continue what was in fact a purely theological piece of research. Similarly in Uganda when there were disturbances in Busoga in 1968 the police brought back into Kampala the two non-national researchers working there.

Non-nationals should perhaps get used to the idea that they will always be under some sort of surveillance and even a courtesy call on their local police chief might be sensible as after all they may well be in danger from time to time just as much as they may be seen as a danger to others. Strangers are always strangers.

The fact that researchers are by their nature and training hostile to overt or covert authoritarian regimes, tends to put them under local suspicion. Their informants are likely to be informants to the police as well and it is prudent to keep notes in duplicate so that they can be posted away against the possibility that the originals might be stolen. It is essential to avoid recording data which can be of danger to those who made these statements.

It is probably impossible to avoid involvement with local politicians as after all permission to do research will be made in a combination of academic and political reasoning. Some research topics are clearly so dangerous that it would not be worthwhile to apply even though all those involved know that corruption or whatever is a national problem. The study of corruption in Tanzania (Green,1981) meant that the writer wisely got academic employment elsewhere before his contract came up for renewal. How far can a researcher into the use of domestic water in Sukumaland go in detailing the corruption in the regional Water Department without first being out of the country and without any plans for returning (Drangert,1993)?

In research terms can non-nationals ever free themselves from being thought of as part of the powerful local political establishment, as the researched community know that they would not be there at all without the concurrence of these politicians. They would not have the semantic skills to twist and turn out of this association and it is doubtful if their assertions that they have no connection with governments would be believed even if they stated this to be the case when much research work by non-nationals is related to smoothing the way for government projects.

10. The foreign researchers' dissatisfaction with their own societies

Western researchers come from university environments which are on the whole critical of the political policies and programmes of their own governments and which may be typical of most younger men and women who are not yet overly involved in personal economic and social responsibilities.

It may be that their emotional involvement with the communities which they are researching and which are so dissimilar to their own, correlates to the dissatisfactions with what they have experienced in their own societies; Redfield living and working in Chicago might have gone into rural Mexico with the idea that his own urban American culture, represented a deterioration of the human condition (Braroe and Hicks,1967). If this is not seen in moral terms, then there is often the feeling that political change is a necessity which at least in the recent past has often had a left-wing if not a Marxist flavour. This becomes a dilemma between appreciating the virtues of this particular small scale society and the inevitable necessity of their incorporation into a larger scale social system based on an outside ideology not of their own creation. The dissatisfaction with their own political and economic system and their approach to the researched community, is based almost inevitably on a partial understanding of how it works and the options open to them which comes from their outsider status. They are often privileged as not only outsiders and guests but as having assumed high status connections because their costs and back-up comes from unknown sources.

It would seem therefore that these dissatisfactions with their own societies may lead to an increase in a form of bias that sees these new communities having virtues which are better than that of their own societies. There may also be dissatisfaction with how the people from their own culture are behaving locally in the exclusiveness of ex-patriate communities. Raymond Firth's wife 'reacted to Singapore society with uninhibited criticism and antagonism to the social life of the European community in spite of their great hospitality to us' (Firth, S.1972).

11. Participant observation and the non-national

It is obvious enough that non-national researchers are going to have more difficulties in getting accepted by communities than a national but there are similar problems with internal expatriates who come from a different part of huge countries so that in practice they are seen as foreigners. There is the general mistrust of strangers, but the national is better able to deflect this from interfering too much with their work because of their own cultural understandings.

Mistrust of others with whom they are not in reciprocal, close kin or affinal relationships is a very general trait in most societies. 'There is a readiness to view people as potentially dangerous and the most characteristic initial reaction to others is suspicion and mistrust. An individual who is obliging without cost to those who seek his aid is understood to have some yet unrevealed plan for capitalizing on his position (Lewis,O,1951:292). This is a very natural reaction by those whose own social and economic life is tied into networks of reciprocities and the ubiquitous use of 'thank you' by non-nationals in societies which may not have such a commonplace concept which can be seen as terminating relationships rather than helping their continuation.

We have seen that full participant observation is not only socially impossible without enormous personal sacrifices which may not be worth the anticipated results; becoming a tramp or a criminal may not further the understandings which already exist from the researchers who have not gone to such lengths of personal commitment before writing thorough reports on what such people think and do. Full-time participant observation such as the Canadian woman who lived with Eskimos (Briggs,1970) will always involve disrupting the routines of the people with whom they live. Was it a study of the Eskimos or of the difficulties of studying them over the long isolating winter. Some non-nationals aim to become totally immersed in some part of the community in which they hope to do research but it has to be stressed again this means a growing knowledge of that part, a small part indeed of a whole society's communalism. In total emersion into another society for a non-national it is likely to be easier in a society which is large enough to have some social experience of accepting foreigners than an isolated one with no such experience; most societies now have had strangers within them for long and short periods as political and religious missionaries, medical teams and passing tourists apart from the influence of television images of strangers who are accepted as showing their human characteristics rather than just as actors.

Whatever the actuality of such immersions it is likely to be for the non-national an exhausting process in their attempts all day and everyday to play an unfamiliar role. There will always be the unconscious as much as conscious need to withdraw into their own corner where they can be, to some extent, themselves. There is always the fact that social science researchers, whatever their protestations of commitment, are simultaneously playing other roles derived from their own culture (Briggs,1970). Their significant others are always there outside these researchers' immediate social horizons; they remain there as someone else's daughter, son, friend and academic subordinate or when the researched people have no awareness. So it is probably best for non-nationals to part-time observe what is going on around them in an unobtrusive a way as possible without pushing themselves into new roles which have deep long-lasting commitments. If they hang about they do not have to explain what they are doing and to do more than that is probably always intrusive and biasing. Behaviour at football, baseball and cricket matches, in public transport and working in the fields and construction sites, worship at Hindu temples and on Christian pilgrimages can provide valuable information if it is not tied to the compulsions of quantitative analysis.

Nevertheless there are two frustrating disabilities at the start of fieldwork for the non-national; not being able to talk or to understand talk without much strain and not being able to understand the system that they need to talk about (Firth,1972). This often means that they take the initiative and thus make themselves more obvious than in their stranger status in the first place. They never fit in because the role which they have taken on is outsider the understandings of that culture Mrs. Firth wrote 'by sitting outside our house near a path I created a situation in which the social initiative passed to the Malay villager. The need to be seen as soon as possible as someone who has something to give as well as something to get. To be dressed alike considerably helps along the game of identification with the observed. But of course it is only a game, played to act participation, it is not the same thing as what comes from long-term socialisation. The complete stranger observing something he knows little about, has no means of assessing its importance (Firth, S.1972).

Chapter 5
The 'Half and Half' Researcher

1. The overseas national

Just as there is economic leakage from national economy when overseas capital is used in local projects, there is also an intellectual leakage. Faced with the low standards of many local universities and their restricted course availability, many parents send their children away to other countries where there are better facilities or where other countries have offered sponsored places for developing society students. However much this is harmful to their natal societies which can ill afford to lose their intellectual elite, this is inevitable when the material returns on the use of their brains is so much higher outside their own cultures and the social and intellectual restrictions are much less adverse. Of course, some never return, marrying locally and taking up positions in these foreign universities and institutions, others sporadically return for family reasons. Some become specialists in the issues confronting developing societies and return to do research in the countries in which they have an almost automatic acceptance.

So there are many researchers who have emigrated and taken up citizenship of the counties in which they have received higher education, who retain first generation connections with their natal homes and neighbourhoods. They are nationals of their second homes while probably retaining citizenship of their natal ones. This allows bilateral visiting for social and research opportunities as they still speak their natal languages and retain kinship connections which can be activated whenever the opportunity requires; they will still be known as the son or daughter of so and so. This means that they will still be slotted into a social network in which they will know or be known to a number of individuals in the older generations. This is a predominantly Asian rather than African community of expatriates who can come and go doing research in the countries from which they came. We need not however concern ourselves with the less educated and lowly status of employed and possibly illegal immigrants who live on the fringes of European and North American societies. At the top end of the social range are those who are part of the brain drain from developing countries looking for better financial opportunities and often greater intellectual freedom. This is not a new social phenomenon as Asians have been within an Indian-English culture since the late 18th century (Dharwadker,2003:254-259).

The potential 'come and go' researchers have deliberately relocated themselves in a Western country and in most cases have no real intention of returning 'home' despite some vaguely expressed wishes to do so. The more successful they become the more difficult it becomes to think of returning; their children are being educated in their new country with growing linguistic differences. No matter who they have married, usually locally and within the same expatriate community, their children will have mixed identities increasingly tied to the new country. Of course, their social relationships would seem to provide almost ideal foundations for doing social science research in their countries of origin, but their affinities there may be only nominal and in the course of time not much more useful than committed non-nationals. These expatriate communities have socially reorganised themselves so that they are both multi-layered and multi-centred more especially in university environments in which it is virtually impossible to maintain a specifically African or Asian cultural environment. They developed what might be called an ambidextrous intellect which allowed them in their research to align more understandings possibly than either the national or non-national ones.

In what ways is their situation different to that of national and non-national researchers? They have ready access to their natal country probably lessened by some official, social and kinship feeling that they are no longer quite 'one of us', some even feeling that they have deserted from what should be their own national interests. They are likely to be sympathetic to political and social ideals which are not those of the ruling elite. They will still retain the ability to wind their way through the manipulations of local bureaucracies which for them would be an acceptable maze rather than a series of obstacles. Since they did not emigrate as children nor are they second generation children of earlier immigrants, they are people of two worlds without being committed psychologically and socially committed to either. They do their research with the intellectual equipment of the Western world while emotionally remaining committed to their natal cultures.

There are two other factors of importance. The bureaucrats of their natal country are well aware of their cross-national connections and it is easy for them to be seen as spies for whom the simplicities of expulsion are not available since they may retain local passports. The authorities in the countries of which they are now citizens are likely to be aware of and sympathetic to the intricacies of bias and conflict within such people because they are an asset. They are thought to have a knowledge of the specialised connections in their natal countries which would be of long-term value to some university departments, and they may have other assets as a source of finance, a centre for immigrant student colonies.

2. Existing in two contrasting worlds

The researchers who have been educated in an intellectual world which is alien to their natal culture and continue to live outside their natal countries become people of two worlds (Lienhardt,1979). Are they in fact any longer hands-on researchers in their own country or do they use their insider knowledge of their natal culture to write as dispassionately as possible about it and parallel problems from a knowledge which is either retrospective or acquired through the updating which comes from reading the reports of others' research there? Certainly they know more about their natal culture than any outsider might be able to acquire except through substantial investment in research.

There is of course also the issue as to whether their émigré status has removed so much of their natal identity other than some basic instinctive sympathy that they do in fact think and write as internal if not external expatriates. But there is a more important contrast between the Western system which socialised people within the ideology of the supremacy of the individual and that of other cultural systems which are communally centred. From the ideological safety of the Western world in which they now live and into the individualism of which they have become accustomed, to return to the natal social environment in order to do research rather than just to attend conferences and visit their relatives, is to put themselves under great personal strain. It may be that such people find it easier to see their natal environments in a more detached light and to write studies which contradict the current political correctness there. Deepak Lal's (2004) conclusion that the economies of British colonies had higher rates of economic development than 'freer' societies would have been such a case coming as it does from someone who had been a socialist adviser to Nehru in the earlier years of Indian Independence. It is not just that all natal developing cultures are complicated and are considered to have a homogeneity which is entirely fictional, but that re-entry puts these 'half and halves' into a continuous series of conceptual dilemmas (Kondo,1990:chapter 1); they look like natives but are culturally inept even if they are known to be highly educated by the standards of the Western world from which they have temporally emerged and this cannot offset their ascribed status. Whatever role they attempt to assume there is an existing role in their natal culture into which they are tied. It is virtually impossible to retain the observer status which is the main Western criteria for evaluating fieldwork or for undertaking fieldwork. For such a person trying to work in their natal culture there are constant threats to their achieved identity, and the invasions of their personal and intellectual privacies.

The ascribed framework of thinking and behaviour is communal and the achieved status is individual. It is impossible to formulate a sentence in Japanese without stating something about the speaker's position relative to the person spoken to. You are never 'I' untouched by particular situations (Smith, R,1983:79), shaped by formalities, kinship, occupation and other people's desires and usages as well as by contextual factors (Suzuki,1978:93). Many other cultures have similar semantic problems for the re-entering intellectual and indeed for all others attempting to return to their natal environments which they have left many years before, complicated in many cases by marrying partners of another natal culture altogether.

The 'half and half' researchers have the advantage or the handicap of being able to negotiate the conditions for their work from two contrasting social and political positions since they are neither nationals nor non-nationals. As we have seen points of view created in one cultural area can be the pre-conditions for those arguments changed by moving into another cultural environment. Such researchers need not necessarily be moving in and out of a Western environment but can be the experiences of an Asian academic in an African university.

A Bengali Hindu woman (Bannerjea,2000) was able by persistence, sincerity and a knowledge of how to manipulate the system to get permission to research a pre-Independence non-violent Pathan Muslim political movement in Pakistan. Her academic advisers would most certainly have suggested that she look for another topic. Only someone who knew their way about both cultures would have been able to wangle her way into such an area of research by good luck and the accidents of meeting people there with a sense of history and their vanity from being approached for their memories, made her position as a 'half and half' researcher advantageous. We do not know the extent of her personal costs for her commitment to this topic for four years. Apart from the fact that she must have had the support of her family for an unmarried woman to embark on such a difficult topic, there was the effort to acquire a good working knowledge of a language totally different to her native Bengali. She had also to submit with grace to Islamic restrictions on her behaviour from which she had escaped in her own community but also by moving to England. Perhaps this 'half and half' experience would be much the same for those who move from one cultural area in an immense country to another. A Bengali moving to do research in Bangalore might have as many low level problems of personal strain as if he had moved to Germany or the Congo.

But it is important to recognise that most of these 'half and half' researchers are moving from one university campus to another which have similar working and social environments. This environment is likely to be more tolerant of legally non-nationals, rather than the situation surrounding fieldwork. These researchers are likely to have problems with their identities even though they may have changed their nationalities and married someone with a similar changed identity or from their new country or even born in the country to which their parents have changed nationality. They may well be still mentally part of their natal cultures and yet the longer their separation, the more idealistic their memories are likely to be even though they may have paid only short visits there to keep in touch with their extended families. Even their ability to speak their natal languages may be impaired as with President Banda of Malawi after living in Britain for many years was not able to speak Chinyanja so that he used English in his public speeches which had to be interpreted.

3. Identity and the 'half and half' researchers

There are four aspects to consider in respect of these social scientists who not only may have divided nationalities but may have dual social and psychological identities. Firstly, there are the individuals themselves who may have a persistent longing for their natal environments which against all their pragmatic knowledge they see in ideal terms; the emotional clash between the pull of their existing realities and what they may remember as situations in which they were more mentally and socially comfortable.

Half their families may still be in their countries of origin and there is a constant flow between these two halves, emotion coming one way and economic assistance going the other way as indeed they are known to be better off than their national relatives. They are often affronted by the behaviour and culture of the country to which they have now become a part. Since they are mostly men Muslim men are affronted by the freedom with which 'their' women are now a legalised part and the Hindu women are upset by the constant ways in which they are forced into states of ritual pollution. They are still so much part of their families of origin even having to take sides in family disputes over inheritance. Then in their fieldwork there are the difficulties between the professional base line from which they are seeing their research which is overloaded with the understandings of the culture in which they are now living and those of the culture from which they have come. This is probably constantly refreshed by the fact that they feel more at home with people in the same situation as themselves; those with whom they share food informally and by choice. An Indian studying his own community in Kashmir was accepted as a rather wayward Pandit now engaged in a special form of social enquiry; they tried to absorb him back into the village life where in their view he properly belonged and he was asked to intervene in family disputes over property (Madan,1975).

Even in cultures which are professionally multi-cultural, their fellow professionals who should perhaps see them in detached terms of competence in their work may have biases in this. The 'half and half' may be seen as having got their jobs from the biases that foreigners have been treated more indulgently in the academic environment than nationals and further, that their built-in knowledge of their own cultures in which they may be specialising gives them not only uneven advantages but prevents the professional detachment which is thought to be the basis of proper social science research work.

Then finally there are the difficulties to which they may be subject should they return to do research in their countries of origin. At the academic level there may be the feeling that they are in some vague way ideological 'traitors'; they should be working for the benefit of their own societies and not for some country which has no real interest in these foreigners of which they are now one. It is often more complicated than this; does such a 'half and half' conform to the social situation in which they find themselves as they are 'one of us' who has come back home? They cannot and indeed should not behave as if they were not, they are returned (Kondo,1990:11-12). 'As a Japanese-American I created a conceptual dilemma for them (the other workers in the factory where I was researching) I was someone who was both Japanese and not Japanese. How can someone who is racially Japanese lack 'cultural competence'; they would be astonished at my linguistic mistakes. I proved a challenge to their sense of identity. In my cultural ineptitude I represented for the people who met me the chaos of meaninglessness'.

These people may feel over optimistic as to the extent that they can be accepted back without fuss into the communities from which they have emigrated and much must depend on the contacts which they have maintained there after they left which was often in early adolescence when they would have had little status to recover except that they were the sons and daughters of people of known status there.

4. Fragmentary nature of their valuable memories

Any "half and half" researchers who return to their homeland in any role which relates directly to their origins has few advantages from attempting to retain their achieved foreign persona. If they become nationals for the purpose of their research project, their non-national half will bring them few advantages except in having provided them with specialised training and a partial non-national perspective. In the national context such researchers are anomalies which to one side of the divide represents breaking away from family responsibilities and on the other envy at those who have been able to get away from the restrictions of local culture which clashes with their ideas of modernity apart from the obvious fact that they are living less problematic lives without political uncertainties and the fear of natural disasters.

The researchers will in their own opinions retain their individualities but to the nationals they encounter, they may represent any number of different roles which are largely out of their control. 'Daughter, guest, young woman, student, prodigal Japanese who had finally seen the light and come home. Overtime my increasingly Japanese behaviour served temporarily to resolve their crises of meaning and to confirm their assumptions about their own identities. That I too came to participate enthusiastically in this recasting of the self is a testimony to their success in acting upon me' (Kondo,1990:13-18). But this was not the end of this researcher's translation of roles, 'in one position my goal had to be the pursuit of knowledge where decisive action, independence and mastery were held in high esteem. In another, independence and mastery of one's own fate were out of the question; rather being a daughter meant duties, responsibilities and interdependence. By moving out of the house of a Japanese family to live on my own I had simply changed the role of daughter for other cultural meaningful positions; those of guest, neighbour, worker, young woman that demanded participation and involvement. I was never allowed to be an autonomous freely operating 'individual'. I was always defined by my obligations to others'. (Kondo,1990:26).

It would seem that these 'half and half' researchers have not only the usual fragmented persona of any person anywhere but are denied the ways that role conflicts can be resolved in these receiving cultures which would conflict with the primary objectives of their fieldwork. They fall between both the advantages and disadvantages of clear-cut national and non-national identities. 'Anthropology highlights the socially constructed features of the concept of 'self' in a particular culture. Inevitably this one calls into question hegemonous American assumptions about identity and selfhood is a founder essence containing inner true feelings' (Kondo,1990:34).

5. Increased insights of 'half and half' researchers

It is correct that any national researcher will have insights into their own culture which are not available to non-nationals without prolonged training and perhaps many return visits to the original area of research. We have already seen the research problems and insights of a returning Japanese woman and now we need to look at those of a returning Indian one.

The intellectual who has been educated abroad within the post-graduate dimensions of another culture, will have augmented their understandings and sensitivities of which a national might not be so readily aware. They are not outsiders and on returning to their natal country, they do not classify either as a national even though they, may retain dual nationality. But however advantageous a particular achieved status may be, there is always a downside, which comes from any form of ascription (Mascarenhas-Keyes,1988) demonstrated perhaps by this researcher adding her husband's name to her own. There was a need to have a multi-native strategy in what was a multi-cultural area for her research. Although tempted to use research assistants she felt that they would have linked to particular social connections and by working alone the difficulties of having a team approach were circumvented. This approach was also helped by living in a house rented form an absentee landlord. Linguistic versatility is a presumed asset but to what extent is this so in a multilingual environment in which each language has its own semantic complications, which might have made using interpreters advisable. This researcher in Goa had to use Konkani with Hindus, Portuguese with the traditional Catholic elite, Swahili for those who had returned from East Africa and English with those who preferred to use it rather than Konkani.

So perhaps there is a heightened sensitivity which allows such researchers to be more adaptable than nationals who might consider such adaptations to be personally unacceptable as lowering status and indeed for non-nationals who might not even consider such fieldwork alternatives as possible. Since these men and women have been significantly trained in the particular fields which they have chosen, we can presuppose that they have intellectually changed in their understandings, very probably at a faster and more significant rate than if they had stayed in their own countries. The point here must surely be that in all three of the researcher categories which we are considering, we have to conclude that it is naïve to imagine that the fieldworker can ever become a self-effacing recording machine (Nash and Wintrop,1972).

The professional development of these men and women would seem to have been a prolonged development of the culture shock which all social scientists experience in entering a fieldwork situation which is socially and psychologically foreign to them. For such people this is a continuing process and in fact a dialectical one in which there are continuing challenges between the patterns acquired through socialisation in their natal culture and those that are forced on them by the resocialisation which is required of them in the societies in which they now live and work more or less full-time.

This is obviously a fruitful stimulus when we read the works of Deepak Lal (1995;2004) and Sen (1992;1999). Indeed there are many others in many fields in and beyond the social sciences who have themselves benefited and mankind has benefited from their prolonged experience of culture shock.

6. Being close to home and the fading of detachment

While by our definition a 'half and half' is a national who has trained abroad and remained there taking on a new nationality, all nationals in respect to that part of their own country become 'half and half' in which they were born and from which they have become culturally removed by prolonged separation from it during their education, and how they live subsequently; at the very least they will have become or thought they have become upwardly socially mobile as part of this separation from their natal communities. Does being an official foreigner deprive such researchers of the particular insights which they might be considered to have by the fact that they were born and brought up in or near to the people and the community to which they were at one time socially connected? To what extent does the pull of the heart towards the place rather than the draw of the people who will have changed with the years even if they do still live in the same localities which must have turnovers of populations proportionate to the years between leaving and returning.

If such researchers return to the country of their birth in any general sense they are just a specialised form of expatriate neither internal or external. The problems arise when such researchers return to research the specific communities in which they were born and are still known there; the extent to which their thinking can be attributed to this particularised past is the issue here. Nadia Serematakis (1991) returned to her natal village in the Mani part of Greece to research women whose identities are constructed through their association with death. Although she described herself as a native anthropologist and may have tried to act as such, everything she did was not considered intrusive as long as it could be seen as an extension of kinship, she was always somebody's daughter. However her involvement in village activities as a woman with a particular kinship space, meant that she could enter the feminine space of death by becoming a member of the mourning chorus with the contractual obligations of a local participant. In a sense she had been taken over by her past and she was no longer a detached social scientist at work. She developed dreams connected with this role which the Maniat women interpreted as warnings (Seremetakis,1991:232-36). Her own body, her unconscious and her consciousness became 'witnesses' that validated women's cultural powers over the status of death. In this case this social scientist's way of knowing separated her from the methodological demands of detachment, led to a sensitivity which constructed a model of a self quite different from the Western rationalised self. This inner voice in the dreams spoke a language for which she had no translation, using symbols she did not control and signs which were both alien and with her, signs that spoke in references to an unknown other that she did not know. Seemingly from her own descriptions she had gone back into the 'native' role and yet was able to describe her involvement in ways that provided primary data on Maniat behaviour. Of course, this data is beyond verification but nevertheless it fits into local patterns and leaves it without the destructiveness of alienating analysis. Nevertheless she had now become a different woman beyond the possible criticisms of her professional colleagues and thus in some ways replicating Castenada and his relationship to Don Juan.

A commentator (Josiphidies,1997) suggests that there was in her case no cultural space left between description and interpretation. She never questions her interpretation but presents it as the way things were and herself as an assured spokeswoman. She reaches back into her own past to recover a bodily memory implanted by the shared substance of commensality and metaphorisation and utterly convincing as a description of personal experience. This is certainly an important issue in which this 'half and half' woman had in fact returned to become a fully involved native, whereas in other cases of individuals returning to research their own communities they have become less involved and better able to slip in and out of kinship and professional roles retaining some distinctions between each.

7. The problems of re-entry for a researcher

Anyone who has been away from their natal environment for a significant length of time cannot return without going through a process of readjustment and indeed if the period is long there will be as much 'reverse culture shock' as if they were entering a socially alien society should they wish to return for research purposes or even just for family visiting. This residence abroad may make it necessary for these researchers to learn again some aspects of their own culture and to understand its implications which they would never have done had they not gone through these higher education processes of obtaining new perspectives (Altorki,1982).

An Indian or Chinese who goes abroad to be educated and stays abroad has the ideas about their natal environments largely stuck at the time of their initial emigration. Since then they have had significant changes in how they think and behave even if they have the personal illusion that they have remained largely the same. They are certainly older and their friends even their partners will be taken from the societies into which they have migrated even if they have been chosen from a group of immigrants similar to themselves from the same country, part or province. Then there will also be changes in the society from which they have emigrated and most of these intellectuals have been away from their home countries for years. Letters and newspapers are no substitute for the parallel experience of growing old within a changing social environment.

Chapter 6
THE NATIONAL RESEARCHER

1. I am one of them therefore I know them

This is certainly the basic feeling of a national social science researcher. Anyone in their own social environment has the same fieldwork status as an informant. They may be well informed about their own community and their own part in it or they may be singularly unaware of the social conditions within which they live because of socialisation and the fact that their education is likely to have dealt with the 'other' rather more than 'us'. But they cease to be one of 'us' by the fact that they are now researchers as they have set themselves apart. By claiming special insights they are in fact attempting to reoccupy roles which they have probably vacated some years previously; they now occupy special roles indeed but not that of anything approaching current memberships of these communities in which they may have had at some time close relationships. In fact it may be almost fallacious to suggest that they have a particular insider form of social knowledge other than that which comes with the ascribed and achieved roles which they currently occupy. The value of such national researchers may be in their contributions to wider ranging studies about particular roles; they have no automatic understandings of how their own society may be functioning.

To what extent then do nationals know 'them' in the general sense of sharing not just the generalities of their culture but its particularities and practicalities involved in fieldwork; they have roles which have to be explained and detailed since this will be a form of bias. On top of this can we be automatically assured that they do have a theme of sympathy in their minds for their fellow nationals apart from some shared parts of nationality and language which gives them significant advantages over non-nationals but not to the extent implied by this heading.

A Chinese anthropologist (Fei Hsiao Tong,1939:xiii) wrote 'if it is true that self-knowledge is the most difficult to gain, then undoubtedly the anthropology of one's own people is the most arduous, but also the most valuable achievement of a fieldworker'. An Afro-American (Szwed,1969:170) wrote 'it is naïve to believe that native anthropologists necessarily will do more satisfactory work than non-natives since the United States black social scientists have brought the same methods and conceptual approaches which rendered their white brothers incapable of dealing with a major portion of lower class Afro-American life and culture'. It is also naïve (Sorbo,1982) not only because the insider's view is only a partial approximation of understandings and also because comparison is the social science's methodological equivalent to experiment and allows them to expose analytical parallels and to falsify hypotheses. But 'he is motivated by a profound desire to improve the life situation of his country's population by academic ambition and monetary reward, but is continuously frustrated by the overwhelming reality of underdevelopment, the policy decisions of his own government and by the fact that he himself is rarely given any power. They come under attack in their own countries from the people they study for their hidden motives or for their interference as do-gooders and scientists in the service of government' (Sorbo,1982).

There is probably no such thing as an isolatable indigenous pattern of thought and behaviour; some identifiable paradigm which is exclusive to some social group. Just as demographically and linguistically they are an amalgam of features which come in from many other outside sources, so will their ideas. The Sukuma quasi-traditional diviners for example, use fire-walking and Coca Cola bottle tops for their diagnoses in which there is the possibility of originality. It is more likely that their thinking mixes the remainders of tradition with media influenced scenarios, advertising pictures and ideas that come from the fringes of all the religions available locally, Christianity, Hinduism and Islam.

There is no evolutionary change from backwardness to capitalist oppression (Warman,1982) but a zigzag between the physical forms which are taken on by literate and non-literate ideologies. Indigenous thought and practice in anywhere in the world is always going to be an amalgam and their model of the nation is unlikely to be that of the current group in power but one much local and personal. Current thinking is always going to be dominantly contemporary however much we maybe conscious of genetic and ethological factors. What is called traditional thinking is just the framework within which current lives are working; an Indian thinker in a village is preoccupied with caste issues, not because of some ancient continuing paradigm but because that is how this community works now; this existential reality does not need any other explanation.

Although it is physiologically possible that nationals may have some immunities to local diseases such as the presence of the sickle cell immunity to malaria in West Africans, they are just as liable as non-nationals to catch others. This is particularly so as their current domestic living conditions are as likely to be different to those in their area of research as might be the case with non-nationals. It is rare for fieldworkers to detail the handicaps which they have undergone and this absence allies itself to the hopes of social scientists that their work would be considered scientific. Professor Srinivas details some of the difficulties which he encountered in his fieldwork (Srinivas,1979). Far from choosing a village on account of its suitability for his research, he ended up in one village because it was the only one in which he could find a place to live, ending up in a cowshed which he shared with animals and a Muslim client of the headman and his family. He was one of the few fieldworkers who have detailed the problems of excreting, and wrote that the headman's son had showed him in which field he could defecate; he commented that this was preferable to the indoor lavatories of the few rich people and made it necessary for him to defecate only at night.

2. The ease of access for the national researcher

A citizen has the legal right to be in the area chosen for research and they do not have to go through any existing procedures to enable research to be started and applications for visas which would apply to non-nationals. They do not come into the country on a tourist visa so their research can usually involve long periods of residence. This comparative simplicity is probably an illusion. The suspicion of the foreign researcher does not mean that there is none for national researchers in the eyes of governments and in the perceptions of those researched and their communities. They are likely to be seen as working directly or indirectly for the government. The ease or difficulty of access must depend on the type of government, the size of the country, the type of research and related cultural issues.

Most countries are only nominally unitary and access for anything more than covert personal projects may require the agreement of different layers of authority. Some nations are federal and whether this is so or not, there are likely to be provincial and even lesser local authorities which would have to be in agreement over anything more than a small community study. There are often frontier areas closed to non-residents and civil disorder limitations in which governments do not want the additional responsibilities over the safety of a well known national researcher from even the most prestigious national university; a researcher whether Hindu or Muslim is not likely to be given access to the India controlled part of Kashmir, for example. Some countries do not have governments which are in control of their own territory whatever their legal status may be. During civil disorders in Uganda, the police returned national researchers to the university for their own safety.

Some countries are so huge that in terms of their characteristics, they have an almost continental scope such as China, India, Russia as well as the string of islands comprising Indonesia. Access to one part may involve as many cultural changes in understandings as might be the case for such researchers moving to Germany or Canada. In these circumstances they are in effect internal expatriates. The study of a Yunnanese community by a Han Chinese researcher would have more or less the same problems as he might have had in going to study a community in urban America (Hua,2001); in fact it would be cheaper, easier and less bureaucratically restricted had he gone to the United States to do fieldwork.

It is unlikely that any national researcher is going to work without the support of some national university and in the background some non-national body which is prepared to provide finance for some projects. ohe national university will be thinking in wider, possibly political terms, in supporting a particular piece of research by a national; they will not want to be involved in any activities which the national government might consider disruptive. How Hua got permission to study a Chinese tribal community in a political climate against any sub-divisions of national unity must have been a marvel of negotiation and covert intentions and indeed ultimately involving his emigration since this study was written in French.; he is unlikely to undertake any more research in the People's Republic of China.

Some research by a respected national will involve enormous amounts of preparatory negotiation with those whose cooperation was necessary. Kochar's study of the excretory habits of rural Bengalis (Kochar.1977.1978.Kochar et al.1976), remains the only scientifically acceptable study of this universal behaviour, if only because of the enormous difficulties of getting the cooperation of sufficient numbers of people to make the study quantitatively valid. One can only marvel at the dedication and persistence of Kochar in this fieldwork which a non-national would never have undertaken in the first place or persisted in the face of so many field difficulties in obtaining this degree of accuracy on a very private and personal activity.

It seems likely that national researchers will always be able to keep in touch with their fieldwork areas of study over much longer periods than most non-nationals; they maybe only a bus ride or rail trip away and thus can keep in touch without the difficulties of writing, indeed they can plan on researching these people in the same community over ten years or more without much expense or the endless search for grants to enable overseas work to be even started. One group of Chinese researchers made a longitudinal study of the social adjustments and functions of Shanghai children

(Chen et al.1995) and a Tanzanian made a long term study of the adjustments to change and poverty in the informal sector of the capital Dar es Salaam (Lugalla.1997) on the door step of the national university.

3. The national researching in their own country

We have to consider the assumed advantages that may lie with social scientists undertaking fieldwork in their own country while remaining resident there. Until recently the number of nationals doing such work has been small in terms of published results but now there are large numbers researching their own cultures for a combination of the demands of personal advancement and national needs. The relationship with informants will not necessarily be simplified by what might be called the nationalisation of social science research since it not only involved the same problems as will be experienced by non-national researchers, but will have new ones added.

The assumption in the minds of some national researchers that they have an automatic advantage over the non-national is of course questionable since this much must depend on the circumstances and topic. The concept of national identity overriding tribal paradigms of thought and bias, new but obvious social class differences and the use of a lingua franca as well as the possibility of the research involving second or even third languages. While the non-national is always an outsider, national ones have always to work themselves into the researched community so that they can come to be treated there as privileged and useful citizens rather than as suspiciou soutsiders and potentially enemies (Naderi.1982).

As national identification is still relatively new in developing society consciousness, it may have been a creation of the struggle for political independence and for use in foreign affairs and for contemporary use in soliciting aid and combating foreign influences. The post-colonial era has not seen a steady progress towards the development of functionally distinct nations with the population thinking and acting in national terms. Most new nations have had a continuous and continue to have serious tribal, religious and linguistically based political conflicts and attempts for parts of a legally defined nation to be subdivided independent units in India, Indonesia, the Sudan and Somalia. The elder who when asked who he was and replied that he was a Somali was laughed at as everyone knew that he always acted in the interests of his clan.

The national researcher is not only primarily a member of such groups or under their influence unless he or she has emigrated from which it might be virtually impossible not to be mentally if not socially associated. So much so that they might prefer not to do fieldwork in an area or on any topic with which they might have some prior personal connections, but to retain a national identity and for research purposes become an internal expatriate. This is what may be happening in practice since student researchers may start their fieldwork within the social environment of their university rather than expensively farther away or confine their research to the university library detached from any serious personal exposure to the political, social and possibly dangerous problems of working elsewhere.

Status factors are bound to intrude since university education leads to a real sense of elitism which intrude on the satisfactory performance of fieldwork. This is no more than a reflection of the non-nationals' built-in sense of superiority which comes from a combination of educational, ethnic and social factors which in their own cultures have a very long history. On the other hand in the Western and North American environments the social sciences in the field of sociology, have always been in the hands of nationals with Americans studying American life and indeed in a growing number of cases urban Americans have been cooperating with tribal Americans on joint projects (Chrisman *et al*,1999;Finlinson *et al*,1999). Anthropology has always been seen more as a field of research beyond national boundaries with American Indians being seen as legally included foreigners and thus targets of such research.

We can conclude that researchers from both African and Asian universities with social science departments, some of which have had long standing and distinguished reputations, may well be experiencing much the same problems as their Western colleagues have long undergone in studying their own cultures as well as those that anthropologists have also experienced in studying foreign cultures, made worse by limited funding and restrictions on intellectual freedom. The main problem is likely to be that of social distance which is almost inevitable when the national as much as non-national researcher come from another cultural environment and are usually very obviously at a higher educational level. The difficulty lies in crossing invisible but very patently existing boundaries which cover every aspect of the researchers and the researched lives.

The national researchers are usually ethnically similar to those researched and thus the first and obvious hurdle of biologically visible difference may not exist in most cases. Nevertheless this does remain a difficulty when physical; appearance is allied to the nuances of behaviour which may distinguish a Brahmin from as member of a lower caste. There are also differences in appearance of southern and northern Indians and similar ones between Chinese from the north and the south. There can be no assumption that physical appearance allows some form of automatic entry into ostensibly ethnically similar communities, apart from the 'half and half' researchers who have returned from overseas. There is nevertheless the feeling of the national researchers that they are one of 'us' rather than one of 'them', and therefore some underlying feeling of affinity. This, beside the persistent illusion of social scientists that they will always be able to lessen social distance by long-term association. What is certain is that the nationals researching within their own country will not have the same frisson of excitement which non-nationals have in entering a totally strange cultural environment.

The non-national knows that he or she is irrevocably different to those that they are researching. It seems likely that the highly educated Dinka may have more of a problem understanding his own culture than a similarly educated Englishman would have in studying a Gipsy encampment in Southern England; the differences in their current experiences are more extreme in the former than in the latter. One can only wonder at the mental convolutions of a Nuer in a restaurant in Syracuse New York who has a series of tribal cuts round his head testifying to his adulthood in that far away Sudanese culture. Deng (1971) however writes from his "half and half" and retrospective situation rather than from any fieldwork since he does not mention any difficulties nor that he had the advantage of conducting his research in his own language.

The non-national social science researchers are often looking at their work in terms of broad long-term theoretical issues in the context of a global environment with the aid of a grant from a non-national source. In doing so they may to some extent avoid the realities of social existence for many people. However, when national researchers look at their own people whose preoccupation is survival in a hostile and overcrowded social and physical environment, they are more likely to relate their work to these realities; it becomes an ethical obligation for them to react as citizens (Dube,1973:21-22).

However, the path of the applied social scientist is usually one which has already been marked out by technicians and politicians; it is rarely a situation in which these professionals are asked what might be the best way to improve the standard of living of the people involved about which they might have as tunnelled vision as any power holder. It is not only that intervention has already been decided upon as both political dogma and slogan but its form has already been laid down. Almost inevitably they are asked how the proposed intervention can become institutionalised while making the minimum disruption to the lives of those involved (Cernea,1982). Unless these changes are entirely staffed with nationals, these technicians will always prefer to work with one of their own kind with whom they can relate on some basis of professional equality. Social scientists are not powerful members of any management committee and they do not usually have their own power base from which to negotiate.

If they ally themselves too closely or too obviously by inference with the nationals who are going to be affected, their suggestions are likely to be seen as challenging the power holders who have assumed the political role of working for the benefit of their constituents of whom those affected will be a small minority. We have seen how the Egyptian sociologist working with Nubians displaced by the Nasser High Dam became disillusioned by their failure to recognise the efforts made on their behalf. So he would seem to have been siding with of the national government rather than that of the Nubians who had lost their homes and the graves of their ancestors to a far away civil power and not from war or famine which they could have understood.

National researchers work from a broader and more knowledgeable background than might be achieved by a non-national imported on the basis of a well-earned reputation for advice from applied work in another country with different social and geographical environments. They may even be mistrusted on this account as it is usually a situation in which 'us' and 'them' are clearly in opposition on many accounts.

4. The limitations imposed by citizenship

National researchers can be considered to belong to only one exact social category and that is their citizenship, but beyond that legal one both national and non-national researchers may belong to as many different categories as may be created in their job descriptions. We have seen that in many cases the national researcher may be considered an internal expatriate, a sort of quasi-foreigner quasi-native who might be considered to be one of them in certain situations but not overall way. A Han Chinese researching Cantonese family life, a Kikuyu researching the Luo port workers in Mombasa, Kenya and a Tamil researching Gujarati Patidars in their trading networks in India and East Africa are to all intents and purposes internal foreigners.

An internal expatriate may have even more trouble in getting accepted as they may be seen as agents of the dominant government in a distant part of the one country. The Lapps of Norway who were waging a campaign for greater legal and ethnic autonomy, saw anthropological researchers as 'foreign interference' providing information to enable their continuing dominance by Norwegian society based in very distant Oslo (Keskitalo,1976). In Spanish speaking South America the bureaucratic processes of the state as part of their colonization made both the Aymara and Quechwa wary of providing any information at all to outsiders which might be to their disadvantage and answers to questionnaires and tests would be part of this ethic of subterfuge (Ortiz,1982). There the national fieldworker has a difficult job to break free from this mist of mistrust even though freedom from Spain was achieved generations ago. This resistance may be much higher in British and French ex-colonial territories which involved far higher rates of interference in the lives of their citizens.

So at one extreme there are issues of great social distance between nationals who are researching and nationals who are being researched. The issues are entirely different when there is some degree of social closeness. A non-national can always pretend that there is some neutrality in how they look on what they are doing in their fieldwork, but a national can never pretend to be quite so neutral. They are ethnically and linguistically committed to a particular culture rather than another because of their socialisation in childhood. There is however a sense in which they are no longer natives unless of course they are researching the academic or professional community in which they work; they have left their natal environments for good and they do not really believe that they will return. They have become different by the nature of their professionalism. Away from this restricted environment they will always have limited social freedom as the nearer they come socially to their natal environments the more their social position becomes related to caste, religion and land holding as well as their membership of a particular family.

Another factor is that they cannot easily escape from the consequences of what they report which is easier for the non-national who has by then left the country and may not wish to return. So that the national researchers are likely to work on matters which have no direct political implications or impose on themselves self-censorship in which they do not publicise matters which reflect on the dignity of their nation or draw attention to criminal or immoral matters connected to powerful individuals.

5. Objections to the 'native' studying 'natives'

There have always been methodological objections to social scientists studying their own societies. This may well belong to a previous era in which it was assumed that an undeveloped society had a social homogeneity so it was assumed that a 'native' was always intimately involved in some part of the community which was being observed. This overlooks the fact that in most social science university courses wherever they are, the only fieldwork practice that is possible for most students is in the surrounding society; cheap, convenient and involves no prior requirements of language or particularised social training. So it seems that most social science researchers start their fieldwork experience in communities which are part of their own existing lives with no large dissimilarities in their mutual understandings. It was this assumed closeness of one to the other regardless of circumstances that made Western social scientists assume that their use would inevitably lead to loss of objectivity. This is not a particularly logical objection because every researcher is aware of themselves as the centres of their fieldwork, just as every research report has to clear away any suggestion of avoidable biases. It is not even sensible to suggest that these possibilities of bias override the economic cost and time wasting needs for learning another language, as well as having to travel long distances in order to ensure that pre-defined biases are as far as possible excluded apart from the fact that other less ostensible but possibly more pervasive biases creep in.

It seems likely that these earlier grounds for objection which were specifically disallowed by Malinowski who asked why it should be necessary to justify the claims of an African to write about his own tribe when a European's competence to study his own society is not questioned (Malinowski,1938). This may have been a combination of Western biases against 'natives' taking over their profession by knowing more about their own cultures from socialization than any non-national could acquire, with which they could not compete. In any case what does a project being run by a 'native' actually mean? By reason of their training and almost certainly many years away from their natal community, one can question the area of research from which they should be excluded because of professionally defined and personal biases. It probably comes down that they should not undertake research involving their own university department.

There are numerous examples of research workers moving short geographical distances and finding themselves in situations in which social distance factors make for intellectual stimulation and as stimulatingly different as if they had crossed national boundaries and indeed moved to another continent. On the grounds of language alone the advantages of researching within their own linguistic group are overwhelming in probably reducing the lead-in period by at least a year and avoiding the inevitable biases which come from the use of interpreters and language teachers.

One suspects also that there is an element of jealousy at the ease with which a 'native' can slip into a research environment within their own country without the inevitable problems of visas, permissions to do this or that and the costs. The social sciences have some notable examples of economically costless research and there appear to be no sustainable objections to research by 'natives' whether in developed or developing societies which have not been bypassed by the huge growth in national research by nationals, apart from the surrounding political and educational realities. The objections may be better founded in the dominances which many national governments have on who researches what, where and when.

6. The dilemmas of national researchers

Research styles vary with the social settings in which they occur and in the middle of each is the research worker. The polarities of this are between the researchers who to some extent work within their 'home' environment and those for whom these understandings are exotic (Wolcott,1981).

While in a Western urban environment researchers are going to come across minority communities which are visibly different to themselves in clothing and ethnicity but show that members are becoming assimilated to most aspects of the surrounding major culture, this is not so in other societies. In India researchers are going to be confronted with differences between Hindus and Muslims and as Hindus differences between castes, sub-castes and Dalits; no amount of manoeuvring by researchers there is going to prevent attachment to some and exclusion from others.

Researchers can be 'at home' studying aspects of the universities in which they have spent most of their adult lives and others may have similar feelings in the socials context of a Hindu joint family in which similarly they may have spent the earlier part of their lives. They will at least have some understandings of prostitution in Manila and Mombasa if they have lived there, or any other problem in cities.

The 'away' setting can involve research which is not only different from what they may have experienced but completely different to anything which they may have experienced at 'home'; cattle thieving in Tanganyika, double cousin marriage in Mombasa or sex workers in New York. The absence of homogeneity except in small exclusive communities, means that in any modern society a researcher is going to be confronted with social behaviour about which they know very little and indeed might well suffer from culture shock from this sudden exposure to these cultures foreign to their own ways of thinking. Of course the converse would be equally true for researchers brought up in so-called isolated nuclear families in the suburbs of Paris or Strasburg confronted with the Hindu joint family or the barrack-like life of the Hutterites in Canada. Similarly, Muslims who marry their cousins for reasons of retaining property within the family and avoiding the difficulties of divorce as in-laws would also be uncles and aunts, would find free choice not only exotic but rather stupid with the unknown marrying each other.

When nationals study some area of research within their own cultures with which they are conversant through socialisation and long-term adult commitment to some particular field of study, the dilemma is over the extent to which they can use themselves as informants and indeed their own reliability in this context. Such researchers are in some ways in over control of their subjects and in broad terms they will have difficulties in deciding what is within and without their topic. In this context rightly or wrongly the nationals in their 'home' environment start with an enormous amount of accumulated information which gives them personal satisfactions and confidence; they do not have to learn and write down so much as it is already in their memories.

In exotic settings in terms of their own socialization and expectations confronting the non-national so much is unknown that their impulse is to try and record too much and then have to cut back probably missing much that is important. This again turns on the length of time that the non-national can stay away from their homes which is not such a constraint on national researchers. The more exotic the setting of research, the greater the personal and material handicaps facing the researchers. The nationals in the 'home' setting of the environment of which they know much about its social foundations, do not however have to be so careful in what they research as they already know the dimensions of acceptable behaviour. They can push and shove themselves into situations which would be impossible for non-nationals who would find themselves almost by definition permanently in exotic settings.

National researchers must have greater confidence in their 'home' environment but there are still the dilemmas of acceptance and understanding within such broad acceptances. The Brahmin researcher is not at 'home' in working with a Dalit community in Banaras any more than he would be in studying the Nagas in Assam despite Gandhi calling the Dalits 'Harijan', children of God. They will always have the linguistic advantages over the non-national but it will be the linguistic code of the educated and probably sexually defined so that there is no social science neutrality in how they use their own languages. Nevertheless, 'only here I feel, do I observe, participate and write under conditions in which I am most likely to understand most of what is going on, with knowing nuances, double entendres, conflicting explanations and so forth all apt to go noticed rather than unnoticed' (Wolcott,1981).

The home and away distinction between national and non-national researchers may have clear cut implications in which the competence of the former is easier to assess because of their ability to function from a less handicapped starting point. They do not have their thinking challenged by the complications of coming from an environment with other cultural understandings but then there is the dilemma that these are challenges to the partial *status quo* of their own thinking as a result of inputs from other cultures. The cultural complacency and possible conceit of national researchers over the adequacies of their own social institutions and systems, can be modified by cross-cultural contacts. 'The overriding anthropological concern for culture and for the relevance of that concern in helping us understand not only the lives of others but our own lives as well' (Wolcott,1981).

Social science researchers, except in unusual and rare individual circumstances, are frequently dissimilar to those they research or their communities. Their social base is elsewhere and they are not permanent members of the communities in which they work. Their level of education is usually above that of the people with whom they work, their speech is a specialised linguistic code and their material life is above that of those with whom they are working. This is apart from their personal practices and ideas which come from prolonged education. They are not as one with those they research and they do not feel that they are, however, much they attempt to disguise these differences.

Although to the observant the social distances between the top and bottom of developed societies may seem extreme, the majority of people are in the middle where the average standard of living in comparison to the averages in less developed societies makes them on the wealthy side of any dividing line between the majority who are subsistence farmers and daily workers and those who are not. Non-national researchers are by definition socially distant from whatever they chose to investigate and thus their connections are the tenuous ones of theoretical training which will be reduced by the feelings which they develop for those they have come to know more intimately; they are always somewhere within the category of 'honorary foreigners', and this may make the detachment which is so much a part of social science fieldwork ideology a slightly more realistic possibility. In the developed world the social distance of most social scientists from the communities into which they were born and in which they are now researching may be less than in developing societies.

A national researcher does not have this automatic almost overwhelming distancing which may well make the research of the non-national free from the dilemmas which are part and parcel of the formers' existence once they start research. There are many similarities in their ways of thinking and behaviour and indeed what they expect to find may not be all that different from what they expect to find from their preparatory reading. Even though they may be internal-expatriates, they are less foreign by most categories of common acceptance. Nevertheless a factor in the national researchers' involvement in this work must be ambition and the hope of bettering their status, and in fact it would be ridiculous to suggest that this is less the case than might have been their situation in business or politics; the abstract idea of seeking for truth, for example, can only apply to a few researchers and even then for only a part of their work. The contexts in which national researchers work have been set by Western professionals both for orthodoxy and the need for publication in an academic market dominated mainly by the English language.

As nationals from less developed societies enter the international professional market by reason of their competence in a European language they provide Western academic life with not only high quality thinking but by reason of their origins different approaches to social science issues. But these non-national posts may no longer be available and the national researchers may be left with the avenue of publication as the only one available for career advancement in which the standards have been set by Western scientific professionalism. The comments of the distinguished Chinese social scientist Fei Xiao tong who suffered extraordinary persecution during the Cultural Revolution are relevant in this context (Pasternak,1989) when he said that 'knowing for the sake of knowing is playing games and more importantly that for the future the basis of control must be factual knowledge not 'isms'., He is suggesting that knowledge at least in China as it was once in Russia is politically controlled and that only in the wider world of academic acceptability can facts be evaluated.

Once it may have become generally understood by national social scientists that their professional futures cannot lie in Western developed societies, there will be a growth in their commitment to work within their own societies within the constraints of globally accepted standards of professionalism. The blocking off of extra-territorial opportunities may have coincided with the recognition of national and local protects for which their own ambitions, understandings and natal abilities are important qualifications. If it is considered essential that aspiring national fieldworkers should be able to evaluate, expose and exploit their knowledge of their own societies they are going to be faced with enormous divergences of behaviour from their own understandings which may be both a shock and a stimulus (Hughes,1960:vi).

On the other hand in these situations of wide social distance national researchers are just as liable as non-national ones to feel uncomfortable in the research environment as well as to develop ethical worries which may start to handicap their efforts to be detached. Extreme poverty, unhygienic behaviour, catastrophic illness and horrendous accidents amongst those to whom these researchers are personally committed can be unbalancing to them despite the social distances involved.

But it is not the problems of collecting data which after all is only the primary stage of any fieldwork research, but the next stage. What happens then is the impact on the research data of what is already in the minds of these national researchers because of their socialisation. One certain fact is that the non-national shares very little with the national researcher except the abstract interest in acquiring new knowledge from their fieldwork. They may have read the same text books and ostensibly taken courses with the same titles and same objectives, but there are always the underlying factors of what they are bringing to their work from the social environments in which they have been brought up and worked.

The national much more than the non-national in coping with research data is dealing with recent history even though this is masquerading as contemporary social science data. An Indian who had researched the views of nationals born after 1980 stated (Kumar,2001), 'the problem for historians arises in the impossibility of distancing themselves from a judgemental involvement in the historical process. What we say will suggest partly by implication, whether we are personally wedded to a one-nation, a two-nation or multi-nation theory, whether we believe in the continuity of history or a nationalistic historiography or not, what content this nationalistic historiography is to have and whether we celebrate the efforts of communities to reproduce these local identities at the expense of a national one or consider it a regrettable attribute of their marginalization'. The national researchers will always be looking at their work through spectacles which are part of their local and personal position.

We can perhaps assume that national researchers may have a degree of objectivity about the work that they have undertaken since they are from the start a functional part of what they are studying and can be less intrusive than a non-national. However it is from this closeness that dilemmas arise. How far can a national researcher go outside their national identity and what does this involve? These researchers unless they can be assured of having a future outside their own countries have to think ahead in terms of their hoped for careers; they and their families remain within their society and this is certainly a restraint on what they research and what they publish afterwards.

Researchers collect what they accept as facts and almost all social science fieldwork results in the researchers learning much about the underside of communal life, for example, dowry extortion, the suicide of wives, and corruption in local government. How far can nationals publish what are certainly facts without endangering themselves and perhaps their families as well as handicapping their future careers? There is certainly much self-censorship in what is published. It is safer professionally to research agricultural methods or the domestic economy of those that live on the streets of Kolkata and Mumbai. The non-national does not have these worries as he or she can publish in the safety of their own country particularly as he may not have to consider whether they need much less want to go there again (Green,1981. Scott,1969).

There is another aspect which may separate the involved national from the uninvolved non-national researcher and this is the blandness of much social science writing. Since the latter may not spend as much time as the national in their research areas, it may be difficult for them to fully appreciate that for most of those in developing societies life is indeed nasty, brutish and short, for which they can only have an imagined understanding. If we start from this assumption that for most of the world their lives are more than occasionally violent and often oppressed and that they live under conditions of periodic privation and persistent shortages, it seems likely that the national researcher must have a more real understanding of these realities. For example, the feelings, of the surviving family members of Hutus who killed or were killed by other Hutus in Ruanda, of women who were abducted or raped during the Partition of India, of the surviving men in families without women after the recent south Asian 'tsunami', of the Nuba surviving in refugee camps after the destruction of their homes in the Sudan, as well as the Sikhs who became refugees within India after the assassination of Indira Gandhi by her Sikh bodyguards in 1984.

7. Being close to home and the fading of detachment

It is often suggested that the nearer the researchers are to their native environments, the more insights they are likely to have about what is going on and the thought behind behaviour. This may well be true but what influence do these insights have on the researchers own conceptions; is the job of researchers to duplicate what they see and hear and leave their analysis to others or are they competent commentators on what is virtually their own behaviour. It is not just the researchers returning to a generalised environment in which they were raised at least a decade before their present return after prolonged higher education. Perhaps Mehta's study of Mumbai (Mehta,2004) was all the better as a result of his prolonged stay in the USA as it made his perceptions all the more acute. Is it just a return paralleling that of a concerned visitor to some part of their past and to what extent do the cultural origins of the researchers take over their perceptions and what is the consequent relationship to what is called truth?

However researchers who return to do fieldwork within the communities in which they were born and had their earliest years are in fact 'half and half' people because they have left their natal communities and they have no intention of ever coming back to live there again even if they are nationals still living in their country of birth; they live and work elsewhere and almost certainly they have not married according to the dictates of these localised cultures.

National researchers more than the non-national have difficulties in putting themselves into a detached framework of thinking when working in their own societies because of their own bourgeois sense of superiority to those beneath them in social ranking. Of course in their own eyes and in those of their families they have deservedly succeeded in becoming upwardly mobile. In doing fieldwork with farmers, Tunisian students treated them as their land-owning fathers would have done, as if they were magistrates conducting an enquiry and at worst as representatives of the Governor on a tour of inspection (Cuisenier,1982). A personal sense of sympathetic understanding for those who are not as fortunate as themselves is difficult to acquire without some imposed introspection.

8. The covert nature of research by nationals

It is obvious enough that research by non-nationals can rarely be covert since they are different by sight and by sound and also very probably by their behaviour not fitting into that of the community in which they have come to work; they are strangers doing strange things for strange reasons. The national fieldworkers theoretically can fit in much more easily into the communities in which they are trying to work. Except in situations when they are specifically asked to research a particular problem of which those involved are aware; factory production problems involving workers, management and union officials or communities which are about to be flooded by a new dam, there is always going to be a covert aspect.

It is difficult to see how any research can be carried out without some degree of deception even if this unfortunate word is rephrased as covert behaviour. The fact that such research is being carried within professionally accepted guidelines is no more than an acceptance of the fact that researchers will always have to act in terms of their professional standards or suffer professional censure. The 'I' is primarily part of the professional 'us' which usually has priority over 'them' who are the objects of research, Even if the research is done for the highest of moral objectives by researchers of great sincerity and objectivity, this may not be shared with those at whom this is aimed.

For research to be without deceit, its covert objectives which may indeed be entirely selfish for career gains, there would have to be a shared identity of interests covering the objectives of both 'I' and 'them'. This is virtually impossible since it is rather like suggesting that in ostensibly ideal marriages that husbands and wives have the same constant interests. In theory the most potentially accurate research would be when the researchers become part of or already are part of the community in which they are proposing to do research, but there are a number of dilemmas in any such positioning. Primarily if the researcher is doing research in this hidden way, it is rather like spying, with parallels to the police informer. Is it morally correct to take advantage of people in this way not just over a single issue but over many months?

The obvious point must be that in such covert work they may well find out matters which these communities keep secret from the authorities and indeed their employers. In the former case there are legal issues since it could be knowledge concerning serious crimes, and not reporting it makes them accessories after the fact; being a care worker and not reporting the abuse of children, and being a member of a gang intending to commit armed robberies or cattle thefts from a neighbouring community. It would be likely that in any research whether within an institution, company or village there will always be those who are breaking the law to some degree since it would be virtually impossible not to do so when it is a basic point in law that not knowing about a particular law is not regarded as any excuse for an offence but it would be a factor in the sentence after a finding of guilt. This of course would not occur in customary law issues and then only when government is largely known through its bureaucratic tentacles.

Perhaps the moral issues here is the magnitude of the deception and whether the assumed moral rightness of the ends justifies the means by which data was obtained. Becoming a casual worker in order, for example, to expose the degree of servitude and indebtedness of such workers, and there are always such issues and whether it might result in the closing down of that work and the workers returning to unemployment. It is widely known that underage children work in large numbers in many industries in India and Indonesia and that this is contrary to national law passed in many such countries as part of their alignment with largely Western trends in social and political thinking. Researchers who disclose these practices do not accept that it is justified by the fact that those people are providing an income for people who would not otherwise have any other social and even more support.

There is another dilemma connected to the degree of covert involvement with which there will be a corresponding decline in the degree of detachment with which social science researchers should observe in their fields of enquiry. They may well identify with those with whom they are now living and see themselves as part of their 'us' and in opposition to outsiders. Even when researchers are overt in their fieldwork it is hard for them not to prefer that community to other ones and to see their own researched community in very personal terms. Margaret Mead's long association with the Manus islanders must have made her think of them in such personal terms.

In fieldwork in which there is a prolonged and major deception there would always be a parallel continuous strain on the researcher maintain identities which are not their own. This is certainly difficult for anyone who is not a professional and well trained actor and indeed in the fact that this deception is not just for an afternoon's performance it may well have the characteristics of schizophrenia. Such researchers have to be prepared to separate themselves from their private personalities and domestic roles. It would seem unlikely that any Indian social scientist would be prepared to take on the public role of as an itinerant *saddhu*, prostitute or casual worker in and around major cities although such work might be of exceptional value. There is however other covert research fieldwork which does not involve any deceit although these researchers are fully involved in what they are observing. It is easy for researchers to undertake fieldwork on rush hour behaviour in Mumbai, pilgrim behaviour in Banaras or one of the four major mountains used for pilgrimages in China, market behaviour in Accra and how restaurants work in Massawa. No one knows that they are being researched and the research work might be more valuable because it is qualitative rather than having the inevitable inaccuracies of quantitative work. We cannot assume that quantitative research is necessarily more accurate than qualitative unless one accepts its superiority over other methods and this surely is a Western bias as a result of the Enlightenment.

Covert behaviour and covert thinking is certainly part of all research which is not narrowly experimental. No fieldworkers are likely to expose themselves to the communities in which they are working by explaining their tentative hypotheses in matters which would be likely to upset their informants. The researchers would not want to upset the working relationships which they have achieved through their considerable investment of time and effort. The hooded nature of all personal rather than just professional relationships is surely part of all social life. This is not very different from the ways in which these informants conduct their own relationships. The villager does not disclose what he thinks of his neighbours and both men and women practice a necessary continuum of deception because that his how village life is maintained between people who have to live closely together for many years and yet are rivals for success in all aspects of their lives. Surely this occurs in most marriages and within families quarrels may occur when the practice of deception has been broken.

So except in extreme cases in which national researchers take on the roles and persona of people other than their own in which there is total deception, some degree of deception is likely to occur in all fieldwork. The point may be that this is not seen as deception as their fieldwork practices are not according to some job specification but rather within the normalities of research as something which helps their work along. It is also perhaps not rational to expect them to be straightforward about what they are doing as they may not know themselves what they have done until they start to write up their fieldwork. Although almost all national and non-national researchers are by their nature outsiders the former have a far better chance of becoming as far as possible insiders; the degree to which the national researchers manage to achieve this may be the degree to which they have been able to manage to role conflicts over being professionally one being and socially another. In other words the control of deception within a framework of immediate social acceptability.

9. The advantages of being a national researcher

Anyone from China, India or Malaysia who comes to do research within their own country has the advantage of knowing much about the overall cultural background before they start; of course, this could be also seen as having some factors in common with the enclosed mind, but surely this does not overlay the advantages (Hennigh,1981).

Such researchers start with some comprehension of the entire system, as well as understanding something of the national culture and the points of tension. Many matters which come up in fieldwork would not have to be explained to such researchers. But, 'my (New Guinea) friends who spent their childhood years away from their village in pursuit of an education found, on returning to their village, that they were incompetent at farming their family gardens because they had missed out on mastering a large body of complex knowledge (Diamond,2005:281). Malinowski (1961:23), wrote that a man who lives in a polyandrous community cannot experience the same feelings of jealousy as a strict monogamist, though he might have the same elements of them. The national researcher is better able to see the feasibility or not of many localised hypotheses in particular situations, and has a distinct advantage over the non-national in this respect.

National researchers may have local commitments from which they are not separated by distance and international law and these remain however much these researchers change their status or domicile. This is bilateral, their communities retain an interest in them as well; there are always exchanges of visits and presents and perhaps even for generations there are watered down reciprocities.

They may have connections in their research area before they start work although it may be easier to carry on research work without disclosing what they are doing but that is not sensible as they will always have to live with and explain what they have written should it be published and even if this is done in a major language not directly known to these people they will find out soon enough what has allegedly been written about them according to the communities' network of gossip. We know that the very public protests about the novels written in English by Salmon Rushdie came from many who had not read the works of which they were complaining.

There may also be clashes between what these researchers see and experience as insiders, and put into print may lead to difficulties with the authorities who are administratively and politically responsible for such situations. It is usually assumed that their conclusions are partisan because there is no such thing as a neutral position in community life; goodness or badness and indeed everything else is always referred to in a context of personal pronouns. But then there is always the difficulty as to whether someone who is a natal member of a community or social group can ever have the necessary detachment to research their own communities. This coincides with the whole issue of bias and probably ends up with the acceptance of such data with what can be assumed to be their known biases and as such is likely to be nearer social reality than the observations of others whose connections with these same communities is both partial and intermittent.

10. National researchers' role ambiguities and adjustments

National researchers can only expect to slide into a fieldwork situation without serious problems if there is some elision between who they are in reality and the situations in which they have placed themselves. This is rarely the case although the problems are not likely to be as severe as those experienced by the non-national.

A Bolivian born social scientist with Spanish as his natal language despite being a national was always looked as a foreigner by Bolivian Indians and indeed came to accept this role (Buechler,1969). He and his wife had decided to research the new land-holders of a former estate which had been redistributed to their former tenants and employees, but neither spoke Aymara and so were dependent on interpreters. Whatever the status of the researchers who may consider themselves to be insiders in the situation which they are going to research, they are either too close so as to undercut the detachment necessary for valid research or have no more than some personal understanding of their own role which may have no bearing at all on how those being researched see them. They have to work out for themselves the type of status and role they can achieve which will further their research objectives.

The Bolivian researcher decided that in the circumstances of these people occupying their newly acquired land, his best role would be for him to act as an interpreter to them of the outside political and social world with which they were having primary dealings for the first time in which the interrelationship was not weighted against them. In this his nationality became an advantage as he adjusted his researched objectives away from any assumption of social affinity with those speaking Aymara.

Another type of insiderness which we have already mentioned was the American researcher who had to take up long term residence in a community in order to study their educational system and who became an insider because he and his wife stayed there so long that they were indistinguishable from other residents who stayed more or less the same length of time before moving on; he became 'our' resident researcher rather than an outsider trying to find out what was going on (Hennigh,1981).

A different type of researcher would be those who are looking into their own work environment, and here much would depend on whether such work is overt or covert. Most doctors will be researching something in which they have a long-term interest and there is no reason for anyone else to know what they are doing. The surroundings of such research would be governed by the ethical rulings of their professional organisation. Such research is covert in the sense that it is not part of an employee's work contract but overt in the sense that they do not conceal their special interests.

The fact that researchers may be insiders in terms of their citizenship is likely to be no more than a sliding scale of nearness and distance. They may be insiders as regards the professional group to which they belong under social association with that urban-industrial culture in which they have spent most of their adult lives, but for a Javanese researcher to assume that he has much in common with rural peasants makes no sense. In such a community he is likely to be given a higher status than their own and seen to be a member of the civil servant class with whom lowly people have to act with circumspection and kept at a distance (Koentjaraningrat,1982), and there would be very substantial language difference in what these researchers would use in talking to him and the language that they would use amongst themselves (Geertz,1973).

So we can conclude that those who are researched make no distinction between the social scientist as researchers and the society or groups which they think that they represent. National researchers in reality are probably just as likely to be classified as outsiders and indeed as potential enemies since it is always assumed that the government has given them permission to be there (Naderi,1982). They can remain outsiders for the duration of their fieldwork however much they might remain personally acceptable to those who see advantages in their presence.

11. Observing and participation

For the national to research some aspects of their own culture rather than the divided intellectual who has to go through a process of re-socialisation on their return whether or not for social research, there are the difficult issues of existing socialisation. There is the initial need to recognise that by taking on a researcher role, they are no longer by any standards 'normal' citizens; they have made themselves different. 'It is not at all obvious or even plausible that a 'native' anthropologist, born and raised in a city,perhaps educated in a foreign school in his or her own country and even more likely trained abroad in Europe or America would be doing indigenous anthropology in the strict sense of the term when studying illiterate peasants in a remote corner of his or her country (Altorki,1982). This quote from a Saudi-Arabian woman academic raises a number of divergent issues which involve any national researcher. Whatever the social origins of the national researcher whether a Mishmi or Baluchi from remote villages, the process of education has urbanised them. Then the process of education is an imposed system of acquiring knowledge which has in all probability no obvious or indeed even intended relationship to what they might have learned had they remained in their natal social environments, and moreover it is a Westernised way of learning. The adult man or woman who has been successful in this hierarchical process of accumulating status by examinations abstracted from other aspects of living has been made socially and psychologically different to those who have not been through this process. If such researchers leave their own current social environments, almost certainly urban and predominantly intellectual in how they see the world around them, and start research outside this charmed circle, it will almost certainly be with people who are socially and economically inferior to themselves. This social situation is not much different to Western nationals researching in their own country; the recipients of their research will have seen the researchers as social aliens and the researchers themselves will not see those with whom they work as their social equals.

On the other hand should they start to do research in the community and at the social level of which they are already a part, there are immediate dilemmas. They may feel that their primary obligations are to completing the research which they have initiated and which their families know they are carrying out. But they are still relatives and no research objectives will ever obliterate the fact that they are still daughters and sons, grandchildren and cousins and they will always be expected to fulfil those roles. Their relatives may trivialise their research even if they know that such work is highly thought of in the outside world in which these relatives now belong professionally.

As a household member by right of birth, such researchers are immediately on call for their primary roles as relatives. They have no immunity which would normally be given to visitors of that nationality or even non-nationals over breaches of acceptable behaviour. Such researchers are automatically slotted into social roles appropriate to their age and parentage which both channels and restricts the possible range of their enquiries. The non-national is always some sort of stranger but the national can never get away from the fact that they are kin, of a certain age with an appropriate marital status.

Access to a society by reason of birth and location does not means access beyond their primary social roles and to obtain data about which such researchers do not immediately have the right to obtain requires careful work. 'It took six to eight months before I could develop friendships that made close observation of their daily life possible to the degree that my presence was more or less ignored' (Altorki,1982). But supposing a researcher and his wife come and live in a community for years and are known from the start as researching into a local activity, the dilemma remains that they are their own best informants because in that community (Hennigh,1981) they were not much different to other married couples who live in a community of couples who have just arrived, stayed for a while and were likely to be on the move again as soon as their jobs changed. It was not a fieldwork situation as they were acceptable residents, and not just suspicious transients. This couple moved from being detached enquirers into an active involvement in community affairs in which the community saw them as a source of information and often advice. The community had an acceptable place for active newcomers and they were treated as any other community resource. This active participation in community affairs was a highly efficient way of gaining information, it was a requirement of local citizenship and it was also a psychological defence against the abrasiveness of long-term fieldwork and counteracted the absence of academic social stimulation.

There was also the factor that by becoming a part of the overt community, life was not much different to that of others there, it was a significant short-cut away from the struggle to gain insights into alien cultural settings even if it was part of their own American public culture. It also reduced the traumas of withdrawal when they eventually had to move house to take up other work. This form of active participation is only possible for nationals in that part of their own society in which their social characteristics are broadly similar to that of the field of study. Once there are social differences and any element of covert observation then the problem is broadly similar to research in a foreign environment.

12. The national's explanatory potential

There can be little doubt that the national researcher in situations approaching their normal social environments, home, village or neighbourhood has a potential research ability advantage over a non-national of even the most persevering and self-effacing personality. An unmarried English woman researcher brought up and educated in the south of Britain would have some understanding of the background of married women in a Northern industrial community centred on factory shift working. Both would have a shared knowledge of English, seen the same advertisements, been exposed to similar media and have some understanding of national and local politics as well as of the workings of the social services as it concerned them relating to health, the police and of course road use behaviour. It may be that they have social differences based on class, speech and dress but even if that is so they are both within a single social system despite a tendency in research to underline the differences as part of the analyses required by each research project; these are usually thought to be relevant but similarities are far less interesting. The researcher may eat, dress and use quite distinct linguistic codes but the linking similarities make for an ease of entry and preliminary understandings which would take a non-national a very long period to acquire if indeed they could ever be acquired in adult life by learning rather than by socialisation.

How far would this apply in India, Malaysia and China, other than nationality? There would indeed be striking differences resulting from education and a rise in social status in comparison to their parents and most of the members of their extended families. Whether these individuals have not travelled overseas or have become 'half and halfs, they are now members of an intellectual class within which they would share a similar basic understandings of what is going on socially, economically and politically in their own culture. More importantly they would be able to sense their way through the intricacies of their local bureaucracies and to know how far and in what direction they can go in their enquiries.

But how far do these somewhat intellectual overviews make an Indian national born, and at least raised in a particular locality which they have undoubtedly left many years previously understand the workings of sub-societies in other parts of India? The answer must surely be with difficulty as they may not even know how much or how little they know.

This author is aware of the example of an ordinary Zigua boy in Tanzania whom a non-national researcher met by accident when they were both sheltering from a rainstorm demonstrated nearly twenty different games which he played with other village children. If that is the 'specialised' knowledge of a ten year old, then any adult who has not been alienated by prolonged education outside his or her natal social environment, will have an unparalleled accumulation of local knowledge. The educated adult researchers may thus be over-confident that they have such useful accumulations of local knowledge from which they themselves may have been alienated by social mobility through prolonged education.

13. The citizen and suspicion of their motives

A non-national undertaking fieldwork is open to any suspicion of their motives which can be thought up as the pursuit of knowledge for anything other than self-interest is not understood by most people in most cultures. On the other hand a national doing fieldwork in any community and for any purpose, unless they have socially accepted connections there already, are going to be the objects of suspicion on two counts. Firstly, they are unlikely to have any concept of disinterested search for knowledge and they suspect quite rightly that they are working for their own advancement and for some outside organisation whose motives are both suspect and unknown. The ethnographer usually confronts the suspicion of having ulterior motives in the immediate encounter with and learns its existence from the people studied; without such motives most people find it hard to understand why only the researcher is paid or wants to spend money to have this experience (Pels,1999:107). Secondly, their experience of outsiders whatever their ostensibly benign purpose or their legal right to be there, are largely seen as intruders whose arrival is followed by trouble or anticipated trouble for the community which has been entered. Any type of intrusion from government officials entails forms of control and hostile reactions to what they are trying to do even if its beneficial effects are ultimately recognised. In this respect the 'not in my back garden' reaction against planned or required changes is an almost universal reaction.

In fact for many if not most people the experience of government is that of inspections connected with policing, taxation, agricultural controls, licensing, education enrolments, public health, road maintenance and building permits. It is within this framework that the arrival of the social science researcher must be viewed. 'This situation is further complicated by the fact that the academic reasons for the researcher's choice of precisely this location for study are, to many of the people studied, incomprehensible (Pels,1999:107), since they are often accidental. Even if the choice is more by accident than planned or made as the result of a civil servant's recommendations that that this area has, for example, a vacant house or indeed someone there who has had some experience of dealing with strangers or internal expatriates, this is unlikely to be accepted locally as the truth.

The whole situation is based not only on the perhaps well justified suspicion of the potential objects of research, and that no good is likely to come to them from these encounters except for the few who may gain temporary status or employment from such projects. It is also the fact that researchers, however well intentioned, are not naturally endowed with the ability to negotiate with people from other cultures over their research requirements.

Any community has both power-holders and those that aspire to power and both might well attempt to make researchers become part of their communal politics. There is almost always the initial dislike of those in power to have a free uncommitted outsider whether national or non-national loose in their socio-political environment.

The national researchers' political ideas may make for problems either in their viewpoints as members of a professional elite or in the fact that their membership or affiliation to a political party may be known in the community or institution being researched. Although there is a clear distinction between being a member of the Communist Party and being a practising Baptist, it may be more significant in guiding thinking than being a member of some national cultural organisation. Perhaps it is easy enough for a non-national whose religious affiliations may not have any deep familial and social background. However for a Hindu of any caste or a Muslim, their religious background is automatically tied into national and local politics and almost impossible to slough off socially or intellectually.

14. The political mistrust of the national researcher

The power holders in any society know very clearly the difference between national and non-national researchers and the fields in which they are working. The latter are quite clearly on sufferance and require the agreement of both the community, institution and the central and local governments for their presence there. The former do not have to go through any such anonymous one-sided process; they are free to negotiate their own fields of enquiry and the personal relationships involved. So far so good but there is a more complicated sub-strata. The non-national can always be expelled or prevented from working by circuitous means if they created difficulties which outbalance the advantages of having a foreigner as a pawn which can be manipulated for the advantage of informants, leading local figures and the community.

The national researchers apart from their natal understandings will each have their own network of connections and the extent of their social and political power will not be known; informants might well be a little wary as to how they should relate to the appearance of national researchers; on whose side are they and to whom are they related by kinship and social obligations. Since they do not go away in the sense of living in another country and may well live within reasonable travelling distance of the researched, they remain a more or less resident pawn in local politics and in the personal lives of important informants. The Egyptian researcher who worked on Nubians affected by the Nasser dam was visited by some of those he knew in that work although in lived in Alexandria (Fahim,1979).

On top of this it is obvious that if a national researcher has long-term relationships with the community which he has researched, there is no longer the same range of possible detachment from the research at which most fieldworkers aim. On the one hand they may become sympathetic to the difficulties of the lives of these people with whom he has become intimately connected; he wants to represent their misfortunes to higher political authorities while at the same time recognising that almost every community within the nation at that level of living has the same problems. The researched community feels that the researcher should represent them as a return for their cooperation in his work. On the other hand the national researcher with his wider perspective of national issues particularly of the shortage of money, may loose sympathy with such people as the Nubians displaced by the flooding of the Nasser dam; their continuous complaints against the government and their unwillingness to cultivate part of the land allocated to them despite the money and efforts spent to meet their needs (Fahim,1979).

The national researcher may also have problems from the fact that with his superior access to what is going on in any community, he is likely to uncover corruption. Although researchers study discrete communities attempting to understand them in terms of their distinctiveness, there is always the state in the background of all that they do. The modern state has many more ways of being involved in local affairs than in previous generations. To omit the state from their work on local communities and their undertakings is almost always an error. The state is always there and the national researchers finding out about widespread corruption and for example, having the Israeli government's censorship and control of research in the social sciences is a concrete moral strain on national researchers which is very much in their minds while it is an abstract issue for non-national workers (Cohen,1982).

In terms of the sensitivities of higher level politics and national dignity there may have been some preference for non-national researchers; their work can be used or ignored according to the local political situation. As foreigners the reasons for the rejection of their information are always ready to hand; they did not know enough to be able to distinguish the realities of their data and their conclusions can be ignored.

The national researcher from an insider position is likely to learn more about what is going on than a non-national. To the extent that they are involved in the immediacies of development, they may well be seen as political adversaries particularly as many social scientists are committed to change (Cernea,1982). Logically enough social scientists are often seen to be spies and their methodological tendency to be participant observers in almost every activity builds up this picture of suspiciously purposeless inquisitiveness which does not make sense to the communities in which they are working particularly if they are non-nationals (Diamond,1964). This is made worse by the non-national's professional obsession with the extraordinary rather than the commonplace (Naderi,1982); it is perhaps ironic that more is known about witchcraft than is known about the disposal of human waste.

Conflicts may break out between the approaches of the social scientist who has no commitment to any local interests unless of course they are researching their own community and as such can be assumed to have a built-in bias; the local power groups who are committed to the *status quo* and would dislike any approach which questions this. 'In a subject where theories and methods are weak, intellectual shifts are likely to arise out of conflicts between approaches rather than conflicts between explanations will be weak and often non-comparable because they are approach specific' (Bernstein,1974:154). So the politically unacceptable will be attacked by the inadequacies of the methods used by the researchers rather than in terms of what they may have disclosed.

Citizen social scientists may come up against trouble not only when their research closes in on subjects which are against official public practices and which have the support of a politically dominant group, but because of the ideological framework with which they approach their work. In this there are two opposing frameworks. The Western world bases its understandings on the rationality of administration in which morality together with religion are essentially private matters. This could be described as the morality of rationality and that certain behaviour does not contribute to the public good; the religiously endorsed beating of wives, subordination of women to men, their circumcision, and public corruption are wrong on these grounds. Western concepts of the public good include behaviour which is in private and which is not excused by religious beliefs.

Thus social scientists may collect facts and reach conclusions which may or may not elide with the policies of those in authority. Researchers may be guided by their own personal moralities which are not channelled by religious beliefs. Whether they get into trouble is often a matter of circumstances since their approach to what they do in the field has considerable flexibility.

At the other extreme there are Islamic societies in which the framework of thinking is that of tradition based on unalterable sacred texts. In this situation social science research and its conclusions have to be within the boundaries of what Muslim theologians lay down as being what the sacred texts seem to say about these issues. This of course is a very restrictive framework within which to work, in which the religious rationality clashes with that of Western social and indeed political science. But national social scientists are not dealing with any overall and effective restrictions on research which will vary between countries. In Saudi Arabia there is a close connection between the hegemony of the Saudi royal family and a divided but still politically influential religious elite and in Iran with its Shia religious dignitaries holding key political positions. On the other hand there are other ostensibly Islamic states with much looser interpretations of what should be done such as in Indonesia where there is a much more tentative connection between the government of a predominantly Muslim state and substantial non-Muslim minorities. So restrictions on social science research are more likely in an Islamic state with its adherence to a single holy text and its derivative interpretations and in any single party or military controlled state such as China or Myanmar, than in Hindu, Buddhist or Christian states in which there are no such structures of authoritarian orthodoxies since their sacred texts are inexact, ambiguous and numerous. In addition to these restrictions there is human ingenuity in the use of alternative interpretations. In such states as Saudi Arabia there is always the avenue of advice 'nasima' rather than criticism 'naqd' but even that has to be proffered in the form of transgressions against God's eternal commands (Asad,1993:200-36).

So in such circumstances much would depend on how the data is presented and to what extent it attracted the attention of political and theological authorities. It is perhaps not that Islamic theologians and political authorities with which they are institutionally interrelated, do not change but that change seeps in without attention being drawn to what is happening by social scientists. Overall it seems that social science research in authoritarian states is limited and dangerous since their association with alternatives defined by those in power as subversion would block such people from further work within their own societies.

15. The national researcher and the underside of society

Whether it is designed or not social science researchers are either by birth members of the professional classes or become so as a result of their prolonged training and current employment; thus they are not necessarily either professionally or personally interested in the under class; those who are deprived of reasonable living conditions.

All societies and not just in the non-Western world have very large numbers of people who are seen as being outsiders and misrepresented as having no attributes, thus making research on them a priority. Afro-Americans are an example of a community which has been consistently denigrated by researchers and politicians as having no culture of their own and of being largely parasitic on the dominant Euro-American community who provide the research supporting such views (Gwaltney,1981). It may well be that the Untouchable community in India is in much the same situation as not having as socially valuable a culture as that of caste Hindus. Both these communities have ways of retaining their own identities and ways of dealing with intrusions from the outside dominant world including researchers.

In such contexts there are particular advantages in being

ethnically similar and a national but this does not provide automatic entry but only a predisposition among the researched community to give them some wider options initially. Such researchers still have to make them realize that they are not just agents of the dominant social groups surrounding them.

16. The national researcher as an employee

Unless the social science researchers have independent means and can thus choose what they work on and how they do it, the researcher always has some characteristics of an employee. Of course, a researcher who is employed as a teacher or nurse can do research on some matters in which they are personally interested but it has to be done covertly if they do not want to loose their jobs or overtly with the permission of their employers. Thus it is possible for teachers to work on some private research programme while still doing their official job just as a doctor can work privately on a personal interest in a particular aspect of say, epidemiology. Since both these undertakings are difficult to accomplish their status as employees must be kept in mind.

Generally speaking social science fieldworkers are employees of a university, a government department involved in agricultural development or social welfare, a non-governmental organisation working with refugees or education or an organisation funding research from private sources such as the Ford Foundation or one dependent on government funding which they distribute to approved projects. As receivers of money from organisations which they cannot control they are to all intents and purposes employees as the source of their funding by their agreement to support a particular project, controls them. There is of course the illusion of academic freedom which is in practice always circumscribed by the factors which we have already detailed. The researchers balance this hope against the practicalities of what they are hoping to do and where they do it and under what contemporary conditions.

Of course there may be public and private codes of conduct but once the researchers have accepted this support they are no longer in any position to question the backing for their own work or the topic other than through covert manipulation, as they are functionally employees. They also start their work knowing the objectives of their employers since they are not likely to be funded for the open ended type of research for which university grants may be provided. Each social scientist has to weigh up the balances between the comparative needs of 'us', the researchers and their employers and the 'them' who are to be researched with a view to some change which is happening or is considered necessary.

An Egyptian anthropologist (Fahim,1979) in studying Nubian resettlement schemes related to the Nasser high dam found it hard to reconcile their continued complaints and unwillingness to cultivate part of the land allocated to them with the amounts of money and effort put into their needs as new communities, when the country as a whole had financial problems. At the same time he was sympathetic towards their reluctance to take any initiatives to solve their own problems. The solution to this ambiguity probably lay ultimately with how necessary it was for that researcher to retain a paying contract from the government.

In major moral issues they can resign from employment and of course suffer the social and economic penalties of breaking their contracts or at least seeming to do so by not completing their research work. In developing societies most employment for social scientists is likely to be in applied rather than theoretical work in which they are employed to look into some problems which are causing anxiety to those in power; they would be sensible to keep this in mind when voicing objections. If social scientists are employed to produce reports on some areas of difficulty then what they report is the property of their employers and copyright rests with them as to what should be published or indeed they may prevent the publication of any part of what the fieldworkers have written; reemployment must depend on some measure of conformity to what these power-holders want.

Von Haimendorf held the view that the information he collected from informants was his property and indeed others have held themselves to hold the copyright on what they have collected

(Fisher,1996:351). Some social scientists have seen their own use of the information that they have collected as needing the protection of law. Horwitz (1993) had her major informants sign release agreements for her to used their information for whatever purposes she needed. Haimendorff claimed that the information that he had collected from a Nepalese assistant who had subsequently become a professor in their national university was his under copyright. This unseemly and politically inept quarrel would have prevented him from doing any further work in Nepal. Both these approaches seem singularly out of place in the circumstances of contemporary research when cross-cultural cooperation in research must be the main way forward apart from the fact that the researchers have both shown an absence of any generosity of spirit.

The national researchers may not have as easy an access to insiders in the areas which they want to research as might be imagined. Any social scientist works hard to acquire and maintain a circle of similar professionals to help them in their work (Madan,1982) who are at the same time rivals; this is a time-consuming process in which their work and their social lives may overlap.

17. The national as an employee of a university

National researchers come from a university and their association with the academic world may be of immediate importance as well as much longer than any with a particular place of research. They may be directly employed so that their research will have to fit into university term times and even if they do research from their homes it will have to be subordinated to what their employers expect and indeed have a right to demand; unless their contract states that it is not so, their primary duty is to teach. As the visiting Professor Bohannan remarked: 'Oxford is a wonderful place in which to work if only the students stayed away'. As the university may have provided the grant to enable the research or approved the topic covered by the researcher in his or her own time, its own status is affected by what they do and therefore affects their further employment (Bohannan,1981).

Whether they progress in the academic world depends not only on the value of their teaching and the quality of their research but on the reputation that they achieve among other academics which at its extremes may promote or block progress. In most cases this reputation is based on the tenuous processes of academic interrelationships and gossip. As university employees they have the distinction of not being employed by the government although the government may in the first place provide the university with its funding in general or for research in certain areas. There is of course the ideology of academic freedom which in the long run probably depends on the extent to which the university will support the public rather than private research activities of their employees.

Many Western universities have accumulated independent funding so that they are able to take a clearer line on the independence of their employee to express opinions and have practices which the wider public would not approve. In this there are limits as, the publicly expressed opinions of senior academic staff for example that non-white peoples may have a slight genetic disadvantage in their cognitive capacities and that similarly women may have the same which might account for their relative absence in high level mathematics. An example of independence was that of an academic losing his job for using marihuana in experiments with students. These problems made by academics to academic audiences receive immediate publicity as being politically unacceptable rather than academically wrong. During the McCarthy period in the United States universities were not particularly prominent in support of their teachers suspected of Communist sympathies. Just as in some universities in developing societies' world they make no public comment about military governments and oppressive regimes. In the former cases except for that concerning the use of soft drugs, they can be seen as moral failings on the part of the university authorities but in the latter it is simply not safe to intellectually oppose the government. No national would investigate black market activities quite so openly as a non-national may do (Green,1981) and then leave the country in safety.

In addition many universities now operate departmental and

university research committees which would have to pass the research intentions of their staff and thus avoid their involvement in matters which would involve public opposition and questions of their judgement; a self-regulatory environment. In the Western universities the boundaries of acceptance for academic work do not coincide with their assumed legality combined with the fact that the universities may not wish to it to be known publicly what they do privately as it might make them liable in law as accessories after the fact.

National researchers in Western societies have certainly participated in projects that have involved law breaking in for example, gang activities, drug taking and corruption. It would probably not be worth the uproar for law enforcement officials to take action against such researchers, as they would gain nothing and they would lose the opportunities of gaining some additional knowledge. Researchers will probably work out their own limits of tolerance to law breaking in terms of dangers to themselves rather than in terms of what their universities might think.

National researchers may operate in countries which have no long established tolerance of academic freedom and more importantly have every reason not to tolerant criticism. In any country which has a single party government and no well established political opposition, criticism of any type is seen as opposition. For informants the fact that researchers are affiliated to a local university does give them a possible separate status to that of those directly sponsored by the government. This distinction may be an important one in which government suspicion of what they are doing aligns them with the opposition (Abu-Lughod,1988).

Nevertheless the relationship with a local university is important is a not so much covert restraint on how far national researchers can go when viewing their own career prospects and ideologies as to whether endangering these is worth any oppositional disclosures. They must always be more circumspect in what they do openly unless their future is assured possibly outside their country and even more so if they are married or born into local families who might suffer from their research activities

18. The national researcher as an employee of the government or a firm

In many situations in developing countries the opportunities for social science research are as the paid employees of the government, it which they have little opportunity to take an approach which is different to that of the government. Any idea of independence must be tied to the possibility of alternative employment, loss of seniority in such a move, pension and status issues. In the situations in which they are not salaried employees of a government department or project, researchers may be hired for a specific purpose and specific period which means that they have no opportunity to go outside the terms of what they are asked to report on; they will want to establish a reputation for usefulness and with this the opportunity for further work.

Since they have been hired for a specific purpose their conditions of employment may be much the same as that of a mechanic who has been hired for repairing a piece of machinery; a short-term preoccupation. The analogy here is deliberate because the contract for the researchers is likely to be quite specific. The contractor is uninterested in paying for or accumulating any disinterested knowledge (Feldman,1981).

So to be employed by a government department or an agency which gets its funds and authority from the government requires a change of mind and heart in those who become applied rather than theoretical social scientists. In this context it is salutary to remember that the possibilities of university appointments for social scientists with higher degrees in the first post colonial years and the expansion of higher education has long gone. If these highly trained social scientists wish to use their potential expertise then they have little choice but to work for employers who are either government departments or agencies selling their research results to their governments. They are in no position to be too consciously critical if they wish to earn a living after their long years of training and idealism.

But there is a snag to this in that they do not have the

specialisations that can easily be transferred from one social environment to another. A soil scientist can be just as useful in Kwangtung and Rajasthan as in Hopei and Assam with no difficulty other than the short period required for personal adjustment. Social scientists may have gained their expertise within particular cultures and this knowledge cannot be easily applied to different circumstances in which there are many unknown factors. The contracting firm which employed an ecological researcher to advise them on their Alaskan developments may have already made up its mind as to what they intend to do which in Feldman's case was that the oil fields in Alaska had to be opened up for the greater good of American society and of course for the financial benefit of the companies involved. The case ended up in the Alaskan High Court in which the researcher became a witness against his employer.

Even where there is no state or private capitalist motive behind a project on which the researchers are contracted to work, they will be tied to the theme of development. In accordance with this theme there is often the imposition of schemes on communities which will not themselves benefit and indeed will involve substantial disruptions on their ways of living. It is the imposition of rapid change and this in real terms is the common situation in which contracted researchers are likely to be involved. And they will be asked to find ways in which this can be mitigated. They are tied into a foregone conclusion and they and those with whom they now have to work with just have to make the best of the situation.

The contractors employing researchers whether private or public, are specialists in their own fields and any interest in human relations will be for the running of their own organisations or departments. They would see community needs whether tribal, ethnic or occupational as standing in the way of progress. A valid viewpoint but not one with which most social scientists would agree. The time factor means that they will not have been asked to undertake community wide background research. In this sense they are narrowly defined as an employee with a job specification and they will have been employed because they already know something about the matters over which the government is concerned.

The most important relationship for such researchers is with their employers and the dimensions of the political, social and economic world in which they are operating and indeed with the few power holders who might have influence on these matters and who might read the research results. The community or issue to be researched is of secondary importance unless they have their own channels of political influence.

Thus in entering this employment or rather the requirements of their employers there will always be factors within which they have to operate such as financial limitations, time within which a project has to be completed, and the ever present limitations of the popular majority who may have no interest at all in minority issues. The researchers will also have to consider that they have been employed as political pawns just to dress up the issue to show some concern for perhaps minority issues, the humanity angle. Then there are always unexpected factors which the researchers can use in support of their assessments such as the *chipko* (tree hug) movement of Garwhali women to protect trees in the foothills of the Himalayas (Berreman,1989) and the Buddhist ordination of trees in Thailand (Tannenbaum,2000) which all caught the popular imagination and in the latter case got royal support.

There are also likely to be the unexpected such as the 2004 Indonesian 'tsunami' which will have stopped major development projects in that area for a decade and the termination of aid by overseas donors because they are exasperated with the extent of corruption. Much may depend on the orderliness of the national structure which enables research to be of value and its freedom from political emergencies but all in all the national researchers are less easy to push aside as they live there and have easier access to issues and supporting views.

What are the advantages of employing a social scientist? Their training and framework of research expertise provides employers, whether government departments or agencies working for government, with people who are able to assess the interrelated functions of a society or institution whether based on occupation, class, ethnicity, region or the importance of dominant figures or power groups. They are also able to criticise the work of other professionals from their own uncommitted positions. Indeed they are useful adjustable examiners of the varying situations confronting all decision makers all the time.

19. Preliminary adjustments in becoming employees

A social scientist wishing for an academic appointment is moving from one establishment to another similar one and this whole field of employment is a mixture of academic knowledge and internal politics in which it is not too difficult to fit in because of the similarities in the social environments. To achieve employment within a government department or an agency has very few similarities with the academic process however. The prospective employee is working within a much broader environment in which the actual matters to be researched involve national and local politics, departmental rivalries and budgets which have to be negotiated amid the claims of other departments for whatever funds are available.

Social scientists by training and the nature of their own research are individualists rather than experienced in team work; it would be rare indeed for anyone to be awarded a post graduate degree for a joint project. So now they would have to learn to fit into a hierarchical decision making structure in which however right they may be, their contribution is only one element and likely to be a peripheral one as well. A principal difficulty is that their research and whole approach is tied into completed events, they are in effect historians. As employees they are part of an on-going process and may have to learn for the first time the constraints of continuous employment and consultation (Almy,1977). Not just to comment on the work of those whom they may consider to be misguided but to live alongside them in harmony with those who are socially near and more important than the distant people who are being researched. On a day-to-day basis they will have to work alongside specialists in other more exact disciplines who have well known and accepted fields of work.

They will also have to adjust to and change the feeling that a scientific fact has to be adapted to but to change human behaviour on the basis of social science evaluations comes within the range of their general perhaps dislike of public relations and politics as manipulative without much regard for the public good. In the uncertain world of politics and development, employers may have a preference for the producers of what they take to be facts. They will need convincing that the social scientists producing analyses of yet more inexact variables have a value for their necessary and immediate decision making.

The successes of hard science in such matters as the increased crop yields of the Green Revolution and the control of some mass diseases are well known but it is much harder to pin-point the successes of social science which has been to contribute to broad cultural understandings; the planning engineer wants statements about what is likely to happen, not just a list of possibilities. Scientists can put their finger on some exact element which can be related to human behaviour; crops or some humans die and they work outwards from a small number of provable facts into social environments in which there are no parallel exactnesses. Social scientists have to explain that in human behaviour there are few such facts that can be pinpointed. They are not in a strong position by saying that they have the ability to assess contributory variables which may in some circumstances have some effects.

Social scientists, to many employers, are an unknown quantity and moreover are seen as having an occupation which does not fit easily into departmental and occupational systems which have soil scientists, engineers, epidemiologists and economists who are well known by what they do and are easily slotted in as such. They have to sell themselves to market orientated employers who are sceptical of their expertise. They may be experts in the understanding of one cultural environment and for their alertness of mind to be of value in an other but this is not likely to be properly utilised in the conditions of short-term employment. Their useful expertise in a particular social environment is cumulative and requires a long term involvement.

They also have to acquire the tact or more often the diplomatic skills which are necessary in making recommendations which in effect are critical of their employer's practices and policies in which they have become entangled by the fact of their employment or because of the accidents of political change. The 'other' with whom social scientists are used to working is no longer as clearly defined segment of society out there as the 'us' of the researcher's is now of primary importance if they want to have influence and further the results of their work.

The close association with institutional attitudes, practices and values which deal with the immediacies of cause and effect can be a shock to those who are used to the more relaxed long-term scheming of a university environment. While university tenure is a cherished goal in which success and value are vague concepts which are not usually directly subject to the crudities of value for money, there may be little security in this world of applied social science. Whereas academic life has its own secretive processes for obtaining advancement and security, the ways of getting and retaining employment in the outside world require different techniques and circles of influence. They will also have to come to terms with the fact that as employees their personal freedoms will be much more restricted since they will have to be seen to work in much the same ways as other employees. They will have to work directly on what their employers require rather than on some rather more abstract issues which they feel should have priority.

20. The national researcher and non-governmental organisations

Social science has developed mainly in academic surroundings and institutions who want to use their expertise and only have this source from which to recruit people with the necessary professional advise and sympathetic understandings to provide information. There can be little doubt that any academically trained social scientist carries with them into the field ideas which they have absorbed in their long years within the academic environment.

There is perhaps an attitude of sceptical tolerance for those who run the countries of which they are nationals; the feeling that those in power do not have the moral understandings and abilities to do the job well while they themselves feel well qualified to provide detached criticism without ever having had the responsibilities which go with public service. Certainly a form of conceit particularly as few have had any experience of wage earning and situations outside their universities. They expect non-governmental organisations to provide much of the necessary criticisms.

In most developing societies these non-governmental organisations whether religious or involved in secular charities, are both linked to outside organisations and as well consider themselves as involved in ethical issues, overriding local political considerations. Their employees or those who do research on their behalf do assume local political roles. Oxfam in Tanzania gave substantial support to the one-party state's villagisation campaign which brought people compulsorily into villages from where they had lived previously in well dispersed agricultural holdings, in anticipation of social benefits which never materialised. This support continued despite the fact that observers on the ground were saying that the damage done to communities by removing them from easy access to their lands. Researchers had warned Oxfam that this would reduce agricultural output, increase the workload of women and overload the land near to these new villages. These fieldworkers who were employees of Oxfam were in no better position to complain than if they had been direct employees of the government. Oxfam was operating in a one-party state and they were in no position to voice these anxieties about this unpopular programme which they had critically supported without jeopardising their whole national operation.

21. The national researcher as an adviser to governments

Advisers have no narrow roles and even if they are asked to look into a particular problem, there is more latitude in what they can suggest and to whom as they are often appointed because of their already established reputations and therefore, have some political leverage. Verrier Elwin (1947) who had become an Indian national and had an international reputation from his publications on Indian tribal affairs certainly had the ear of Prime Minister Nehru. As nationals they will be working from a position of already knowing the political, linguistic and social background to what they are asked to discuss.

A difficult and different development for applied social scientists is having the knowledge which they have gained in research used in litigation. In Australia aboriginal groups had taken the Federal government to court since it has been ruled that the land was not free of people with any title to the land when the first Europeans had arrived (French,1997). Aboriginal claimants have used national anthropologists as expert witnesses and their field notes have been subpoenaed and the evidence in them about informant's memories of tribal behaviour dismissed as inadmissible evidence as it was second hand. The tribunal accepted only participant observation as evidence even though memory data is accepted as valid under tribal custom and an accepted part of social science methodology (Morton and Shaw,2003).

22. The advantages of being employed for social science research

As employees of power holders and decision makers in their societies, national researchers have the opportunities to contribute to the long range good of entire communities with which they have become involved and they are often surprised that their recommendations are put into effect. It is difficult to adjust to this fact of having to take personal responsibility for what they recommend rather than seeing what they write as distant suggestions for the generalised good of society.

National social scientists in developing societies who have invested much time, effort and money as well as a separation from their families, need to get employment, particularly as non-national opportunities may be diminishing. In contemporary times this is more likely to be available in the world outside academia in the institutions which for better or worse will always want assessments of how they are working or should work. They are more likely now to recognise this need and to employ researchers to look at their own working. This is an opportunistic world in which social scientists are well qualified to participate. As employees of on-going and powerful organisations they are in a better position to help those who should be helped than possibly as teachers or as the writers of articles in little read professional journals. They are able to edge their suggestions into action since they are now part of the decision making process; they are now part of the designers of change.

Chapter 7

National and Non-national Researchers Issues in Common

1. The dangers of fieldwork

There are always dangers in social science researchers going outside the protections of their own social environment where they are protected by their own communal systems of safety. There have been well publicized events which have highlighted these dangers, for example, Goodall, the chimpanzee specialist of international fame was protected from would-be kidnappers by her own staff, the finders of the pre-human bipedal hominid remains of Lucy in Ethiopia had to move abruptly because their national co-workers and guards feared their camp was about to be attacked and the women field biologist who was murdered in Kenya. Overall their standard of living is a constant invitation to theft, not just envy.

Such terrifying events have occurred in areas which their governments have considered to be generally safe but, even there, a national wanting to research his own community can receive death threats. Research and personal danger are not usually associated with individuals in their choice of subject and usually they learn about urban gangs and cattle raiding from informants just as they do about drug addiction and AIDS enabling behaviour. Now, just as in colonial times, no researchers would have been allowed into disturbed areas and even if they gain access to such communities under the cloak of membership of Oxfam or Medicin Sans Frontiers they would hesitate to make this known for fea of compromising the work of these international charities. There can be no doubt that research work on such dangerous fields is valuable but sponsoring organisations and individual prudence keeps researchers away from pre-determinable dangers.

Certainly governments would dislike reports on behaviour which brings disrepute to their administrations but since the researchers have long left the areas of their fieldwork by the time that their work achieves publication, the civil servants enabling such work are likely to be blamed for exposing their responsibilities to such publicity.

But today with mobile phones and e-mail facilities from battery driven computers, authorities are even more nervous of allowing non-nationals into areas and access to topics which might put them into danger and all the mass medias publicity which would come from their injury or kidnapping. So at any time globally there are always going to be areas and topics which will not be approached by social science researchers whether they are nationals or non-nationals; Algeria, Liberia, Afghanistan, Kashmir and Colombia. In these areas and elsewhere there will always be research topics such as organised crime, corruption and voting fraud which carry with them the risk of personal injury such as the United States Watergate scandal and the more recent postal voting fraud in Britain.

Social science fieldwork is a patchwork in space and time and tends to be concentrated within areas of overt social passivity. Important areas of research such as the Moplah revolt in India and the Saya San rebellion in Burma can only be approached historically from official records and the reminiscences of participants. Just as now the Gaza strip and Baghdad will not be chosen as places for research by non-nationals despite their importance for the social sciences.

Whatever may have been the realities in many social systems, this imposed bias against research in disturbed environments may have led to seeing many societies as having stable social systems and an aura of quietude which may have been far from typical. Evans-Pritchard wrote about the social system of the Sudanese Nuer without mentioning that the British military presence and punitive expeditions with their ideas as to what constituted good government had in fact destroyed their ability to govern themselves;

in another guise these same principles may appear to govern the current government's prolonged war against the southern tribes and the current civil war in Darfur.

Of course there are dangers in fieldwork particularly when the researchers tend to look different ethnically or by costume and by what they are carrying; thieving and mugging occur everywhere. Informants will tell researchers that there are certain places which they should not visit and when it is inadvisable to go there. In many cultures women never go anywhere alone and for anyone to walk about at night invites suspicion of criminality, adultery and witchcraft. There are dangers from seriously breaking local laws and customs The social science researchers are particularly subject to suspicion because it is never immediately obvious what they are doing and why; they are not just doctors, archaeologists or agriculturalists. They do not carry about any obvious tools of a known profession.

However nationals would not experience many of these difficulties because there may be an element of physiological immunity in their ability to endure nationally prevalent diseases such as malaria and parasitical infestations. The national would also probably not experience the same stresses from isolation as members of their community with a shared dialect may exist everywhere; they can often better the conditions of their environment without affecting the objectives of their research.

In many cases social scientists have to decide what is safe to write about particularly when they are not writing about people who are unlikely to see whatever they publish and who are of lower social status to themselves. Thus particular prudence has to be shown when they are writing about people of greater importance than themselves, then dangers are present at several levels. The researchers are usually of junior academic status and to pitch their conclusions at a level which conflicts with the known views of their superiors is likely to blight their future opportunities. Non-nationals are often assumed to have post-imperialist attitudes and intentions so conclusions would have to be turned to avoiding any such accusations.

There is also the need to protect the identity of key informants. If the parallel is not too extreme the key informant to the reporters of the Washington Post about President Nixon's involvement in the Watergate affair, could not possibly have been disclosed even if they had known who it was because that person might have been in deadly danger. There are always dangers in writing about a well established and self-protected local or national oligarchy providing relayed information about their possible corruption and rewards. A fieldworker wrote, 'I am acutely aware that I have less freedom to write than the average newspaper columnist who chooses to imply or to state outright information that I must suppress' (Sheehan,1993).

In many developing societies and even in the subordinate parts of developed states such as Greenland with Denmark and Lapland with Norway, there are no groups of experts who have an assured and influential status independent of the national power-holders. The access to power and promotion lies in conferring with these power holders (Petersen,1982). In what are in practice one party states those that provide different opinions on policy are in effect seen as an opposition; public non-conformity with those who wield power and who indeed may have come to power quite legitimately is likely to be a dangerous practice where no concept of an open public opinion has had time to develop and become part of political life.

2. Dangers to the minds of researchers

In thinking about the dangers involved in fieldwork there is a concentration on what may happen physically from criminals and political disturbances. But there is a far more damaging danger to their minds. It is necessary to think about what may happen to the minds of researchers whose work is confined to the consequences of violence and extreme deprivation. They can become deadened to these terrible events or they end up being unable to separate themselves from what they are recording in their notebooks.

In the former case as Bob Geldorf the founder of the now

enormous charity Band Aid, stated that while he was originally upset and impelled to act by the horrors which he saw, put in the end he became deadened to this and it just became numbers. In the 1947 Partition disturbances the Calcutta Police collected over 3000 anonymous bodies from the streets and we can presume that the only way they could cope was by deadening the mind to what they were seeing and doing.

One can study murder without being personally horrified by its prevalence because each case is more or less separate and unconnected and the assessment is in terms of whether it is commoner in one area than another. Even when a disturbance in Karamajong involved forty deaths the concern of the researcher was more over how this event should be coded and not about the human cost. The researcher remained detached because the individual misery did not impinge on his consciousness or his conscience (Tanner, 1970). On the other hand some social science researchers have worked for so long with events of such frightening proportions that it is bound to affect them as persons; they are contaminated by the massive nature of the evil which they have seen or recorded. The genocide in Ruanda and the destruction of European Jewry in the Holocaust and certainly in the case of Bettelheim, contributing to his suicide (Bettelheim, 1986).

It seems likely that most researchers who have investigated any major tragedies in which they have heard the personal stories of survivors will be deeply affected. Iris Chang researched the prolonged barbarism of the Japanese army in 1937. Nanking where hundreds of thousands of civilian Chinese were killed and raped (Chang,1996). She went onto research further into Japanese war-time atrocities. This accumulation of evil depicted in photographs and the stories she heard from survivors led to her suicide.

Shail Mayaram (1996) wrote'I had been researching for a few years before I realized the enormity of the Partition experience. As I investigated the 'record', its silence and what is deliberately silent, the problematic character of 'history writing' was exposed. At a personal level this work has shaken to the core of my being premises and beliefs I have grown up with and to which I know some members of my family adhere. Time and again I was confronted with my own positionality; an inner struggle to rent a silence, a repeated refusal to recognise oral testimony until it had been 'verified by authoritative' sources. I write this account with the growing realisation of how we are all participants in the making of silences.' What can have been the affect on Butalia who worked for years on the recollections of those who survived the killings, abductions, rapes and arson which accompanied the 1947 Partition of India which provided the cause for the instant migration of millions to the safety of the other side of these new borders (Butalia,1998).

In such cases it is irrelevant to look at psychological explanations to account for the reactions of such researchers working on recent and still living history. It is not so much that they were mentally damaged by what they found out such as might have been the case for the imaginative who worked on the Black Death in Europe, but their discovery of such unparalleled evil perpetrated by people who are still alive and amply recorded in national and non-national records with which their minds could not cope. As the Chinese Professor Sun Zhaiwei of Nanking stated 'Nuclear scientists wear protective clothes and have their health checked by doctors. Perhaps the historians of the extreme need similar measures. Yet for now we have to take care of ourselves' (Augost, 2005).

Another aspect of the dangers which may affect fieldworkers are the cumulative stresses of isolation not only living alone in the social sense of not being in the centre of known associates and associations but being alone in a strange environment separated from friends and relatives and known food and fun, which is limited for their culturally understood comfort and compounded with language difficulties. Above all they are separated from people with whom companionship can be shared.

Few individuals in the social sciences are trained to or have the personal abilities to approach strangers; they are not door to door sales people and those that they do approach in the sense of 'chatting them up' are likely to be people of similar age and backgrounds or are grouped together on a package holiday. It is rare for fieldworkers to carry out research with people who mirror their own characteristics even in quite general terms. Then there are the strains on the researchers who have undertaken this work for their own advancement; there is always the very real possibility of failure in going to work with strange people in strange surroundings, none of whom have any reason to help these intruders except for reasons of their own which are certainly nct shared with the researchers. There are parallels here to pre-examination stresses with the big difference that there are no syllabus which can be worked through in advance. It is a venture into the unknown and if it was not so, it would not be research in the first place.

Most researchers approach their research with some design not only in their own minds but also as approved by their supervisors. There is the quite irrational hope that these strangers will conform to this outline and often enough they will be faced with situations on the ground which will put them under stress even if imaginatively it could have been expected; the culturally unexpected are always intruding such as the blood sacrificing of animals, the police beating up suspects and extreme poverty. Planning to research a fishing community only to find that their economy is based on coconuts and rice growing. Most researchers come from patrilineally minded cultures and to be faced with matrilineality in which their social behaviour is the wrong way round is fundamentally stressful.

While research is designed as an abstract process, most researchers are ideologically committed to human betterment for which their projected work is a part and this is often augmented by their own personal sense of intense involvement with the people they are studying. They may often not understand the workings of reciprocity and think that their presence there is appreciated within the senses of common humanity. British researchers may feel little personal commitment to a slum community in Britain which they are studying but an illogical one to a foreign one whether it be a community of cattle herders, or subsistence farmers for which there can be no real reason to be emotionally committed; even if they do not actually say so they often think in terms of 'my' village or community.

Finally, there is the stress in the change not only from the certainties of the lives which they have previously experienced but more widely from the supposed certainties of international life prior to World War Two to the widespread major uncertainties of contemporary life in which almost everything is politicised without the relative certainties of stable social systems (Saberwal and Henry,1969).

3. Researchers are always social aliens; a shared outsider status

Hard science disallows any form of attributable and quantifiable influence over the assessment of its data and presumably they are not too pleased with the idea of fashions in scientific endeavours (Kuhn,1962). Doctors are sometimes criticised for their lack of empathy with their patients and this aspect of medicine is left to nurses. How does empathy affect the social science researchers? There is a distinction between the psychologist to whom people come and whose experimental work is in-house or not too far from their professional location and the sociologist or anthropologist who almost always works away from home.

These professionals leave environments to which they are accustomed and go and live often for quite long periods in alien environments with which they have no connections except the initial approval of the government for non-nationals. There are thus social distance factors affecting their conscious or unconscious sense of well-being even before they have opened their mouths which must affect their ability to be detached observers. What are some of these unsettling factors which are shared by both national and non-national researchers? The university professional environment from which they have come promotes a special way of thinking and individualism which does not help such people to relate to outsiders, even those in other professions. As regards their own work they are if anything taught to be selfish as ideas cannot be copyrighted, and their colleagues are rivals. Most people, even researchers, come from an environment which they classify as home and in which they are more or less comfortable. Entering a social science fieldwork situation has few if any parallels to going somewhere as a tourist for a short time. It is a radical disruption of physical and psychological comfort which takes considerable effort to overcome as it involves the 'little' things in behaviour not the major differences which may appear in carefully prepared text-books on culture and custom; dirt, disease and loneliness.

Few researchers realise the extent to which their behaviour is governed by the verbal and body languages in which they have grown up. Non-nationals are of course quite incompetent in this even if they have stayed for long periods in their research areas and even the national may have difficulties if they have moved over substantial cultural barriers involving ethnic and religious differences; most have experienced the difficulties over having friends to stay for longish periods or indeed staying with friends and moving into a totally strange social and physical environment is immeasurably more stressful.

A major handicap in fieldwork is the age and sex of the researchers. Most researchers are in their early twenties and yet they want to get information from people who are predominantly much older than themselves in both age and experience and who already have achieved status in their own societies. By the time that researchers have achieved age status in their profession and no longer do fieldwork because of family and academic commitments or have left the social sciences altogether; perhaps age and theory are allied. Much the same applies to women researchers with the additional handicap or advantage that they have access to fields barred to men while at the same time have wider restrictions on their potential areas of enquiry.

Most people think of religious differences as being related to major theological issues but these may be of little concern to the people with whom the researchers have come to live and for whom religion is a matter of ritual practices conditioned by habit and social conformity; researchers have to come to terms with their importance which cannot be diminished by the researcher's biased conceptions of magic and for their important avoidances based on pollution which do not occur in the university environment. Because of the level of their thinking researchers may see much of what they experience in macro-national terms while those researched in micro-ethnic ones and an almost inversion of values. Finally, researchers are not in practice classless ideologues and for them to move from one comfortable social slot into another may well make for psychological discomfort.

Social science researchers are usually social aliens inserting themselves like some bacteria into some social body which usually has had no previous experience of such peculiar intrusions; admittedly some communities have been comparatively well researched such as the Coorgs of India and the Zuni of the United States. Researchers have to emotionally support themselves in situations which by definition make them lonely people supported by their professional needs. This loneliness has not been accepted as an occupational hazard requiring both human and professional attention.

All the manifold differences between what these researchers are used to, whether domestic or institutional, amount to considerable strain. No matter who the researchers are and how carefully they have prepared themselves there are bound to be social and psychological difficulties. Because as a matter of current political and social notice it is often assumed that these differences and the related tensions are ethnic as might well be the case of a Nigerian researcher in the United States (Ogbu,1974) just as it might be for a Sri Lankan in East Africa (Wijeyewardene,1961) or a Dutchman in Tanzania (Wijsen,2001).

People look different and they know that they look different and no legislation will avoid the fact that appearances predefine the attitudes of both researchers and the researched. Punjabis look different to Tamils just as in Kenya, the Luo look different to the Kikuyu. This is not to suggest that there is an insuperable barrier both within and between nationals and non-nationals or within nationalities, but that in such situations the researchers will have a longer period in which they have to work themselves into where they are working and in this a longer period of loneliness. The difficulties are far more complex than the simplicities of visible ethnic differences. It is no longer a white on non-white research environment with its particular strains which may be as much on the side of the researched as the researchers. The Yugoslav civil servant who stated prior to the break-up of that country that 'we are all Yugoslavs now' was living under an illusion. Ethnicity in the form of tribalism is as common in Europe as in Africa or Asia and movement across the religious-cum-national divide in India and Pakistan may well have to be considered an insuperable divide and one not worth attempting to cross except possibly in terms of cricket.

The basic ethnic differences in early anthropological fieldwork has overlaid those which have always existed in any form of field-work within national and over national borders. Booth's study of poverty in London, Britain (Booth,1889) and that of Srinivas and Majumdar's prolonged work in India (Majumdar, 1962), were all upper-middle class intellectuals and it seems likely that they had very little in common with those they observed other than nationality and perhaps some shared but partisan interest in their fields of study. These researchers whether they are national or non-national will have experienced tensions stemming from the same sources of differences between what they are used to and what they have to endure in the field; at its crudest filth is always filth, and post-graduate researchers will not appreciate it in terms of interesting social differences.

The assumption of a researcher's detachment is a tenuous one. It may exist to an acceptable degree in working through statistical records; there is not much bias that a researcher can put into calculating birth dates, the areas of fields and the data recorded in a medical clinic. However in the field in the interrelationships which they have with others, who they are as persons is a vital part of every piece of social science research.

Since researchers are always by definition outsiders and usually very obvious ones at that, unless they are doing covert fieldwork, this is often seen as an insuperable barrier against getting valuable information. Firstly, however, obvious the researcher maybe, it may not take long for the 'social alien' to be more or less forgotten; they become part of the social furniture. After a time working in a Kenyan prison, this author came to be classified as a civil debtor. Secondly, just sitting and watching can provide valuable information as people are not going to change their daily routines if at all because an outsider happens to be watching if indeed they are seen to be watching rather than just doing nothing in public.

The watching of Bushmen children showed that in a subsistence society, they become accepted members of adult society quite early in their lives and have some sort of relationship with everyone they meet . East London children on the other hand have more restricted and formalised relationships particularly for sex segregation (Burton-Jones and Konner,1973). Thus sitting and watching is a viable methodology whether from the path-side, railway platform or from a window overlooking a school play ground or market except that as always they can only watch one part of the visible scene at a time.

4. The researcher as a solitary person

Except for the rare cases in which the researcher is accompanied into the field by a companion, they start work as single individuals whatever may be their social and kinship relationships in the places from which they have come.

Humans are not geared to being solitary and wherever we are we try and join up with others; in its various dimensions we want to be loved and it is hard to enter a community with the best of intentions only to find that we are isolated by every single social category about which we care to think. Further we inevitably find that the people we hope to research in their well-established ways of doing things, have no particular wish to bridge this gap except for their entertainment and personal gain. So researchers are usually lone individuals whether they are nationals or non-nationals and it is certainly an assumption that the foreigner is by definition more socially alone. They have similar difficulties in working their ways in but they probably all share the contrast from the university environment and its stimulus, and their own domestic surroundings. It may well be that this is the first occasion that they have been so socially alone while being surrounded by people of whose behaviour they have no knowledge so that they do not know what to expect. This feeling of isolation is likely to be worsened by language difficulties; a non-national may have some knowledge of the major language as a lingua franca but not at the level of its vernacular use with people who are not used to speaking to foreigners. The national may have little knowledge of the local form of a major language or a derivative form used by many in their more private lives.

Field workers may not have the resources to fill up the spaces in their daily routines which were previously filled up with social activities and inter-relationships; the need perhaps for a distracting hobby vaguely related to the community in which they are living such as plant collecting. It may also be sensible for researchers to keep in touch with their own cultural backgrounds by re-reading favourite books (Hau'Ofa,1991). In the societies from which they came they will have been included in many social activities without having to make any effort because their existing roles made their inclusion automatic. In the research situation they are not automatically a member of any shared social groupings.

In their new surroundings then they are unknowns about whom nothing is known except their strangeness until they are introduced by another outsider of some status or some see advantages in some limited contact with them. In such situations nationals who do not live too far away from their home environment, may be able to develop some tenuous social connections based on tribe, caste, clan or distant affinal relationships, even similar names or membership of a minority religion. These strains are likely to be more noticeable at weekends, public holidays and their own celebratory dates when they would not expect to be alone. The importance of mobile phone use by fieldworkers can only be a partial antidote to this or a further unsettling influence.

5. The contrast in living conditions

Any individual undertaking fieldwork as opposed to a project which conforms to office hours based on a normal place of work, experiences a radical change in their total life environment with quite distinctive characteristics. It has no similarity with being a tourist or even with the mobile back-packer. It is not the same as a transfer from one country or part of a country to another within a single organisation or by a change of job in which there will always be similarities. The only continuity is an intellectual one, a self-chosen wish to enlarge or use a particular fund of knowledge which is not overtly connected to their social and economic needs. Taking up fieldwork dislocates these researchers unless they are working on projects in which they already have recognisable roles such as nurse, teacher, social worker in which their research is a by product of these basic roles in which they and the community in which they are researching, know about in advance of their arrival.

Researchers who move out of their established social environments undergo as a matter of course not of choice, a number of radical physical, psychological, social, economic and even religious dislocations.

(a) *Physical:* There is a change of diet in general and a restriction of choice to what is available and since much research is done in communities lower down on the social scale, so that what is available may be experienced as tasteless and unsophisticated. Since all researchers may have been warned that to refuse offered food is a serious insult, there will be health worries over both food and water. There will be changes in physical comfort and worries about where and how to defecate and how to clean themselves both then and in washing generally. In visiting friends within the same level of culture there are not these difficulties which immediately arise in what may seem to be overcrowded living conditions.

(b) *Psychological*: The general feeling of being out of place and not knowing how one should behave is always a strain but this is far wider since it applies to every aspect of life from how and where one should sit and how one should behave to people older and younger than oneself, of unknown status and of the opposite sex. The correct behaviour for a researcher is not known to either party.

(c) *Social*: There is the inability to communicate in the micro-terms of interrelationships which only comes from detailed social knowledge particularly when the basic constituents of the language are alien to the researcher's own in the use of tones, sex related verb use and the semantic uses of language which are alien to the researchers own social understandings such as the reluctance to make direct statements or indeed to talk to people older or younger than themselves; Kumar writes of the fact that many people in Banaras talked indistinctly because they had no reason to be distinct since they assumed that the people they were talking to could pick up the threads of what was being said from the clues they heard (Kumar,1992:163). Language teachers speak clearly but socially there is no particular need to do this nor is there always the social need to look at the person to whom one is speaking.

(d) *Economic*: Those with whom the researchers come into contact know that these strangers are better off than themselves by dress, equipment and speech and they may want to take direct or indirect advantage of these differences, indeed it would be unusual if they did not since they must know that it costs a lot of money to get there and to stay there. On the researcher's side they may well feel embarrassed at their relative affluence and feel under some strain to provide help.

(e) *Political*: Whether they like it or not researchers are seen to have influence or they would not be there in the first place. But these strangers are often asked to intervene with authorities with whom they are assumed to have influence. Whatever their own origins they will be met by uncomfortable contrasts and the tensions from feeling that they ought to interfere but know that they should not.

6. The personal issues involving domesticity

Fieldworkers recognise as much as any other worker that too much social and geographical space for too long is going to be a strain on any relationship and perhaps even more for non-nationals from Western societies who put so much emotional investment in these hopes based as they are on the predominant value of individualism, but then for nationals in non-Western societies there may be equal strain in their separation from their families as sustaining social groups. If husband and wife are in the field together the active relationship may solve some problems while creating others since the relationship requires the giving and receiving of support on a daily basis. Whose field situation is it and whose needs predominate and which of the two runs the household and does what at other times. Margaret Mead's marriages to social scientists did not survive their competitive fieldwork situation of competing interests and Victor Turner's wife had to put her own career on hold not only until the children were grown, but until her husband had died (Turmer. E. 2000).

It may be that the partner provides personal support for the researcher and the research but the partner has no reason for being there other than providing this companionship; she and it is usually a woman has the double load of isolation from her partner during the day in an environment in which she has no interests and from the social life which she had created for herself in her own society. In this situation it is far worse than having a partner away as an airline pilot or on a daily commute to work which leaves her at home in an environment largely of her own choosing within the cultural comfort of nearby shops, television and a mobile phone. In the fieldwork situation she is likely to be alone for long periods in a situation of a very different standard of living and without the incentive or the means to communicate with the people outside their temporary home.

There are exceptions to these built-in difficulties in which a different form of culturally acceptable marriage exists in a national setting or when a non-nationals are married to nationals and *vice versa* in which the national one acts as a valuable lead into the researched community. Not so much as a key informant but as explainers of their husbands' motivations and of course as a more culturally efficient manager of the household. We can suppose that Verrier Elwin's Indian wife performed such multiple roles for many years and that she was of more 'use' to him as an anthropologist specialising in tribal affairs than any possible English one.

There is a possible typology with national and non-national couples; both in the field as potential rivals or cooperating in joint projects, one supporting the other in the field, one in the field and the other at home and one in which the relationships are kept alive by infrequent visits to the other in the field. The only difference might be that the national couples stand a better chance of keeping their relationship going and in adding to the ability of one partner to do more effective fieldwork. In national-non-national partnerships whether there is research co-operation or not, the researcher is bound to benefit from the additional insights which will be passed on over the cultural boundaries on a day-to-day basis.

Domesticity has another important aspect regarding the status and respectability of both men and women researchers. Few societies other than highly developed Western ones and even there not across the board for all social groupings, have a less respectable role for the unmarried adult; they are somewhat suspect. In all such societies there will be derogatory speculation as to why they are not married. The unmarried male researcher may be seen as a potential predator on the women in the researched community. The unmarried woman researcher is in an even more ambiguous position as to why she is not married but allowed by her family to be in an unprotected social situation; this might suggest to some that her behaviour in the past has been so bad that her family have disowned her. Her unmarried status as a desirable and educated adult woman requires an explanation. She may also be seen by the married women in the researched community as a potential threat to their own marriages as a predatory woman tempting their husbands. It is politic then in many situations to be either visibly married whatever the actual status of the partner may be or to show in pictures and conversation that the researchers are already married in their own societies.

7. The shared differences over language

Once social scientists go beyond the descriptions of what they see which is likely to be as much personal as professional, the languages used are both a matter of choice and interpretation. Firstly, within the national lingua franca language which may be shared between the observer and the observed but with all the variations which relate to age, sex, education and status; in both Japanese and Burmese, there are substantial differences between these languages as used by men and women. Whether the researchers are nationals or non-nationals there will be substantial difficulties in thinking within cultural idioms which may be against their own patterns of thought. Western educated social scientists ostensibly are lineal thinkers and would find it difficult to speak much less think in forms of speech which avoid any such straightforwardness.

Then there are the linguistic codes within a single language group such as the restricted and elaborated codes in English (Bernstein,1975), the patois of convicts in Kenyan prisons and the business and professional English of educated Indians as well as the differences between Burmese spoken in Yangon, Mandalay and Myitkyina. A further example is the problem of whether Bhojpuri language spoken in Banaras is separate to or just a derivative form of Hindi. Finally, the use of languages which are totally distinct from the languages of the observed and which may be a necessity for professional usages; the personal linguistic framework of mind in which the original impressions were recorded and which are subsequently described and elaborated.

In all these situations there are complexities far beyond the relatively simple ones of the non-national learning a totally new language just for the purposes of fieldwork; a boxed-in type of understanding which does not involve the use of that new language again when they return home or for nationals to similarly return to their homes in another part of their country in which another language is spoken. This last is perhaps the most difficult because there is a process of translation involved not from a static text which can be carefully considered, but from fleeting impressions recorded in memory, on tapes, in field notes or in digital media. In these cases the writers for that is what they have now become, can use any language which is consistent with their intentions and the requirements of their professional status.

Thus Indian or Sri Lankan social scientists who have worked in East Africa will have had to go through three linguistic shifts in recording their findings, their natal languages, English and then kiSwahili or whatever is the language of the communities observed. A Han Chinese from the north working in Yunnan would only have to make a single shift from Mandarin to Yunnanese but such a researcher would have to make another leap into English as the commonest language for the publication of social science data if he wanted to get international recognition for his work (Hua, 2001).

While we tend to think of the time consuming difficulties of the non-national Westerner learning a language for their research, it would seem that this is a comparatively simple single step change in comparison to the multiple linguistic changes which are forced onto those national social scientists who work within large multi-language nations whose lingua franca are of no use in the international and professional world of social science. Westerners do not realise the extent of language changes that social mobility forces on the intellectuals of developing countries. A Tanzanian Roman Catholic bishop went through five distinct languages before ordination; his natal tribal language kiHangaza, national lingua franca kiSwahili, English in secondary school and first seminary, Latin, and French in North African senior seminary; much the same would be the experience of any national social scientist.

Thus all social scientists come up against not only the complexity of the language used in the researched communities and the language used by the researchers who will rarely be from the same community and social strata. In what language do they think and perhaps dream and how far is that a logical rather than imaginative use. It seems likely that imagination and thought patterns are those controlled by the dominant culture and that it is probably impossible for anyone to change those that come into their minds with their natal languages. However when social scientists come to write in another language then to some extent what they write will have to be channelled into the structures rather than the thought patterns of another language; in this respect we may be forced to recognise in some respects the poverty of English to express what was common enough in the kiSukuma language spoken by five million Tanzanians with its tones (Richardson,1959) and twenty four tenses (Batibo,1985).

8. The feelings of guilt

Apart from the personal political opinions of both national and non-national researchers which are unlikely to coincide with those of the power holders in or above the community which they are researching, they are faced from almost the first days of their research with the plain fact that there is a 'have-have not' dichotomy in where they are now staying and that they have to see themselves as being among the 'haves'. Of course, there is poverty in some parts of the societies from which they come and they know of it as part of their training, but it is obscured by their professional lives and domestic arrangements. They may well have a feeling of the immorality of the global distribution of wealth for which they have no direct responsibility, but now they are face to face with people from whom they are separated by all the categories which are used in assessing standards of living; higher incomes, better clothes, back-up health care and better equipment.

So this comparison is often a sensitive issue on both sides of the divide which is as much economic as social. It would seem difficult for them not to have some feelings of guilt over their own comfort and their access to back up services, particularly when it involves them personally and specifically when it involves the illnesses of an informant or a close member of his family or when they are appealed to as someone who has a vehicle or private stock of medicine, some special knowledge or more importantly some assumed political and social connections outside the community which could help these people who have been, are and will be so helpful to the researchers. These feelings are not just ones experienced by Western researchers but also Indian ones in East Africa were noted as having the same problems but related them to getting their fieldwork accepted (Saberwal,1969).

So in many if not most research situations the researchers come from relatively affluent and assured backgrounds in comparison to the situations of those who are the objects of research. This may bring about a feeling of pity so that the researchers do what they can to help individuals but a broader feeling that the overall situation is wrong and that their work should be directed to trying to reduce these gross inequalities which so often make their lives both brutish and short.

Their clothes are better and they have a range of personal equipment of which most if not all of the surrounding community are envious; even a Parker ball point can be the envy of an Egyptian policeman outside a Cairo museum and Ethiopian children at the rock churches of Lali Bela prefer ordinary ball points to money for wordless guiding round these sacred sites. These feelings of guilt accompany the fieldworkers all the more when they begin to see the people with whom they are working as part of a more personal cooperative effort in which they are much more professionally and personally dependent than they realised.

It is hard to see that there are any differences in this except in personal sensitivity between non-nationals and nationals and it seems likely that the latter will experience these contrasts, not only more directly but more continuously since there is no break from contact by leaving the country in which research is being carried out; a continuing oppressive feeling of partial responsibility for the situation of widespread poverty. From this develops the researcher's feelings of obligation to those with whom they work; a conscious struggle between their personal and fieldwork needs and those of the surrounding people which impinge on these primary needs.

There is no such thing as a neutral role in social science fieldwork even though it is their professional aim and researchers have to live within the roles which they are given by those whom they are working with. Whatever this role is, it will have to be tied to the appropriate inter-personal obligations which are not likely to be specified in advance of the actual situations which arise. Apart from the fact that the role of research is likely to be an unknown to those in the researched community, these obligations are going to be tied to the fact that these researchers are outsiders with outside connections and certainly wealthier than they are in their actual standards of living.

Certainly researchers are going to feel that they have obligations to the people with whom they are living. These would include the use of the researcher's transport according to the local evaluations of how transport should be used so perhaps it is better to have no more than a bicycle which can hardly be used for funerals, weddings and important business deals. Certainly they will be called on to provide medicine as it is often assumed that they have better ones than those which may or may not exist in the local dispensary. Then they will often be asked for personal loans which in the nature of social practices, will be seen as gifts and of course they will be asked to provide letters of recommendation to ease the ways through the mazes of bureaucracy. Researchers apply the standards of their own societies to these requests. Only national researchers know and can use without giving offence the circuitous ways of the people themselves for avoiding giving a clear answer to such requests. Initially researchers are seen as guests but once this short period is passed it may be felt that as semi-permanent members of the community of a certain obvious or imagined affluence they should contribute to its welfare and some at least of its members. This is likely to be felt especially by those who have been influential in getting the researchers accepted enough in the community to get their work started.

9. The feelings of affinity

Almost all fieldworkers whether they are nationals or non-nationals come to have some affection for the people with whom they have spent so much time and effort; it is as if some part of their personalities had been caught by that of those whom they have been trying so hard to understand.

Against this is the fact that fieldworkers remain strangers retaining the physical and psychological characteristics of the cultures from which they have come and there will always be limits to how far they will consciously or unconsciously go to show what they consider to be cross-cultural affection or affinity: how far indeed will those researched see this behaviour in the same terms, and not just as a rather odd aspect of outsider behaviour of which they might be able to take advantage.

If fieldworkers have developed this affection or affinity in which the former may characterise the non-national and the latter to be associated with national researchers, then this detracts from their professional ability to see these people in neutral terms. Fieldworkers often refer to their principal and long standing informants as friends rather than as acquaintances or work associates and assistants. Kumar working for many years on and off among the artisans of Banaras refers to only one man as a friend because the relationship was undemanding and gave pleasure to both of them; the old man and the young mother (Kumar,1992:73).

Fieldworkers certainly have long and close relationships with some of the people with whom they are working but if friendship is primarily defined as 'one joined to another in mutual benevolence and intimacy (Oxford English Dictionary,1956:752), then the social relationships of social scientists and the principal people on whom they depend for their research cannot be classed as friends. Such relationships between two totally disparate persons would be unlikely in any cultural setting, certainly not in situations of great economic, social and intellectual differences. Any form of personalised commitment during fieldwork is thus a move away from neutrality. Whether with men and indeed more particularly with women or any person of the opposite sex. The relationships with informants are in essence not much different to that with servants upon whom the fieldworkers often depend for keeping them in good running order.

Friendship in most cultures has the specific content of equality in material reciprocity as well as openness about personal issues; it just does not fit in with the superior material and social status of the researcher nor indeed with their role as collectors of information. Fieldworkers always take by the nature of their preoccupations, more than they could ever return; it could not be otherwise. Perhaps those researchers who refer to the people with whom they have worked as their friends are coping with their conscious awareness of how much they have got from them for which they have not given any equivalent returns.

There is a difference between the affinity that occurs with nationals sharing a country to which the researcher's long term obligations and the strange and highly personal form which often develops between the non-national researchers and the peoples with whom they spend their initial field trips. The emotional impact is often so strong and so long lasting that for the remainder of their professional lives whether they express it or not, they are subconsciously thought of as 'my' tribe, village or community. But these feelings would seem to apply only to rural communities and tribal groups who are distinct. It does not apply so clearly to fieldwork in the slums of some of the major cities in developing societies where there may be a high turnover of population. It is possible to see that these attitudes are hangovers from Rousseau's 'noble savage', however irrational this may be to their professional thinking. This is probably combined with the feeling that they are researching a disappearing way of life and that this is usually a decline in moral standards irrespective of material improvements.

Researchers from Western developed societies who go to work in so-called tribal societies and communities to which they have no cultural affinities at all often suffer from a form of cultural shock which leads them to find aspects of their social behaviour which are assessed as better than the ones in their own societies. An assumption often based on relatively short-term observations that these people have virtues of simplicity and openness which is absent in their own societies. Of course, this is questionable since the social behaviour of these so-called tribal peoples is always as complex as that of developed societies despite the absence of material wealth. This is perhaps the commonplace experience of anyone entering a totally different social environment of which they have had no previous experience. A cross-cultural phenomenon and it seems unlikely that any national or non-national researching parts of their own societies would develop such proprietary as much as emotional attitudes of mind.

This feeling of affinity does not seem to occur in fieldwork within developed nations except for some ideological preference for rural rather than urban ways of living and when they have some shared interests. An understandable sympathy for individuals caught in situations which may be beyond their personal control such as drug users and single mothers but no feelings of affinity for the school or prison researched.

This affinity may be shown in the frame of mind within which they write their reports and of course primarily they write for their own career advancement. Some write (Sheehan,1993) for those with whom they most identified as well as fighting for the political and social causes which they supported and in this she 'had made the astounding ethnographic error of not realizing that I was living part of the dark experience of the average Irish woman. I had internalised and personalized their second class status'.

There is another aspect to affinity which relates to their professional associations in which the distinction between national and non-national researchers may not be as distinct as their legal status might suggest. With post-graduate status may come a parallel pattern of living; they have self-segregated themselves from the way of living of most of their fellow nationals with whom they are likely to do fieldwork. They may value their privacy and feel a need for relief from the all-embracing nature of fieldwork in the company of their fellow professionals. The fact that a national researcher is a Hindu among Hindus or a Han among other Hans does not avoid the oppressive nature of unvarying round-the-clock unavoidable association with those who may feel that the researchers are people who can be dropped in on at any time and in a sense kept under observation for the safety of the researched community (Srinivas,1979).

10. Dispassion as a form of bias: the moral dilemma

One of the major difficulties in fieldwork is the acceptance of emotion in others and particularly in the 'performance' of emotion in cultures other than one's own. There are differences in attitudes to the suffering of others and of animals which is seen in psychological and pathological terms rather than as acceptable ways of behaving in that culture. We worry about emotional differences and probably assess them in terms of our own cultures which we disguise under the cloak of alleged dispassion. Thus to view the emotions or seemingly the absence of what is no more than our emotions in the behaviour of other cultures in dispassionate terms is not an abstract approach 'The dispassionate observer cannot exist since such an observer distorts the data by behaving in an inhuman manner. The solution would seem to be to participate but to try to limit one's passion and involvement to the minimum expected by the community' (Beals,1970).

In this it would seem that the national researcher would be better able to assess emotions more or less in terms of the contexts in which they appear and to control their own feelings which would not have such covert cultural contrasts. It is not a lack of the 'right' emotions that is often said by Westerners to exist in Eastern cultures, but a response to different paradigms which includes repetitive suffering from high mortality rates, natural disasters from earthquakes, famine and the recent Indonesian 'tsunami' for example, and the religious convictions of largely uncontrollable spirits and repetitive ongoing lives.

There is another aspect to dispassion which is the obverse of being unable to appreciate the significance and nature of emotion in peoples of cultures other than one's own. The spread of the social scientists own emotions into their fieldwork; a clash in which their own emotions become an out-of-place disadvantage. Laura Bohannan, wife of Paul Bohannan wrote (Bowen,1964:99,156,184-5) movingly about watching a friend die because her relatives would not take her to hospital. 'One cannot make friends with a community. One has to make friends with individual people. I had allowed myself to become too absorbed in the personal affairs of the few households I knew best. But what is one to do when one can collect one's data only by forming personal relationships. It is hard enough to think of a friend as a case history. Was I to stand aloof observing the course of events? There could be no professional hesitation I might otherwise never see the ceremonies connected with death and childbirth.'

Here again the inculturated national researcher may be better equipped to balance the personal entanglements of compassion with the professional needs for dispassion. Even the 'half and half' migrant researcher would know better than the non-national the extent to which the social scientist can become emotionally involved and the appropriate ways to remain both attached and detached without upsetting the community and the key people with whom they are necessarily involved.

11. Likes and dislikes and a common humanity

It is an impossibility for any research worker who lives in the field for any length of time not to develop the same range of likes and dislikes as they would have done in their own societies. However it is more likely that they will develop more of these perhaps quite justified biases as they meet people behaving in ways that they must find quite unacceptable by their own standards of behaviour and that of the societies from which they came. This is very likely to occur in the treatment of women, children, animals and the sick. Kumar speaks with feeling of her disgust at witnessing the beating of a suspect by a police inspector and by her unpleasant manhandling in an excitable crowd (Kumar,1992).

These may be the likes and dislikes which come from surface irritations but in cross-cultural situations there are likely to be situations which are deeply upsetting to the outsider who has been socialised to use other not necessarily scientific solutions to misfortune. In general this is a difficult issue in community studies as it means taking sides for or against people even when the researchers know well enough what they are doing will almost certainly shorten their lives. These tensions may be less so in narrower topics or institutions in which the researchers have to relate to specific individuals whose behavioural style is already known in advance (Gans,1982).

However it seems to be a feature of fieldwork that researchers have benign relationships with their informants and at least find the community or institution in which they are working not so much benign as bearable. The fact that Colin Turnbull disliked the Ik for the selfish crudity of the ways in which individuals survived must be a rarity, in that he applied Western sensitivities to that project into which he had gone because he could not go to research his first choice. The fieldwork arena is taken as a job lot and this would seem to be atypical to 'normal' inter-relationships in which there is a grading of acceptability.

Fieldwork, therefore, requires the creation of an attitude of mutual acceptability which does not exist in reality; an artificial network of benignity which each researcher tries to weave round their activities. Here are researchers who come in from a different social level and who are trying to put into practice an abstract way of thinking about the human behaviour of those with whom they are now associated, they are being required to be inhuman; in a state of pastoral neutrality. In any area of research there will inevitably be individuals who are disliked because of their personal behaviour or because they are associated with what the researchers consider to be immoral behaviour. In their outside life researchers can usually pick and choose their associations but this is not so in fieldwork.

Researchers can appreciate and indeed understand the psychological and social situation in which individuals behave outrageously by the standards of the intellectualised environments from which they come; their body language alone as well as their pattern of speech would show that they disliked having to talk with mass murderers or child molesters. It may be necessary to study a distasteful topic as to leave it out of any coverage would leave an obvious gap. Straight researchers might well find it difficult to appreciate the practice of homosexuality in prisons but it would have to be covered. Researchers who study deviances from socially acceptable behaviour must presumably have come to terms with this difficulty before they start work on prostitution, criminal gangs, paedophilia or the social behaviour initiating AIDS. What researchers cannot adjust themselves to in advance is the unexpected unpleasantness which can erupt in the study once it has been started.

Almost inevitably because of the artificiality of the conditions under which most research is carried out, the relationships with informants must be methodologically suspect; they want this relationship with the outsider which may not have much to do with the research objectives. Bilaterally there is a disguised benignity.

In any area of research there will always be people who are disliked for personal reasons and more importantly those who are hostile to the researcher as a process, as an element in the topic researched and as an outsider. By definition researchers have to work round these difficulties without having the social room to manoeuvre which they would have in their own personai lives. It does seem that in fieldwork there is the creation of an artificial situation which the researcher has to like which does not coincide with much reality; they are more or less forced to go through the motions of liking rather than to admitting to dislike.

12. The illogicality of human behaviour

Social scientists in their research have had to transfer training obtained through book knowledge with their theme of detached irrationality about other cultures to field situations in which these behaviours are real and to a great extent unimaginable by these outsiders who observe them. It takes a great leap of understanding and perhaps sympathy to experience and to accept as a fact that in a Yao matrilineal community, the biological father of children is not responsible for them but for the children of his sisters because most researchers come from patrilineal societies in which relationships between the sexes is individualised and personal. It is not easy for researchers to assimilate and accept that such people are rational in behaving in this seemingly irrational way. This is an example of behaviour which is in the grey area of customs which are understandable and on the fringes of rationality since there must be elsewhere in the world examples in which because of divorce or death, brothers are doing just that for the children of their sisters.

But what about behaviour which is deeply imbedded in a culture and which researchers would never agree as rationally acceptable? The Amazonian Wari until about 1960 practised exocannibalism by eating the bodies of their enemies and social outsiders in disrespectful ways as well as endocannibalism by eating substantial parts of the bodies of dead relatives. This latter was simply the norm and the most respectful way to treat a human body and part of the process of achieving emotional detachment from memories of their dead (Conklin,1995). Such complex semantic issues suggest that cannibalism may symbolize not only evil and chaos but also social order and the regeneration of life-giving forces (Sanday,1986). Of course, the behaviour of the Wari is extreme even by academic research standards and usually fieldwork stays well clear of matters which are not easily assimilatable into the understandings of field-workers. By and large they stay within the rational framework of their own cultures; there is nothing likely to be particularly upsetting or culturally stressful in industrial relations, farming and slum conditions.

However fieldworkers are always likely to come upon cultural practices outside the range of psychopathology, which are quite plausible in an academic sense, but it is altogether another thing to accept them as practices which are based on rational understandings given their cultural premises. There must be innumerable examples of what must be for the most experienced fieldworkers behaviour which provides them with some form of culture shock. There is also the fact that what may be shocking for the fieldworker has the other side of the dyad to consider, the people being researched can be shocked at the behaviour of the fieldworkers. An English researcher eating a tin of snails to celebrate his birthday, produced the comment from his cook that he must be really poor, if not mentally deranged to eat such filth.

Even exchange visits between university students and cultural professionals never prepares fieldworkers for outlandish behaviour for which they may not be able to create a positivist explanation which contains some sharing of logicality with their own understandings. Thus whatever happens in their fieldwork, researchers are not only on their own, a psychologically and physically challenging situation in itself for those who have become used to association with friends and relatives, but are faced with behaviour for which they may be unable to provide a context by inference and analogy.

There will always be difficulties between what researchers are told is the right way to behave in a matrilineal African or polyandrous northern Indian society and the realities of who actually behaves in these ways when there are no mother's brothers in a particular family and when polyandry may be a reflection of poverty rather than social ambition. Then if this was not difficult enough fieldworkers find that much of the information which they receive comes within the frame of qualitative data and that it has few characteristics which would make it acceptable as either legal or scientific evidence. Researchers have to decide when they can move over from guesswork understandings to providing a logical qualitative assessment from long association with the people whom they are researching. Perhaps we have to accept that researchers can never do more than that.

Through no fault of their own, the production of these contexts by researchers can be assessed as works of trained or self-taught imagination developed in the field without any supervision or criticism by those about whom it is written. All this comes latter when these fieldworkers write up their work and are subject to professional criticism which must lead them to vigorously defend their material on which they have expended so much time, effort, money and social capital.

So the provision of a context is likely to be an imaginative pastiche which is found acceptable because there is little or no comparative material against to which it can be compared. They could have got it all wrong as was the case with Margaret Mead's early work (Mead,1928) but have produced a work which was internally cohesive and academically acceptable for many years (Freeman,1983). No one knows what may have been in the minds of Tagore and Shakespeare when they wrote their poetry despite the fact that there are definitive texts for which there are no parallels in assessing human constantly varying behaviour. We have to accept that in most fieldwork researchers are basing their formative opinions on visual and auditory impressions and that probably the only verifiable fact is that these men and women were there in a particular social setting for a known period; all the rest is in effect speculation.

13. The unreliability of informants

Fieldworkers take a great deal of trouble in finding and retaining what they consider to be reliable informants but there is little reason to suppose that either of the participants in these relationships have the same objective or indeed have much idea of what is going on in the other's mind. It is not that either are deliberately deceitful but it has to be accepted that neither are really comfortable with the others perceived objectives. There is no real reason why informants should be any more reliable than informants for the police; they have to give themselves a role and that need not be one that readily allies with that of the researchers. Unless researchers already have some social links with people in the community which they are researching, they must of necessity have a 'bag carrier'. Someone who makes it possible for these researchers to start fieldwork; where to live, how to eat, where to wash and excrete and so on. Whoever these people are, the researchers have started to be under obligations to people about whom they know nothing. The process by which they are under surveillance has already started.

Even if they know the language in the grammatical sense they are unlikely to know it semantically. The people they are talking to are passing on information on their linguistic terms not in that of the researchers' more academic understanding. These fieldworkers have almost a moral belief in the probable if not possible honesty of those providing them with information. The providers of information are doing so for their own purposes in the ways that they have been socialised to do for people with whom they are socially connected and more specifically with strangers, even those who are close to them both socially and ethnically.

In any situation the control of information passed on to a researcher for example by Japanese townspeople and Ethiopian immigrants to Israel was filtered by monitoring his activities via social networks in which he was peripherally embedded and by prevarication, procrastination and rewards for perceived 'proper' behaviour. This is the usual way information which is passed on is expressed in terms of the dyad's, researcher's and the informant's different cultures and the results must be viewed as a composite between two information sets (Ashkenazi,1997).

There is also the almost universal cross-cultural trait of not giving direct answers even to the most direct questions. Many languages are tools for the social expression of ambiguities and not for straightforwardness which is supposed to exist in Western cultures. Ambiguity may be a social necessity in small communities in which the same people who are in continuous association and is certainly the way in which intrusive strangers have to be treated for the social safety of the speakers (Weil,1995).

The last thing that informants are doing is to supply fieldworkers with information in the detached sense of social 'truth' within which they are working. Certainly it may be correct that informants may come to like the researchers with whom they are working and indeed it would be strange if this was not so, but they are also getting returns on this association which they value and which will influence them to keep such associations going. They may get prestige from association with this stranger from a different and richer world, avenues to advancement, economic returns out of all proportion to the researchers understanding of values and always the opportunity to bend the information towards local social ends of which the researchers will often not be aware. Researchers out of human weaknesses from being alone in an alien environment come to need such tenuous relationships as well as the ever-present push to get on with their research within the time limits of their official permission to stay or as allowed by university terms. Researchers rarely wait and see people who can just sit around hoping that something will happen. Informants are seen as a necessity for getting on with the research but are rarely seen with the scepticism which their multifaceted roles make necessary.

14. Failure in research projects

Any research project in the social sciences which is based on scientific methodologies and as such subject to the rigorous scrutiny of those working in the hard sciences would be bound to conclude that the work produced while attempting to be scientific, has failed to meet its requirements. Human behaviour is too varied both in its individualism and in its different environments. Thus the social sciences have either to create their own accuracies by rigidly restricting and defining their fields of operation; the general failure of quantitative methods except under these conditions and thus the researcher is more inclined to qualitative fieldwork. It could be suggested that there is a greater personal return from the latter than the former.

With the more general failure of scientific methods in the social sciences there is not only the questioning of the methodology, the first line of attack in any criticism of fieldwork, but its alliance with the personal failures of the researchers themselves. What might be the result of any evaluation of fieldwork in cost benefit terms to the institutions concerned? In the period 1963-67 about one-third of the grant aided non-nationals affiliated to the East African Institute of Social Research in Uganda's Makerere University failed to produce any written evidence of their research endeavours. These researchers were all working on abstract issues of no immediate benefit to their host country. It is possible that they may have written up their work subsequently and that it did not pass editorial acceptance. Others stayed for long periods with minimalist productions related only to their obtaining a higher degree. This may not be a problem of any significance since the matching of personnel, projects and productivity can never have the functional tightness of an organisation and its employees based on the need for profit rather than loss, or even of an approved programme related to political needs.

A great deal of attention has been paid to the errors of Margaret Mead rather than the methodological weaknesses of her first fieldwork in Samoa. It seems likely than virtually every non-national social scientist doing fieldwork for the first time outside their own immediate cultural understandings, will have made similar mistakes. In private they may admit to difficulties rather than errors but this finds no place in their published work. Mead's error in Samoa, without any adequate knowledge of their language, was to assume that her informants were telling her the truth; every Samoan who knew about her during that fieldwork must have known that she was being hoodwinked and yet she was surrounded by a benign wall of silence.

Failure is of course a complex and relative term in the social sciences. It cannot be paralleled to a failure in the pragmatic sciences when the initiator of an experiment fails to complete it or fails to produce the anticipated and hoped for results, but the latter failures contribute to the store of knowledge. In the hard sciences there will always be drop-outs for a variety of physical, psychological, economic and social reasons, but there is a difference in that Chinese, Indian and British scientists are assessed by similar standards which are largely independent of culture.

Failure in fieldwork has not only its immediate reasons but it can be redeemed by the fact that information and experience can be stored and retrieved latter. So there are failures which centre on the individual, a personal inadequacy to cope with fieldwork and failure to find out what they want through shortage of time. There is a narrow dividing line between the different forms of failure; the drop out rate for those involved in any research project is a combination of hurdles and ambition but for the non-national there is the lure of fieldwork in a strange and hopefully exotic community in which the non-national is at a disadvantage.

15. Researchers efforts to do research being rejected

All research in the form of some individuals inserting themselves into communities in which they do not have any understandable part, is an intrusion and in the case of non-nationals a very obvious one. Clear cut rejection would only occur when the community feels that the behaviour of the researcher cannot be tolerated, but most would feel that it would be imprudent to reject such strangers since they must be there with the agreement of higher authorities whom it would be unwise to offend. It may be that then non-national can obtain some varieties of information simply because they are outsiders and thus can be trusted not to use what they are told for local political and social advantages. Turner's relationship with the Ndembu healer Muchona the Hornet (Turner,1967:131-51) illustrates this cooperation.

There is always the distinction between a largely unstructured community in which the researcher's relationship with those in the community is usually a matter of ongoing negotiation. This cannot be done so easily in researching an institution which has a bureaucratic structure as defined by Weber; fixed hierarchy of authority, written rules of conduct, full-time officials, home and work roles separated and the workers do not own what they operate (Weber,1978:956-63).

One suspects that almost all researchers underestimate the difficulties of what they want to undertake and they are not likely to detail such thoughts in any application for a grant or in reply to an enquiry from an agency which might want to employ them on an applied project. In a sense all research are studies of some degree of failure, an underachievement in terms of their immediate hopes and longer term ambitions.

16. The need to adjust to the unexpected

Although it is reasonable to assume that the national researcher is proportionately less likely to be confronted with the unexpected, both they and the non-national are going to be constantly confronted with the unexpected not only in the cross-cultural dimension but also in the social life of those with whom they are working. A reactive dimension which is difficult to predict and disturbing when it does occur if only because the researcher is already in an uncertain role situation.

A certain resilience is always necessary in fieldwork because it is virtually impossible for everything to go smoothly; notes are lost, informants do not materialise, ill health is always just round the corner when researchers are working outside the environments to which they are accustomed and the social situation is always potentially unstable. It might always be the case that researchers should have the ability to readjust their projects to new situations without feeling that their work has lost purpose; the stability which they may have hoped to depict can be turned into a study of instability.

17. The researcher as a person— age

Unless research is undertaken from the researchers own home on a commuting basis or as part of their daily work, it involves difficulties and strains on whoever does fieldwork. The vast majority of fieldworkers are young with few social commitments in their own social world and probably male. This means that they can engage in work which is based on a person with few social ties. It is not work that can be easily undertaken by married men or women nor ones in which they can bring their partners into the field with them since it compounds difficulties of adjustment to the society in which fieldwork is undertaken and probably brings about strains in the relationship unless both are researchers. A married women in the field unless she has her own research interests is under much more strain that she would have in her own society where she would have known social connections and the opportunities of employment.

So the majority of fieldworkers are both young and in practice may have only a single fieldwork experience either because they find less stressful and better paying employment or they find that family commitments and their teaching schedules do not allow further long periods away from their principle work. This initial fieldwork provides the foundation for all their subsequent work and thinking which pivots round their fieldwork acquired knowledge. The majority of social scientists have very few fieldwork experiences unless they can be carried out during work or from their homes.

The problem thus is that fieldworkers are young and see their research through the eyes of the socially inexperienced, and this is not to denigrate their understandings but these still remain that of the younger generation. Moreover their relative youth means that older people who are part of the researched community will see these researchers as young and socially inexperienced and probably in their minds not really worthy of the serious attention of those older than themselves. Through the tolerance of the researched community towards strangers and those with unknown but possibly dangerous or valuable connections, they can be treated as older than they are, honorary middle aged men and women, but they are still in practice socially young. In the language of the Nilo-Hamitic pastoralists of eastern Africa than they would be graded as junior warriors who do not have much say in the governance of their community except as cattle guards not yet of a suitable age to marry or to have their own household and cattle.

On the other hand there are a few social scientists who have made repeated visits to the same locality such as Abrahams with the Nyamwezi of Tanzania (Abrahams,1981;1994) and Mead with the Manus island people who have aged as their informants have aged. Such people have an accumulation of insights of the communities which they have researched and at the same time will be proportionately more influenced by them. As they become better known in a field area and as their own process of becoming older becomes self-evident to the people they are studying, so they acquire a reciprocal identities; a sharper social identity on both sides of the research work. Such people move more obviously in the researched community's social world and they become part of the accepted social scene so that it begins to be assumed that any relationships are permanent (Strathern,1981).

Ageing as a useful factor in fieldwork is probably commoner with nationals who do not have the same problems of keeping in touch with those who were researched as well as the internal politics of their university while non-nationals have the additional problems of absence and finance. Overall fieldwork is a young man's activity who sees society from that age related viewpoint. So fieldwork is by definition a physical activity which can be extremely tiring and for one reason or another is an activity confined to people in their late twenties and a predominantly male activity. If the researchers are older, then what they attempt to achieve and do in fact achieve involves less physical effort in what and how they are observing.

We can also assume that as the researchers get older that their state of mental alertness in their fieldwork will be in some measure different to that of younger researchers. There may well be a proportional corresponding increase in overall understanding which have hitherto been restricted by age and sex. Since they will be less physically active there may be an increase in their reliance on what people say they do rather than in observing what they actually do. Would the national researcher have advantages over non-nationals in these age related processes? It may be that they will be better able to integrate ageing into fieldwork since they may be working nearer to home without the need for exhausting travel and be less dependent on grants which tend to go to the younger researchers.

18. The researcher as a person— their appearance

Apart from any passage of letters introducing a researcher or requesting permission to research in a country or area to which and in which they have no legal or social connections, they are seen by those that they come to do research, before they are heard. There is no reason to suppose that many minds are not made up by what they see as the researcher enters their social and indeed their world of eye contact.

First, they see age and sex and this divides up social relationship into the sexual dyad and its binary oppositions and to grading by age. This is done almost instantaneously and has both biological and ethological foundations; the researchers cannot pretend to be what they are not by sex and age. Then costume is noticed and most researchers approach the community in which they hope to work dressed in the clothing appropriate to their own social backgrounds. The Goan woman researcher working in India was very careful to wear clothes appropriate to the social venue in which she was visiting and thus had five changes of costume (Mascarenhas-Keyes,1988).

Whatever costume they wear it indicates the researchers own position in their own society and how they feel appropriate for approaching people with whom they have no other social connections. Thus on the whole they dress in ways which are indistinguishable from that of tourists, and their behaviour may well be judged by how they dress. They may think that this informality is no more than showing their social neutrality but it is not and indeed never can be seen as socially ambivalent.

Sometimes their costume is the same as government servants and assumed to be informal but they are thought to be wearing the informal uniform of their employment and it was an additional handicap on their acceptance as researchers free from any government connections (Nakhleh,1979). A Nigerian anthropologist who visited her father's village after living some time in the modern developed ambience of a major city and its university, once wore jeans, but she was laughed at and men called her 'the master who does not have a penis'(Amadiume,1993:188). Shorts and t-shirts are similarly seen as appropriate for teen-agers. Even if the researched community knows full well that the fieldworker is a non-national or an internal expatriate, they would probably appreciate that the attempt has been made to dress appropriately to their age and sex.

19. The unmarried researcher

In almost all societies for an adult man or woman to be without a husband or wife is seen as contrary to commonsense and in many communities the unmarried adult man is socially treated as inferior and the adult unmarried woman is rarely accepted except as biologically incapable of sexual intercourse or deliberately flouting her parents intentions for her future welfare. Women who have become Christian nuns in Africa have often reported that they have been subject to beatings and abduction by their male relatives. The American woman researching in Mexico was frequently told by local women that she should marry a local man and her in-laws would ensure that she would be able to continue her research without being troubled with domestic duties (Landis,1970). Other women researchers have found it better to state that they had a husband at home solved some issues but the mention of their absent children reimposed prejudice against them as unfeeling and of course there would be a lack of understanding that a woman could want and achieve some measure of independence on her own away from male dominance. The Coorgs of the Nilgiri hills in southern India tried to provide Srinivas with a wife (Srinivas,1979) but he does point out that having one's wife in the field cuts down on loneliness but at the same time cuts down on the amount of time that could be spent with the villagers whom he was studying as well as reducing the time that he could spend with informants.

20. The researcher as a woman

Fieldwork has always been essentially a male activity in which women been an important minority not only because of their relative rarity but because of their contributions to understandings which have tended to be from a male point of view, male researchers talking to men about the activities of women seen from that social position. The works of Blackwood, Richards and Mead make this obvious point. Now there are numerous women fieldworkers in the social sciences but they nevertheless have problems since they come to work in societies which pivot round the social power of men.

Although in the Western university environment equality of the sexes in fulfilment of legal and what are now seen there as ethical obligations has achieved some success, away from this these women whether nationals or non-nationals are going to have to live within paradigms of male dominated social systems except for certain 'home' environments. In some cases, as in Saudi Arabia or Iran, it would be virtually impossible to work round or away from restrictions imposed on women by social convention and nationalised interpretations of Islamic law.

Maleness within what might be called their global dominances, is able to transfer itself with some adjustments into most national and non-national social environments. Femaleness on the other hand is always subject to some facets of social control initiated and maintained by men. The failure of women researchers to conform is usually followed by social sanctions, while a similar range of male behaviour would be more likely to be subject to ridicule and excused as no more than the ignorance of a temporary 'licensed' outsider. For offences against proper Tiv behaviour for an adult a woman anthropologist was classified as a witch (Bowen,1964); though it is not clear to what extent this modified her fieldwork or her relationships with similar women there of independent minds and individualistic behaviour.

A single unaccompanied woman is often the subject of sexual advances by men in societies which do not have an acceptable role for unmarried women except as prostitutes, concubines or nuns. Negotiating through this web of male entanglements may be a serious handicap for a woman researcher wanting to concentrate on research that goes beyond women's roles (Landis,1970). 'The local Mexican villagers found it hard to understand how, I obviously attractive in their eyes, could still be single. The obvious remedy was to suggest that I marry in the village. The important thing was to get me safely married so I would cease to be an anomaly...even if I had not been warned that intimate relations would result in the loss of the capacity for objectivity…my very inability to predict the short and long-term repercussions of such relations, either for myself or for the people, would have acted to break any impulses I might have had in moving closer (Golde,1986:85).

It seems likely that national women researchers may be much better equipped to accept these restrictions and to work round them than to take up positions in which they would feel aggrieved psychologically if not socially by being controlled in ways which would no longer be accepted in the societies of non-national women researchers. Women as social scientists hesitate to speak out about the violence to and abuse of women for fear that their freedom of action would be restricted by their sponsors or by the civil authorities in the area in which they are working (Howell,1990). For the women fieldworker to be alone, tired and perhaps disoriented seems to be one of the factors to be assessed as part of the dangers of fieldwork.

21. The researcher and family commitments

Social scientists as professionals committed to field research lead lives which are divided between the public requirements of their profession and what they would consider to be their private lives. There has to be a balance between these two distinct forms of commitment and any over emphasis on one sphere detracts from the possibilities of functioning appropriately in the other. This is a fact of life but the balance between the two is guided by cultural factors which are partially independent of these private and public categories. It is also a fact of their professional lives that fieldwork involves researchers in going away and doing; a clear distinction and the implication of sacrificing one for the benefit of the other.

There are a number of factors which have to be considered. To what extent does the culture think in terms of 'I' or 'we', age centred concepts of appropriate behaviour, physiological and psychological personal needs, acceptance of economic necessities, the requirements of social and geographical mobility, ambition, the inequality of resources and the acceptance of dependency?

Many of these factors as difficulties are avoided by the fact that most researchers in the field do not have close personal commitments since they are in their early twenties and their mental and physical vigour makes separation from 'home' a not particularly serious issue because they are of an age when such separations are normal. Once social scientists take on social commitments to another individual which cannot be based in reality on a 'take it or leave it' foundation, there are complications in any attempts of the researchers to cope with both spheres on something approaching equality. Firstly, there is the issue of whether the partner is a fellow professional or someone with whom the relationship is purely personal and emotional.

If we have men and women social scientists of equal potential competence, then there is tension between them as to whether they are carrying out fieldwork as a complementary team or as individuals. In most cases the woman puts her career on hold until the children grow up or the husband dies as was the case with the Turners and in the field she runs the household so that her husband can work without domestic obligations and necessities. The house-husband role was performed by Colin Turnbull's partner or by men who put their research needs in the field behind that of their wives, or have independent careers which are carried on by agreement indifferent locations. The basic dilemma is shown by the wife of Raymond Firth during his fieldwork in Malaya (Grinker,2000). 'In the loneliness of the field situation, the presence of another can be a real source of comfort, but the obligation to give support to another may be a serious additional burden on a fieldworker to begin with' (Firth, S.1972). The researcher must inevitably be concentrated towards the partner rather than towards the surrounding community being researched so that the partner can be a handicap on the use of time and more so if children are involved; partners are a support for fieldworkers but probably only to the extent that they are not too demanding. For the researchers to have their children with them is probably an asset to family and community studies since they make their own connections and the researched often like their presence as showing that the researchers are 'real' human beings.

This issue should be assessed in time and motion terms. The unmarried or rather the uncommitted fieldworker is always faced with the restrictions of time; entering the fieldwork areas and leaving it are finite events certainly affecting the non-national more than the national researcher. Even allowing for full time working there is never enough time for what is planned and more importantly to carry through with what develops from the work itself. Is the pay-off in having a partner in the field worth the further rationing of time that can be used for the fieldwork itself and the further time consumption of keeping that partner happy enough to remain there as they are often in greater isolation than that of the researchers themselves.

Nevertheless these relationships are important for two overriding reasons. They present a relationship which is seen by the surrounding community in terms of their own relationships; the appropriate status for a man or woman with regard to their age. In this context lesbian and homosexual relationships which would be socially unnoticed in their own communities are often inappropriate or even dangerous in some fieldwork situations. Secondly, the researchers are always in the field for finite periods as obviously they come and go but within the time available for research, there are substantial deductions for sleep, eating and self-maintenance. Any personal relationship has to be evaluated in terms of whether or not their presence aids or detracts from the amount of efficient time spent on the work itself; their contributions to physical and mental well-being, housekeeping and the prevention of ill-health. On the other hand these providers have their own agendas and they are not servants who can be hired and fired; personal attention has to be paid to them so they are consumers of the researchers' time. Any long-term relationship requires reciprocity. There is always a possible disruptive factor to consider.

22. Women researchers and fieldwork

Any researcher entering fieldwork is going to be faced with obstacles which have to be overcome or at least worked round. We can take it that as most societies and institutions are organised through various forms of male domination that being a woman researcher is going to be a basic handicap but not necessarily an insurmountable one, but there are other connected handicaps. Apart from the fact of being unmarried which can be lied about, a woman may have or have had obligations which detract from her total commitment to research. She may have been a housewife for some years and have a family and much may depend on the support of her husband. There would usually be family objections compounded by the fears that almost any chosen area would be considered unsafe. In one case the researcher was a Muslim married to a Bengali Hindu which was not seen as bridging two religious cultures but as apostasy by the Muslims and polluting by the Hindus. This basic domestic fact made her avoid religious affairs although it was an important element in most activities in the small market town she was researching (Gupta,1979).

An aspect of research by women social scientists is their choice of objectives in whatever field they choose to investigate and which are almost always going to be dominated to some extent by men. Thus they are likely to experience difficulties which they would not have in social environments which might accept women professionals as the equals of men.

In all societies there are single sex groupings in which there are minimal difficulties for researchers of the same sex as well as groupings in which the sexes mix in well understood circumstances; there is never total segregation across the social spectrum, as in contemporary societies women are invading long-standing male monopolies in economics and politics. Researchers of either sex will have to work in contested social environments in which there will be multi-facetted biases built into their fieldwork.

Of course, in most Western social environments women are protected by law from unequal treatment in employment and indeed from inappropriate behaviour in those settings. However, it may often seem that in their choice of assignments they are not so much doing important research into previously uncovered aspects of behaviour but making the point that they are entitled to equal treatment. In all but a few societies women have entered the professions as doctors, lawyers and politicians and even Muslim states such as Bangladesh and Pakistan have had women prime ministers, but at the lower levels segregation may predominate. Thus it is that men do not try to research women's activities nor women those of men though both may work in shared areas but with cross-sex difficulties. On the other hand women in societies in which women have segregated roles are able to do valuable research within those fields closed to men and in this national researchers would be better able to do unique work within these locally understandable restrictions than a non-national who would be inclined to see such research in terms of making points about inequality.

If in the opinion of men in male dominated societies women researchers are intruding into their preserves of privilege, then it is not surprising that there will be obstacles put in their way. However there are public activities which have often been dominated by women, such as markets, in which they predominate as both buyers and sellers and in which valuable research can be done without women researchers appearing to be nudging these societies into social changes imported from the more modern parts of the societies surrounding them (Barua and Devi,2004) as would be studies of maternal beliefs and their perceptions of their children's pre-school competence (Sharma *et al*,2004). Even in Saudi Arabia there are important businesses run by women who are able to run them electronically without any personal contact with men outside their domestic roles.

23. Failure to complete fieldwork projects

We know that there are successes in what researchers have achieved as there are external standards which show this clearly enough. Some research reports are published since they are felt to be of value to a wider reading public than the university world and some indeed become classics with a general readership and are often reprinted. Many aspects of fieldwork successes are the articles published in specialist journals with limited readership and their principal use is their appearance in bibliographies attached to applications for jobs or further financial support. Other projects which are commissioned by government departments or non-governmental organisations are not likely to have a high failure rate since the work is produced on a cash and carry basis but they are often confidential and not available for general readership. The majority of successes are theses produced for getting higher degrees and university libraries have large numbers of such works available on microfiche to others working in similar fields. It is perhaps these which are a truer measure of success since they are examined by specialists in those fields and have little to gain by passing inadequate work.

No initiator of research whether donor or university is going to publicize failure because it means wasted money and even lack of judgement by those promoting such work. Furthering research is always a gamble since they are manifestly attempts to find out about the unknown by people who are also unknown in their potential abilities to work in social and physical environments which are radically different to those to which they have been socialised from birth and in subsequent social environments such as the vibrant university communities in which there will be many like minded individuals and much personal companionship. So failure is a touchy subject since it demonstrates a lack of judgment of those who have encouraged and who have promoted its initiation.

There are failures which have nothing to do with either the individual or those who have encouraged the topic. Political situations are always volatile and disturbances may make it impossible for researchers to continue their work where they are known to be internal or external expatriates, foreigners particularly of nations which have become unpopular with the public for political and religious reasons. The national government at the local level may pull out researchers as they do not want to be responsible for their injury and their own governments may advice that they leave.

National disasters such as the recent 'tsunami' in south-east Asia may make it impossible to continue research, and researchers in such areas are more likely to be involving themselves in relief work than in their stated fieldwork. However, it seems likely that in the vast aid industry there are those in the field who are actively involved in assessing the expenditure of vast sums and employing large numbers of helpers in terms of attainable profit and loss since they run as businesses. The Red Cross and Oxfam must have their own commissioned researchers who are working in political and environmental situations where local governments would disallow uncommitted outsiders from continuing to work there.

There are also situations in which the research materials on which reports have to be based are lost through no fault of the researcher; notebooks can be stolen and baggage can be lost or stolen by the police on an internal flight, film spools can be ruined by children and a valuable genealogy chart which did not look like writing can be used to clean a baby's bottom (Varadachar,1979). It may also be that some researchers would find it personally impossible to continue working in areas in which so many of those with whom they are involved are suffering deprivation from which the researchers are protected by their relative affluence and outside support.

It seems likely that most of the failures to complete research are due in some degree to the inabilities of the researchers to adapt to the situations in which they find themselves. It is known that many find university life difficult because this is the first time that they have lived away from home even if home is only a short train or car journey away.

Research may often involve live-in communities and under conditions of which they have no experience at all; being brought up and educated in Mumbai or Manchester is no preparation for working among farmers in Mahalipuram or for studying the decaying marital systems of matrilineal societies in central Africa. The understanding of the theoretically alien may be of very little use in cushioning the contacts with the totally alien nature of their social practices and intellectual framework of thinking.

Above all very few young people are prepared for social and intellectual isolation and they come from social environments in which they are already part of many groupings and have an enormous numbers of other groupings which they can join for sport and stimulus. They miss the stimulus and protection of friends and even their families over whose ideas they may have been in constant tension.

The discomforts of camping and back-packing or participating in packaged tourism to famous places, are not the same thing as having to live with little water, no sanitation and moreover always in the public eye of the curious so that they never have the privacies to which they are accustomed. Regardless of what their research objectives may be there is no relief from this day to day isolation from what they are accustomed to as people who have usually been cushioned from difficulties by their membership of groups of like minded people.

So there will be those who are unable to cope with the living conditions, the sights, sounds and the smells to which they are unaccustomed. Few realise that hygiene is as much social as medical; Hindus and Muslims working with Christians are going to be shocked by their ignorance of what is necessary behaviour for them just as the Christian researcher in reverse is going to find their behaviour inexplicably restrictive if not almost primitive. Most researchers may be able to relate more easily to the overtly traditional than to urban behaviour of groups other than their own; squatting down to eat in the house of an Angami Naga may be mentally easier than doing the same thing in a family housing themselves in a concrete cylinder in Calcutta which will eventually be used for a new water supply system.

Whatever the mental and intellectual pressures on the field-worker there will be more purely physical difficulties. Ill-health is certainly going to account for a high proportion of failures despite the fact that they are likely to be well protected by pre-emptive immunisation. They are unable to cope with the change of diet since most of their eating will have to be in what is locally available rather than in what they are used to eating; it is rare for a researcher not to have stomach illnesses and to pick up infections from environmental and social sources from which they cannot separate themselves without causing misunderstanding; to refuse food or water on sensible hygienic grounds, will often be interpreted as mistrust of the donor even a suspicion that they are suspected of being witches, on the frequent observations of this author. It is not surprising therefore that missionaries in some ways parallel researchers, and carry back to their own countries large proportions of parasites as indeed researchers must do (McQuay,1967). Also, researchers just as missionaries are likely to bring new diseases as well as ideas into the communities they research (Guppy,1958). Hookworm and the resulting anaemia is common enough and it is likely that some researchers will leave the field because of a combination of illness and lassitude produced by poor diet and discouragement.

Researchers are also professionally optimistic that they will be able to achieve what they set out to do not only relatively easily but in the time for which they have funding and leave from their universities; they may also have a feeling that they have 'right' to receive information from the people they approach as has been seen in the behaviour of students from Indian universities approaching informants whom they consider to be their social inferiors. Researchers may come to realise the multiple barriers which they have to overcome in order to get at their chosen topic.

There may also be the dislike of the researchers for the community in which they are committed to working which may be particularly strong for women who have been accustomed to some degree of social equality are now tied to traditional roles of which they no longer approve or accept. Since interpersonal distaste is always a factor in social interactions it must surely account for fieldwork difficulties. This is likely to happen even if researchers steer themselves away from work which they suspect they will find unpleasant however important and necessary it maybe. A more commonplace problem is that expressed in a letter to a London newspaper (*Daily Telegraph*,3 March,2003) 'I am a volunteer in a project in a village in Asia. My host's brother often hits his younger sister and female cousins. His grandmother says this is all right he is older, is a boy and bad girls deserve to be hit. I disagree.,

Research projects are created out of a need to know some important aspects of the behaviour of people and many possible projects are beyond the capacities of anyone to research however well funded and competent they may be such as the sexual lives of married clergy and such a list is endless. Researchers may find that their chosen topic once they get into the field is inaccessible to them in the time that they have available to do the proposed work. The community may be just as suspicious of the national as for the non-national and close ranks against their enquiries.

Some researchers find out soon enough that they cannot work on the project on which they had set their minds because the community does not have these functions. To go to the East African coastal village in the anticipation that it will be principally devoted to fishing only to find that they are almost entirely subsistence farmers, needs rapid rethinking of what they can do and they may be unprepared or ill equipped to make such changes. The original research topic may have some overtones of wider significance and the fieldworker may not wish or indeed be temperamentally unable to change to some topic of more commonplace significance.

There are always likely to be cases in which the community refuses to accept the researcher such as the pueblo American Indians who carefully restrict the entry of strangers into their communities and the Egyptian police who carefully steer researchers away from topics which are likely in their opinion to cause trouble regardless of their personal acceptability and linguistic competence. Lastly, there are the possibilities that once they are in the field it is taking much longer than they anticipated to learn the language and to become at least partially accepted into the community and the money runs out before this state is reached. Overall there are likely to be different failure rates for restricted topic and open ended research and personality differences which make some people more suited to abstract library research and others who are good in relating to strangers and strange situations.

Chapter 8

Issues in Difference between Nationals and Non-nationals

1. Entering a new field of understandings

No matter how well prepared individuals may be intellectually moving into a new social and geographical environment is always going to be a stimulus for comparisons; a new picture is flashing into the mind and of course the more extreme the change the greater the impact. The newcomer to India from a Western environment would notice at once the social activities which they would not have seen in their own societies such as caste marks on the foreheads of men on station platforms and the differences in the costumes of women in their brighter colours. New researchers are bound to be affected by the actualities of what they are experiencing. No book however well written can evoke situations in which all the senses are involved at the same time.

To what extent are nationals prevented from having such striking contrasts or is it that they do not notice because being in their own countries blinds them to many very striking social differences? A Dutch researcher living in Maarstricht is living in a cultural area shared with Belgians and Germans and with the abolition of borders, the only differences within a few kilometres are legal ones.

Only researching within a small social environment to which such individuals have been socialized from birth, there might be little excitement or stimulus from entering a research situation. Within large societies the differences between one part and another will be as great as crossing any number of national boundaries; a

Cantonese working in Beijing or a Tamil researching Bengali behaviour would have social differences forced on their understandings.

There are likely to be many differences about which the non-national researchers are not only likely to misunderstand but to make social mistakes in their inter-relationships. The acceptance of auspicious times for certain activities and the fact that time for subsistence farmers is not a constant since it runs more quickly in the cultivating season as well as the lunar timings in both Buddhism and Islam and the fact that timings are different on sunny and cloudy days.

The citizen potentially has the same stimuli from entering new environments except for the fact that shared citizenship may bring with it a sense of affinity. But is there not a more basic difference that the non-nationals are there because they seek to research environments which are strange to them and indeed the stranger the better since funding may well be more readily available. The national on the other hand with more limited resources is likely to seek researcher opportunities nearer to home; if nationals want higher degrees or applied research opportunities there is little dividend in going to the opposite end of the country seeking social contrasts and making everything more difficult from start to finish.

The Western trained social scientists or those thinking along these theoretical lines are used to finding splits in any community but there is the pragmatic feeling that these divisions are always in the process of changing and that in so far as they are structured into a society they are only temporary features of those communities. In some Asian communities researchers are faced with permanent divisions that are not restricted; caste Hindus are divided from Dalits and Muslims from Hindus and that is apart from ethnic differences.

While in Western communities researchers can move backwards and forwards over whatever social divisions exist without too much methodological difficulty, in the Asian setting it is not possible to move in ways parallel to this. These divisions are set into forms of social concrete which makes the researcher's involvement in one group prevent access to another. The relationships between these dichotomous groupings has been laid out in contemporary violence as much as in history (Naidu,1980).There is the further complication that these minorities which number in millions often have extra-territorial political and religious loyalties. Researchers have to accept the violence built into these binary situations (Bourgois,2002) which cannot be glossed over as one of the bases on which social science research has to be built 'an unusually polarized understanding of politics and practice'.

2. Networks of influence

Any social scientist as they grow older in their profession accumulates people who can help them in their careers and fieldwork and in their own countries they know ways of getting things done which are unknown to foreigners. Thus English researchers working in Germany, for example, would not know their way about and vice versa, unless they had a German partner, were bilingual and had lived their for long periods.

We can accept that in most if not all traditional societies and their ways of thinking, that it would be rare for any act to be carried out without some reciprocities in mind. Thus we can better understand the reluctance of contemporary Chinese and Indian bureaucrats to carry out their obligations without exchanging an administrative act for some return in terms of prestige or indeed the partial humiliation of the person asking. Their bureaucratic position may be the only power that they have which gives them the status which they may not have in the outside world. In this type of situation it would be quite normal to have to get permission from each and, very official at whatever level by some sort of bureaucratic negotiations since the authority of the head of a firm or department does not ensure the automatic compliance of their subordinates (Sheth,1979).These are networks which are based on the manipulation of paper with their arteries of print. It is not for nothing that Mehta from his experiences in returning to live in Bombay after living in the United States for some years concluded that India is a 'no' culture in which this is the standard answer to any request (Mehta,2004).

In developed societies it can often be assumed that a private or public bureaucrat will function more or less without having to have a personal connection with the person asking them to do the job which they hold. This is not so in many countries such as India or China where to get what the researchers want may only be possible without delay if there is some personal connection; both come from the same village, clan, sub-caste or having some distant classificatory or affinal connections. Mehta writes (Mehta.2004) of his relatives visiting Britain and finding that all their arrangements can be made in a single day, each of which would have taken prolonged and irritating personalised negotiations in India.

So national researchers knows their way around and would have all their personal connections ready to use in order to get what they want and probably not even by bringing them into obvious play. Kumar's father was a senior policeman and another had a father who was a senior civil servant. Researchers from the Western world would feel that the use of such connections would compromise their hope of professional detachment, while someone from an Eastern society would feel that they had no alternative but to use such connections if they wanted their fieldwork to get going at all.

National researchers would know how to get themselves accepted into the social networks of the people with whom they are working; they would have the social awareness to know just how much they should contribute to the wedding expenses of an informant's relative and how much help to give to give students with their tests. There is a fine dividing line between fitting in with such donations in cash or kind and becoming the object of some forms of social extortion. Those who are being researched know that these researchers are well off and the national knows far better than any non-national how demands on them can be diverted without giving offence and without breaking useful and on-going relationships. There are many ways, for example, of saying 'no' in Japanese without actually saying 'no'.

There is another aspect of language which separates the national from the non-national publishing their results in languages which do not have an international readership from which they might otherwise collect important networks of continuing influence. The importance of publishing in a major international language is shown by the work of Zhong Wei Chen, the innovative Chinese micro-surgeon who has published 92 papers in Mandarin Chinese of which only 33 have been republished in English so that his influence in the western world has not been as great as it would have been if his techniques had been published in English at the same time as Chinese with its very limited overseas readership (*British Medical J*,2004).

3. Personal involvements with those researched

There are differences between nationals and non-nationals in their senses of personal involvement with the communities and institutions which they are researching which go beyond their attachment to the professionalism of the social sciences. The fieldwork of the national researcher involves 'us' while the same work by non-nationals involves 'them'. With the former there is an overall sharing of ideas and concern about the society in which they are both working. On the one hand there may be an acceptance of the deficiencies of their society as part of its culture. It seems unlikely that a non-national could have written with such efficiency and understanding of the negative 'no' nature of Indian society from such a very personal point of view which is a major part of Mehta's qualitative study of Bombay (Mehta,2004).

Non-national researchers may tend to have an emotional liking for the communities which they have researched often referred to as 'my' tribe which is not extended to urban slums or deviant communities. For them the sheer enormity of the contrasts between 'us' and 'them' becomes an attraction without any full realisation that they may have been used by its members for their own purposes.

There are differences in how women researchers can go about their fieldwork. Foreign women do not fit into national patterns of what should be the proper behaviour of adult women whether married or unmarried who are not nuns, nurses, teachers or market traders and they may also be part of the fantasy world of national men; they just do not fit in. She may be sending out all sorts of signals unconsciously which the local people are bound to misunderstand particularly when they might wear what is generally regarded as male costume.

Despite the anxieties of their families, national women researchers have almost always been treated with paternal concern once their marital status has been established and understood in local terms. On the other hand the unmarried one is suspected of misbehaviour if not currently at least in the community from which she comes which would account for her unmarried status. But there are difficulties over matters of personal hygiene. Male researchers are more used to being in the public eye and thus do not have the same problems over excreting as women. Dua(1979) left her research area in order to live with relatives because of being unable to perform such intimate activities in public or restrict them to the night.

4. The acceptance of fatalism

An additional strain on researchers are the wide-ranging differences in patterns of thinking. Whatever their personal attitudes to progress researchers are part and parcel of professional, political and scientific thinking which is activist; that they can have an active influence on the societies in which they live. And even if they are not personally active they are active in their thinking. This is always pushed into their thinking because of the contrasts or similarities between their lives as researchers and their own societies to which they can return at will. There is thus a difference between non-national researchers from Western societies in which the environment from which they come, despite its contrasts, is activist in its thinking about the changes which they can organise. The national researchers from developing societies are faced with the contrasts of political uncertainties, widespread disease and malnourishment as well as restricted opportunities from which the people they work with are unlikely to escape; whatever activism there is related to short-term survival.

Researchers come to work in communities which are not only

different in how they carry on their daily lives and this is relatively easy to accept and understand and may form the bases of their fieldwork. What non-national researchers may find difficult to understand is the fatalism of much of Eastern approaches to the understanding of life. Many communities of subsistence farmers and daily paid workers feel that there is not much they can do to offset fate which for them has some metaphysical connections explaining their swings from poverty to plenty (Joshi,1979). Many such people feel that there was not much they can do to offset this except for the resources of religion which they were able to use from their own understandings. The non-nationals from a different cultural environment would tend to think that this was pointless ritualism and they ignore the fact that not only do these people not think it is pointless since statistically it is often as successful as scientific methods which try to improve their lot pragmatically.

The approach to life of those who think in terms of a single life and those who think in terms of an endless succession of lives is bound to be difficult for those coming from the previous Western cultural environment to appreciate as having widespread significance. One has only to think of the Western world's lack of comprehension of Islamic suicide bombers and Japanese Shinto kamikaze pilots anticipating paradise from their deaths to realise that these differences are deep seated and perhaps are unbridgeable.

5. Women researchers

A community which is approached by a researcher may have little knowledge of what that means as an occupation but they will have models in their minds as to what outsiders represent irrespective of their explanations and connections. The British popular model of the male or female researcher may not be much related to reality, but it will have been based on some knowledge of what they are doing. In less developed societies there may be some realistic models of male outsiders and what they are doing, but not for women beyond the roles in which they are already known as teachers, social workers and perhaps nuns. Indians are very fond of the cinema and television serials and the models on which outsider women may be allocated are likely to come from American and Bollywood productions (Varadachar,1979), which makes their roles as researchers more easily misunderstood.

6. The use of interpreters

While the national may have a knowledge of the main language of the community they are studying and some ideas of their cultural background, they have to recognise that their informants are in fact interpreters; they are giving their views of their own community and their involvements there in terms which they think the foreign researcher might understand. This will be even more so when they are in practice internal ex-patriates as would be the case when a Gujarati goes to do research in Calcutta or amongst the Coorgs or Muria. A Spanish speaking mestizo in South America will not speak Aymara or Quechwa and a Burmese will not speak Jinghpaw the language of the Kachins who live in their northern mountains.

The use of interpreters those who translate what is said to them into another language which is that of the researcher must certainly be a difficulty for non-national researchers and nationals ones who similarly do not have the time to learn a new language for research purposes. There are also those who just do not have the ability to learn a new language possibly because they do not like becoming a pupil again or do not have the capacity to relate to a foreigner in such a personal way. There are also those who research into observable practices and they require informants rather than interpreters for activities such as time and motion studies of farming or domestic work.

Let us presuppose that the researcher's natal language is English and of course they will be graduates. It will be rare indeed for the interpreter to have a similar knowledge of English so it is almost inevitable that the interpreter's English will not be of a very high quality. It would in fact be best if the researcher accepted that what they are hearing is a poor translation by a national of what he thinks the researcher wants to know. What is probably being used is a basic English vocabulary of perhaps five hundred words with very little use of any range of tenses and idiom.

If the researchers are French, German or Japanese then the situation is even more complicated because they are hearing limited English spoken by a national in terms of a grammatical English which they themselves have learnt in school. So it must be the case that complex ideas in the interpreter's language will be translated into far simpler words as they change to English. It is also the case that the national interpreter will have learnt basic 'English' but have no knowledge of the sociological concepts that are being asked.

The most important point is that interpreters in the field interpret what is said to them as quickly as possible; they are probably not professional translators who are using one language as far as possible to replicate the communication of meaning and thought in another. So these interpreters will be reproducing what they hear not necessarily in good faith motivated by the same professional interests as the person to whom they are talking. Translators have to work on two levels at the same time; the passage of immediate information while at the same time interpreting its ideological meaning. The court interpreter for Kikuyu into English at the trial of Jomo Kenyatta for sedition was told that he could translate at the first level and that the second level would require appropriate specialised witnesses.

They will have their own understandings and biases. When working on the genealogies of Swahili families in Mombasa this author took several months before informants started to put the maternal lined into their diagrams. What the interpreter passes on is a replication of how they see the situation which they are interpreting; they can only use their own specially trained eyes and ears. In their relationship with these external or internal foreigners they are also their part-time teachers of language and customs. The researchers recognise their value in helping them behave in the correct way in situations which are strange for them.

By using these men as interpreters researchers are in effect buying their way into a certain segment of the society or community being researched and even if the researcher is a woman much of her information will be got from men as few women have an adequate knowledge of the outsider's language or to be socially accepted by the community as an interpreter. Using a Brahmin interpreter makes for stressless conversation with Brahmins (Berreman,1962) but excludes the lower castes from anything approaching tolerant access and vice versa (Beteille,1965).

Since most research has to be carried out with the agreement and connivance of the local political and social power-holders, they are likely to provide or arrange for interpreters in agreement with their views. Those with other views are not likely to accept their interpretations as unbiased and possibly not even acceptable. Even the most acceptable and supposedly well informed researcher using interpreters will get a circumscribed knowledge of the community in which they are working. They can pass on information about their own extended families from which the researchers benefit as well as inflating their own family status. They are unlikely to feel any obligation to pass on the views of those opposed to their own family and status (Buechler,1969). There are additional problems from the fact that interpreters are employees so that it is in their interests to keep on talking, providing a flow of information and explanation. It would seem likely in largely non-literate situations that a complex language will be trivialised. Overall interpreters must be seen as dubious necessities.

7. The minimalist possibilities in the grandness of ideas

The social scientist of whatever nationality or natal culture, is attached to ideas which have no boundaries; by academic fact if not by disposition, they are internationalists since abstract ideas have no close relationship to particular social situations. They think rather than live in a world of broad understandings. The belief systems of academia are possibly as pervasive as those of religious systems.

No doubt that the grand old men and women of the social sciences can knit together the manifold research experiences of perhaps fifty years within the circumscribed world of academia. Nevertheless they as are all social scientists, are basically minimalists as they have all done fieldwork in circumscribed fields of enquiry or communities. These patchworks of understandings are indeed cumulative but they remain patchworks.

For the non-national these are accumulations from occasional visits to earlier research venues as outsiders and in the main tend to be theoretical enlargements based on earlier minimalist experiences and conclusions. This is not to denigrate what is perhaps the inevitable minimalist nature of all fieldwork but to recognise they are the building blocks of any possible enlargements of understandings.

Clearly the national researcher is better able to accumulate fieldwork in support of their original experiences. The researcher living in Bengal can keep in touch with her original area of research in Banaras as it is only a railway trip away (Kumar,1992), just as the Egyptian researcher living in Alexandria is visited by his Sudanese informants with whom he had worked earlier in Assuan (Fahim,1979). The national social scientist has the same minimalist fieldwork accumulation of information and also has a touch and go relationship with their areas of work and specialisation but by the fact that they live in the country of their research experience, and it is always being worked over in their minds in terms of that particular country as a concrete rather than abstract form of reality.

8. The predominance of applied fieldwork

In applying social science understandings to specific situations in which governments or institutions want to put right situations not of their own choosing or to initiate changes for the betterment of the people for whom they are concerned, researchers are likely to be dealing with situations which probably replicate in outline situations which they have previously been studied; slum clearance, new housing estates, displacement of populations because of dam construction, famine relief and refugees from civil war. The only replication can be in the titles of the research reports as the actual on the ground situations are not and can never be duplicated. All the researchers may get are pointers which they may apply and may be entirely inappropriate.

On what basis can social science researchers make recommendations from which decision makers can take action because they are not professionally geared to make clear statements about what other people should do in specific situations? What was the quality of the evidence already available from studies of similar situations in which social scientists have been involved? It seems likely that the social and political planners for the Yangtse High Dam will have studied the research reports on the planning and subsequent removal of populations caused by the flooding of the Nasser High Dam in Egypt rather than earlier population movements in societies under colonial domination. Much the same would apply to refugee work which has expanded enormously in Asia and Europe whereas previously information would have been largely related to Second World War in Western Europe.

But even in this researchers have to be careful. Oxfam without adequate research of the implications of the proposed program, supported the Tanzanian national villagisation campaign which proved disastrous in its effects on social behaviour and agricultural production. It is prudent for social science researchers to remember that with human social and environmental differences, it is usually unwise to make recommendations concerning anything more than small and relatively static communities. Grandiose political planning and social science understandings do not fit easily together.

What outcomes are critical to a decision? In many cases the outcome has already been decided on political and economic grounds. The dam is going to be built regardless of the human dislocations which will be caused and refugees have to be moved yet again from the political borders where they are a continuing liability. Thus much applied research is fenced in by preconditions over which the researchers have no influence at all; they do not have and perhaps would not wish to have sufficient political influence to change this.

What is the quality of evidence accompanying these cultural

outcomes? Research whether it is applied or not, is usually restricted in the amount of time available and applied research particularly is almost always tied to a dead-line. In these circumstances the quality of evidence must inevitably be poor and in some cases very little data is available as in refuge camps and the use of resources available to their administrators. They may not know until too late that the public dominances in a refugee camp may duplicate those of the civil war situation from which these people have fled as in the Ruanda refugee camps in Tanzania.

There is always the need to balance benefits and harm which may result from decisions which have to be made and this is often in the form of immediate distress to those involved only to be followed at some much later date by some possible benefits to almost everyone other than those primarily affected. The inhabitants of Wadi Halfa lost their homes, livelihoods and even their family graves and indeed it would be hard to quantify in human terms the effects of the millions moved because of the Yangtse High Dam.

Social scientists may not be competent to make these evaluations which go beyond their own narrower specialisations; they are not likely to be trained beyond their own microcosmic evaluations in the context of the macrocosmic world in which they are now being required to work. What indeed is likely to be the strength of their recommendations without putting their personal feelings and indeed their own careers in jeopardy when national rather than non-national researchers realise that they are not particularly powerful in the first place. These applied researchers who have been trained in the academic environment of individual thought, are now having to work in bureaucratic committee situations to which they are usually no more than adjuncts and junior ones at that; they are not likely to be influential in their own right.

Of course, as non-nationals they can just put their recommendations on the table and leave it at that because they will leave country as soon as they have done this, the fulfilment of their contracts. The national on the other hand does not leave and to the extent that they hold strong views on what they have researched, they may well become involved in internal politics. Social science researchers in applied work should realise that 'it is perfectly reasonable for governments to balance a number of interests in forming policies. Scientific evidence...between exposure and risk is only one consideration. Others include analysis of costs and benefits, risk analysis and appreciation of the degree to which policies fit with public values' (Marmot,2004).

9. The range of imaginative correlations

Outsiders undertaking social science research are reaching out from their own culture into circumstances and understandings which are largely unknown. If they are westerners, their imaginations are related to and within the genre of Shakespeare, Tolstoy and Tennessee Williams plays and so on. Onto these cultural understandings into which they have been socialised by their education, comes professional training and institutionalised correlations which are inevitably based on Western ideas, the paradigm of Western social science.

Once they are working in a culture which has definitions alien to those into which they have been brought up, they have no such understandings. They have to acquire this learning and probably only do so when it seems necessary to acquire such broader cultural understandings but that is not the same as being brought up with these ideas. The acquisitions of such understandings have to be inserted into the time and motion restrictions and the immediacies of fieldwork and the idioms of local languages.

It seems likely that they apply imaginatively the ideas of their own culture to these alien environments whether these come from Freud or Durkheim. May be then they are able to construct correlations from the centrality of Western thought onto other cultural realities such as applying Freudian ideas of Oedipus to West African relationships. This is not to do more than suggest that these correlations are right or wrong coming as they do from an established body of Western thinking, but that researchers from one culture do not have the capacity to imagine outside their own cultural understandings.

Outsider researchers in India working in a community would not feel much obligation to read the *Mahabharata* though the fascination of villagers for viewing all its twenty-six episodes on television might make them watch a few if the serial happened to be showing in the community in which they were working. They would view it but not empathize with the watchers who feel themselves to be one with what is happening on the screen. The outsiders may well feel that this is no more than some fairy tale nonsense.

How many would read Naguib Mafuz in Arabic for an understanding of Egyptian urban life or feel a need to see Oscar Wilde's 'A woman of no importance' in Arabic with the hero as an airline pilot. The use of Western television soap operas with subtitles in Mandarin, Hindi and Arabic are quite likely to be seen as influential because the outsider sees them through their own intellectual understandings. A discussion of film censorship among the Swahili in Tanzania showed that they disapproved very generally of sexually provocative costume on the grounds that it was a sign of poverty.

'The situation facing a social anthropologist who wishes to interpret a concept, assertion or doctrine in an alien culture, is basically simple. He is faced with an assertion in the local's language. He has at his disposal the large and infinite set of possible sentences in his own language and he may not be wholly happy about this situation, but he cannot avoid it. There is no third language which could mediate between the native language and his own in which equivalence could be stated and which could avoid the pitfalls arising from the fact that his own language has its own ways of handling the word, which may not be those of the native language studied and which consequently are liable to distort that which is being translated. Naively people sometimes think that reality itself could be this kind of mediator and 'third language for a variety of professional reasons this of course is no good' (Gellner,1970:24-5). It would seem that the national is going to have a better chance of accepting and understanding the thinking of those within their own cultures than the outsider who however much they may have liberal understandings are always going to see the thinking of those in other cultures as somewhat weird and nonsensical.

10. Differences in financial support

Western societies not only have long standing universities with reputations attracting financial support from donors but they often have their own financial reserves from which social science fieldwork can be supported. There are nations with viable economies which means that they have financial surpluses which can be used for research purposes. Some have government supported research committees making funds available and others have enormous charitable funds such as those of the Ford Foundation which support hundreds of research projects in the social sciences. So in these societies we have a number of social science research projects which are thought to be worthwhile, supported by viable economies which are well able to support such fieldwork. These funds are made available for research which is related to current social problems such as the study of drug addiction, juvenile crime, homelessness, and the breakdown of marriages. They also support projects such as the use of language and the nature of religious belief the results of which do not have any immediate applications. For the researchers who hope to get support for their projects it is more a question of convincing donors that their ideas are valuable; the money is there once the donors are convinced. In these countries the prospective social science researcher is starting from being a citizen in an affluent environment which subsidizes their work as part of knowledge and education being a 'good' thing even if it has no immediate relevance to contemporary issues.

In developing societies there are huge populations and their economies are often maintained in an uneasy balance between political demands for improvements in the voters standards of living and national incomes which only remain in credit with the help of international aid and the ignoring of debt repayments. Some of these nations have national support for research such as China in which the funds available for social science research are less than the amount that appears to be spent on votive candles by pilgrims to their four sacred mountains. Others such as India have charitable organisations for example the Tata Foundation who support research into social problems. But these countries have populations many times larger than any of the rich Western nations and presumably many more problems which merit the attention of social science research so that money for any project is in a poor competitive position unless it shows some relationship to immediate political as much as social needs.

In developing societies national social scientists wanting to do research have three options. Firstly, they can enter into the national need for applied work and be paid by the departments or projects which want information about how to further their predefined ends. Secondly, they can apply to their own universities for support in which these prospective donors will both have very limited funds and have an eye on the possible implications of supporting projects which will be politically safe rather than likely to earn them a reputation for irritating political power holders or finally apply to an overseas or national charitable foundation donor to provide such funds. There is always likely to be an applied edge to any project that is supported locally.

Alternatively they can always do research as a spin-off from what they are already employed to do; teachers can research school systems and doctors can do the same for the workings of hospitals which will involve a mixture of covert and overt approaches. However if they want to research fields like religion and symbolic systems which the power holders will see as having little immediate relevancy, then they will have to find a university capable of providing adequate support. Thus it would seem to be a situation in which part-time research in small scale projects is likely to be the easiest to carry out and most profitable in contributing to national understandings of behaviour.

11. The researchers and their escorts

However carefully researchers inject themselves or are injected into a research project in which they have no business to be except for their research, they are not in the first instance dependent on their own abilities; they have to have human keys to insert into the lock of communal understandings through which they gain admission to the workings of the community. In this there are differences between the nationals abilities to start working and that of the non-national who has no cultural or other connections with the communities in which they have chosen to try and work. The people who volunteer to help principally non-nationals are called informants, interpreters and research assistants but they could also be seen as escorts. They do not just materialise and they have their own reasons for doing so but they are the filter through which the community will decide their degree of cooperation.

National researchers are much better able to go their own way and seek their own helpers if indeed they need them. Non-nationals on the other hand need the help of local sponsors who have local status and an assured position. With such people on their side researchers are more likely to be tolerated by the community at large and to have access to some types of information to which otherwise would be blocked to them. Not only would such escorts smooth the way in general but provide some support for the assumed roles which the researchers have chosen for themselves. These escorts will gather informants and often point out the mistakes and shortcomings of such informants. In a society divided on religious grounds yet unified on another cultural level, these escorts are essential if the non-national researcher is not going to be pulled into segmented views while missing the overview of cultural unity (Laitin,1986).

Although escorts are usually useful in a general sense, they are particularly useful when researchers want access to a particular social group to which there is no open entry and in this the national will usually have the edge over the non-national; a worshipping group of adherents in a particular temple, church or mosque, a sports team, the drinkers in a particular bar or a school. This is particularly important when information is wanted from a street society on the fringes of legality (Whyte,1981) in which the key role of 'Doc.' was the entree to that society.

Thus there are key informants but this is not their correct classification as they have taken the researchers under their protective and status enhancing wings and thus guaranteeing to the community that both the escort and the researcher are in their own ways working for the latter's benefit. It seems likely that most non-national researchers who have achieved a measure of success in their work have had these escorts. Of course, these escorts received benefits for the help they provided, but it is better for the non-national to have their help than to attempt to work without them.

The length of time in which researchers are involved with particular people and communities makes for differences. Non-nationals have 'come and go' relationships in which emotional and social involvements are always reduced by distance; they are only of peripheral importance to most researchers kept alive by a few not particularly well written letters. For national researchers, interrelationships can be kept going much more easily and we have seen that in one case they became indistinguishable from the people they were researching (Hennigh,1981). These national researchers become so involved in what they are researching that they become in effect their own principle informants

12. The researchers representing dimensions of outside power

Most social scientists have liberal, ethical and political views or at least consider that they have them. They probably view those that they research with some tolerance and without particularly worrying about the illiberal aspects of their domestic practices, even if they consider that their behaviour and ideas are unsuitable outside that particular environment. It is not quite that they want to be loved but they certainly approach those that they research in a benign frame of mind; they would not be there if they expected to dislike the community which they were researching. This is not likely to be an important factor in how they are viewed by those whom they are researching.

Unless there is a coming home element in which the researcher is known as a particular person with roots in that community, they are always strangers and particularly peculiar ones at that. Many traditional communities accept strangers as brief transients who are expected to move on. But no community accepts strangers who stay on whether the community likes it or not.

Certainly no community has a process for accepting intrusive strangers who ask questions and behave in ways for which there is no existing model. Those researched want to provide for themselves an explanation for their presence and the easiest one to hand is likely to be that of a plain-clothes detective or some sort of government agent for undisclosed programmes which are automatically assumed will disturb them. Generally, they represent membership of an ill-defined outside power which might possibly be manipulated, but they are often best avoided. Researchers represent some part of the outside world with which there are ambiguous relationships, since they want to be part of it; an outside life which strangers show with their higher standards of living, visible skills and their assumed connections.

There is always the assumption that these researchers have political and social connections with the outside world in which the nationals are living and they are potentially more dangerous to the *status quo* with their linguistic skills and ability to move in and out of several roles at different levels of political influence. Thus there are even less reasons to antagonise national researchers but rather to try and manipulate them, whereas with foreign researchers the manipulations may be more for more personal and individual reasons.

A further difficulty separating national from non-national social scientists is their feeling of being patronized by and alienated from their Western colleagues. Although they start from different premises as professionals in their own countries, they may feel later that there is little to talk about and the nationals are almost always forced to use a learnt language, whereas the non-nationals are talking in their own natal languages (Cheater,1987).

13. The dangers of fieldwork by non-nationals

Non-nationals doing fieldwork are operating away from the cultural environment into which they have been physically and mentally adjusted and whether they like it or not any deviations from these practices are going to cause some degree of tension. The most serious of these is the long-term isolation from people speaking the same linguistic code and behaving in both known and predictable ways; it has few parallels to boarding school or backpacking with friends and the well-known gap year after completing secondary school before going to the university in which they are again likely to be with friends, doing a known job and seen as directly benefiting the people with whom they are working rather than observing. This loneliness of the non-national researcher is a most erosive factor in their well-being and a major cause of failure to complete projects.

Most of this research work is done by ethnic European non-nationals and in remote and underdeveloped areas. It is not necessarily much better for married researchers as 'they became so needy for human contacts after months alone together in the Kalahari desert that they could barely carry on a normal conversation with town residents (Owens and Owens,1984:101).

A Canadian woman researching an Eskimo community assumed that her efforts to establish rapport would eventually be rewarded but ended up with increasing irritability, resentment and failed attempts to understand either her own behaviour or that of the family members who had provided her with a home (Briggs,1992). This is no doubt an extreme case of sharing a small room with an Eskimo family throughout the Arctic winter. But the point must be that for non-nationals attempting to leap over very wide cultural divides will usually have been a leap into the dark.

For American social scientists working overseas it is reported to be a dangerous profession. In 1964 and 1976-88 for those under 65 years of age cardio-vascular deaths halved but motor vehicle deaths rose eight times and deaths from other forms of violence rose four times. In work overseas the rates of ill-health are likely to be at least the same and probably higher than that of missionaries who high rates of parasitic infections and disease (McQuay,1967; Kendrick,1974). Researchers do not have disease preventative behaviour built into their ways of doing fieldwork and they may even feel that to behave differently to those of the people they are researching adds to the already serious social distance handicaps. Kumar on finding an American researcher in Banaras living at the same level of squalor as the people surrounding him, wondered why he was endangering himself by not taking even the simplest health precautions (Kumar,1992). The higher standards of both public and private living of Western and North American professionals may make them in practical terms oblivious to the lurking presence of health problems. This may make them aware that they are ill and attribute lassitude to psychosomatic aspects of their lack of progress in their research rather than to, for example, hook-worm infestations.

14. The non-national researcher as the focus for anxieties

As the focus of social science fieldwork is on some aspect of participation centring on the researchers injecting themselves into social environments of which they are not a part, the non-national is always at a disadvantage in comparison to the national. The discussion is usually in terms of the researchers taking or being given a local role by a largely passive or not actively interested community. The researchers are self-defined as the active element in the inter-relationship and the local community is thought to be largely passive on which rather than with which they work. They can limit cooperation in the researchers' objectives in which case these strangers have to find some other ways of getting nearer to their objectives.

There is a certain innocence in the ways that these non-nationals enter the social environments which they want to research and perhaps forgetting that their presence will not only have social consequences but they expect their moral credentials to be above reproach ; they do not expect their presence to be upsetting in ways that are well outside the boundaries of their professional job description. The anthropologist who went to study the Arusha of Tanzania (Gulliver,1957) wanted to live among them in an area of high population density below the barrenness of the higher and lower levels in which population pressure had forced some to move. With the agreement of a small community, he pitched his tent on a bare patch of grass and in doing so upset the pattern of their livestock grazing. We can thus imagine that in such matters and presumably in every other possible ways the non-national and their distinctively different ways of living forced down a community is always going to be to some extent disruptive and the cause for anxiety. In this minor case of disruption the Warusha will have slotted the anthropologist's behaviour into their wider folk experience of Europeans in terms of political power rather than their social behaviour about which they would have been curious but otherwise in different. The infringements on their society of the German and British colonial administrations and the settlers who were occupying parts of their ancestral land will have been at the forefront of their minds in encountering any non-national other than a missionary. The picture then may well have been a passive if not sullen acceptance of the anthropologist as just another aspect of a wide variety of intrusions. He was not in physical danger because of his personal behaviour or because of any wider range of hostility.

The situation on the Rai coast, a remote part of Papua New Guinea into which a researcher came was totally different. He entered a community in which their perception of 'whites' was quite specific and quite outside the range of what he could achieve by personal influence and professional persistence (Englund and Leach,J,2000), since it was assumed that Europeans had magical ways of getting money from the graves of the dead, one anthropologist was frequently asked to achieve this in ways which were part of the secret knowledge of white people (Wood,1995). He could do little to avoid this as the appearance of the dead and white people appeared analogous to or linked with their affines as sources of wealth for men. These people had very little understanding of the impersonal role of money. There was also the widespread belief that 'whites' were killing local people to remove their internal organs for sale overseas. People in one village scattered on his arrival and a previously welcoming informant asked 'what are your people up to. We are afraid now., This indeed is a dangerous situation into which this well-meaning researcher came and over which he had no control and in which he became personally implicated. We have to accept that there are situations which have no rational basis but that their rationality has to be accepted as a reality and not just some part of another culture. It seems clear that the non-national can walk into situations of potential danger which they could not anticipate and which they would not necessarily be warned about by informants. The culture based clash over rationality is beyond social resolution.

East Africa has had similar problems when the practices of modernity clash with much longer standing magico-religious concepts. Many of the people in Mwanza town in Tanzania did not believe that parts of living human bodies removed by surgeons in the hospital were burnt. Doctors from the Filariasis Research Institute who wanted to collect blood at night when the parasites were present, were always accompanied by a bilingual administrative officer. The Haya on the western side of Lake Victoria had similar worries (Weiss,1998).

There are other situations in which the behaviour of non-national researchers can be understood in terms of what they are working on but not in the roles that in the minds of the researched go with such enquiries. Enquiries about witchcraft is a recognition that it exists as a fact not just as something which has functions regardless of its reality; in their minds there cannot be any abstract discussion of a reality which they see as a real not just a hypothetical danger in the lives of those with whom it is discussed. Education in itself does not get rid of the conviction that witchcraft exists and is a pragmatic known and accepted way of dealing with enemies and rivals; perhaps only those educated in the social sciences have taken on the scepticism of Western social science (Jahoda,1969). The abstract and the ability to see matters in abstract terms may be a matter of a particular form of personal cognitive development and may have no general application to a culture; it may be grammatically expressed in a different way which is not in common use.

The non-national social scientist researching the Pentecostal

Church in Malawi had the uncomfortable experience of being observed by the Pentecostal congregations rather than be accepted as a participant observer. No Pentecostal Christian he met would accept that an impartial anthropological interest drove him to participate in Pentecostal services or to have casual conversations on questions of belief. He had to have a personal thirst for 'the Words of God' . Of course, when he became ill it was stated that he would only be cured if he admitted that the Holy Spirit guided his life (Englund,1996).

These cases suggest that non-nationals through no fault of their own become involved in situations which they could not be expected to anticipate and that this resulted in additional anxieties. In these situations a national would not have had the same difficulties and to some extent he would be on the same wavelength of understandings. In Malawi a national would have known the pattern of Pentecostal thought and thus might have found it easier to integrate his or her behaviour into that of the Pentecostals thus getting the wanted information more indirectly.

15. Failure factors affecting non-national and national researchers

It is perhaps short sighted to see failure in such clear terms that can be used with non-nationals in the sense that some researchers achieve nothing for their efforts. There are always going to be failures which can be attributable to the researchers themselves and to the circumstances beyond their control in the areas in which they work. Failure to achieve what they set out to do is common enough; they may have received a grant for a specific project and found for example that it was too complex to be worked through in the time allowed. It may be that national researchers working in their own society may underestimate the difficulties in studying what may be a totally strange culture (Messerschmidt,1981) with political undercurrents which they do not wish to be associated with as nationals. Issues of conscience are more difficult to deal with for nationals than for non-nationals who can escape and publish their anxieties from the safety of their own countries. There are various forms of ascription over which a national may have difficulties. A Bengali researching in Bengal might be an example of failure because he or she would not be given the same latitude as a non-national. They would be expected to be in the know as to how to behave so they would be given little or no time to learn (Messerschmidt, 1981).

There are also difficulties for a national in researching their own or an allied community in which they will always be seen as the son or daughter of a particular family or clan when researchers want to work in their home environments. Sometimes there will be no alternative to accepting this ascribed status as a Saudi woman doing research in Saudi Arabia where there could be little or no possibility of her working independently and none at all for a non-national.

The chances of failure are higher in researching institutions which have a defined corporate identity and relatively rigid structures which are defended by their hierarchical occupants than by communities. The non-national will always be a very noticeable outsider but nationals can find that members are not prepared or allowed to discuss their own institution, and that they do not have access to documents or meetings. The researcher cannot pretend to be a member of the institution with an acceptable functional role and could not become classified as a useful friend. It is easy for the researcher to be classed as disruptive and frozen out on the grounds that no good can come of their results even when the institution itself has brought in the researcher. The national may be able to proceed by more indirect means.

In communities there will always be factions which look at the researchers in different ways and the failure to gain one form of understanding can be turned into a study of for example factionalism from a feminist point of view. Of course, researchers can write about failed projects in the same way as they would write about projects which appear to be adequately completed. However the national researcher is always contributing some understandings of their own societies which must have cumulative value.

Chapter 9

Issues in Understandings : The Problem of Language

1. The sharing of a language

The sharing of a language does not mean that there are automatically shared meanings for words or the ways that they are used and there may be several and indeed continuing misunderstandings between people according to the nature of their relationships; the sending and receiving of a verbal message is rarely if ever between people of equal understandings because of their different social and psychological environments. Even between nationals speaking the same language there are social differences and then there are both internal and external expatriates who are using languages which they have learned and finally, there is always the widespread use of interpreters, few of whom are professionally trained in the second languages which they are using. Only in the first case is there any possibility of equivalent vocabularies and in all other cases the richness of each language is rarely shown clearly in their translations. In these commonplace situations there are likely to be restricted vocabularies so that a fan of additional meanings becomes attached to each word. This is the inevitable consequence of the use and understandings of words and grammatically forms over the boundaries of particular language groups but the background of difficult concepts complicates understanding further. We can presuppose that the speakers of a Bantu language will have general thought patterns associated with that enormous group of languages from southern Africa to central Kenya. There is no reason to suppose that such Bantu speakers will have any easier understandings of a Nilotic language spoken in the southern Sudan than a Bengali Indian.

This is not the problem of simple misunderstandings which could occur in day to day relationships but the more complex ones that can go wrong within tonal languages which are particularly easy for outsiders to misunderstand and misuse difficult. The effort of the speaker to make a correctly pronounced statement can go completely wrong and if the correct message is communicated, this may be due to the listener getting the point from other clues which may be available if indeed they can be spotted at all over cultural barriers. The serious difficulties come when researchers are digging into the more important aspects of the lives of the people they are researching using the words that are heavily soaked in cultural, ethical and personal understandings (Tanner,1993); what both parties like, dislike, believe in are implicated in their understandings. The reasons for the silence of informants on key events by which they black out the horrors of their pasts (Mayaram,1996), is too often attributed to lack of intelligence.

There is always the fundamental fact that except for a large range of shared objects, there is an absence of sharable meanings so when researchers and informants talk, and there is a process of mutual interpretation rather than just processes of translation. The deeper the researchers penetrate into another culture, the greater the possibilities of getting wires crossed (Nida,1964). The listeners to anything spoken in a language other than their own is always committed primarily to their own cultural understandings of the sounds which are entering their minds. Even the simplest of nouns such as the word for 'bed' brings up the image of a bed in their own cultural settings and there is little agreement on the cross-cultural understandings of colour. Marriage has one meaning in English but among the Sukuma there are twelve different types of marriage with a different word for each (Cory,1953). When it comes to abstract words, there is little reason to suppose that listeners and speakers are even marginally aligned and great care needs to be used in using abstract words without attaching examples of their meanings in that context.

Many languages have no heritage of common meanings

stabilised through detached abstract scholarship and printing. There is little reason to suppose that a small amount of schooling over a couple of generations will change the meanings of key abstract words which are no part of the educational syllabus. Once words are put into print it is very unusual outside scientific writing and legal drafting for an author to use words with exactly the same meaning throughout their writings.

Since researchers come out of cultures and institutions with complex literary traditions, there is always the feeling that their own literate languages are superior to the non-literate ones which are being used by their informants. It is wiser to remember that the use of all languages is overwhelmingly verbal and that only a very small part of human inter-communion is ever written down, much less printed. So in listening to informants or communicating to them, it is always two verbalists at work trying to understand each other.

There is certainly no reason to suppose that any language is superior to any other within its own cultural environment and outside this there is much that any language cannot express whether we are thinking of English, kiSwahili or Hindi. Any opinion on linguistic status is probably no more than the expression of cultural superiority, a form of intellectual racism. In fact the complexity of the verb forms of kiSwahili shows up the relative poverty of English in this respect which has to use a phrase to express the Christian concept of loving one another which in that language is included in a collective verb form 'wanapendana'.

2. The categories of interpersonal language communication

The most important issues, perhaps the most fundamental, are those raised by language in social science fieldwork and the variations in what is used in language. A language can be divided into four schemas of which only the last is the most important for researchers. Firstly, there is the language of the informants. No matter how many people use a particular language, there is no overall homogeneity and every person uses that language in such a way that it is both individual and polyphonic so each informant is speaking to the researcher in a form of their language which is founded in a particular cultural environment and individualised (Aarts and Aarts,1982). Then there is the linguistic-cum-cultural understandings of the researcher and in this the national researcher with any degree of shared language and cultural understandings has an advantage over almost all non-nationals regardless of their degree of absorption and intellectual understandings of their host society. However theoretically possible it may be to become a full member of a culture other than their natal one as an adult, it has to be accepted that no one can make this psychological and social transfer from one culture to thinking and behaving correctly in another. In the uses of language the odds are heavily load against the non-national.

A national can often establish a reciprocal relationship by using a vernacular relationship term. Dua called other women by the term for brother's wife rather than sister as this relationship was regarded as sacred and to be suspicious of her behaviour would then rebound on them (Dua,1979). Finally, there is the language in which the researchers communicate their findings to their colleagues and students and in the end the abbreviated and probably foreign language in which the findings are published.

In any research the moment an informant speaks whether they and the researcher see themselves culturally as one or not, the question of language and meaning arises. A prominent feature of rapidly changing societies and the creation there of provincial, federal or unitary states, is that these people grow up into and use layers of different languages, not just the local and professional differences in how spoken English is used which exist in Britain. We have already described how an East African intellectual may have had to move himself through five languages. It seems likely that a Tanzanian who wishes to be upwardly mobile will learn three languages or indeed those who have commercial and political relationships at the simplest level at least two. So at whatever level of understanding information is being passed whether from the researched to the researchers or in reverse, there are complex issues which need to be discussed.

3. The layers of language and the national researcher

There can be little doubt that in the use of language the national fieldworker has advantages over the non-national and that this advantage continues until the results are written and presented in one of the major languages. It is at this stage that the advantages may in some degree turn against the national who may now be using a language into which he was not socialised.

The nominal sharing of a single language does not necessarily mean a semantic understanding across the whole range of social and philosophical issues, but it does make for an ease of entry into any community. The researchers' speech will however be the linguistic code of the educated. In talking and writing an educated Chinese Mandarin speaker might have the ability to recognise and use ten thousand ideograms, while a working man might not know as many as two thousand. In their natal languages the educated may always tend to overextend their linguistic understandings and usages. Thus the most important issues in fieldwork arise around the use of language, perhaps even the most fundamental one; in practice there are always languages involved rather than a single language so that it is an immensely varied scenario into which the researcher comes with their own natal language fixed in their minds.

Scientifically there are four ways of analysing any language, the sound level in phonetics, the morphology in the structure of its words, its set rules in its syntax and finally, the meaning of individual words and whole sentences as used socially. Of these four aspects only the last is part of any formal understandings and of concern to the researcher. The natal speaker is unaware of syntax, they know how a word should be pronounced and they have no interest at all in how a word is constructed. In the fieldwork situation there is the language used by informants but no matter how many people are alleged to use a particular language and this is usually a political allegation, it has no homogeneity. Every individual uses that language in an individualised way so it is both idiolect and polyphonic.

The researching listener has to understand the phonetics, the sounds which are broadcast into their ears, how the words are constructed and the set rules which have been found to exist in the formal understandings of the language which may be known only thought the formalities of learning a language; the non-national in learning a language which is not their own learns these things. The national who has spoken this language from birth will have no knowledge of the formalities of syntax and certainly does not possess or feel that he has to buy a grammar (Aarts and Aarts, 1982:6-7). Nevertheless under almost every social category the language of one natal speaker is different to that of another.

Then there is the linguistic-cum-cultural understandings of the researcher and in this the national researcher with any degree of shared language and culture has an advantage over almost all non-nationals regardless of their degree of absorption in and intellectual understandings of the people with whom they are working. Lafcadio Hearn (1894) became to all intents and purposes a Japanese with citizenship and the name Koizumi Yakumo but he did not become one any more than Cushing became a Zuni Indian (Cushing. 1978). However, theoretically possible it may be to become a member of a culture which is not one's own as an adult, it has to be accepted that they cannot make this social and psychological transfer from one culture to thinking and behaving correctly in another as adults. The odds must be always heavily loaded against the non-national.

4. The absence of a single language

Most if not all societies are polyphonic. Goa the ex-Portuguese enclave on the west coast of India has four languages in current use; Portuguese, English, Konkani and Swahili used by returning emigrants from East Africa and each is related to status and within a single language there are linguistic codes which similarly are related to status. A researcher investigating historic events within the living memories of villagers (Amin,1995:5) found that his familiarity with the local Bhojpuri dialect 'may paradoxically have accentuated the use of status markers in his conversations while at the same time have helped forge a community of purpose in his work'.

In some cases the users of an official national language are seen as outsiders or thought to be supporting an outsider political viewpoint. A Chinese who is not a Han but who speaks Mandarin may see this as a political necessity associated with control from Beijing and would rather use his own language which is not the medium though which political control is exercised. Swahili is the national language of Tanzania but it is not in the same linguistic group as that of the Maasai. In large states there are minorities with their own languages such as Romany among Travellers in Britain, while India has states with their own languages such as Tamil linked to strong feelings of sub-national nationalisms.

In some countries composed of many different small community languages such as Papua New Guinea, the international and national language is Tok Pisin (Pigeon English) which is also spoken in Melanesia generally. It has an unknown semantic content because it is assessed contemptuously by outsiders as not worthy of serious study and is often the subject of jokes about its 'quaint' phrases.

How much in fact do national researchers know of their own culture as it is expressed in language? In many field situations they have to learn to use a language in a new way as they may be only conversant with the language and behaviour appropriate to their own social circle. It is indeed difficult for the Westerner to stop saying 'thank you' repeatedly in cultures where such a phrase may not exist or if it is, it is used sparingly as it seems to terminate a relationship rather than enabling continued reciprocity.

Language in fieldwork is always related to behaviour and only independent of it in book learning. It may be more difficult for a national to learn the social intricacies of a dialect as they may think they know it already and for non-national to put into practice the fact that there are in social terms no single languages. In Ireland a woman researcher who was not living in a family may have been in some danger from not knowing the appropriate behaviour to be shown by a married or unavailable woman. Some Protestant missionaries in pre-Communist China leant their Mandarin from low status people and thus came to illustrate that low status within their own fluency. To learn another language so fluently as to avoid making a long series of social mistakes is probably an impossibility. While a non-national and indeed a national may obviously be assumed to be of higher status economically than those they are researching, their social and political status is ambiguous and constantly illustrated by their use of language and the language which is used to them.

5. Learning a new language by total emersion

During Second World War in India it proved possible to produce Japanese speakers in eighteen months under school conditions but more recently the same language has been taught by total emersion at some expense and in six weeks; it thus proved possible to learn reasonable commercial Japanese by using it ten hours a day on the basis of no Japanese spoken, no food; a harsh process. Most non-nationals learning the language of the area in which they hope to do research do not put themselves through such a process. All their work is against the clock; the length of their grant in financial terms, the amount of time for which teachers are available and their cost.

In the time available what is learnt is likely to be idiomatically restricted and probably reflects the social status of the teacher. This is particularly so with language use which reflects social status and the teacher being older than the researcher so that the phraseology and grammatical forms would be inappropriate for a non-national in their mid-twenties. Also, language learning is often at first restricted to everyday affairs rather than its idiomatic use for the specialised needs of the researcher working on religion, land use and sexual relationships; it is probably more valuable and more impressive for a non-national to be able to say with confidence 'when was the last time that you propitiated your patrilateral ancestors' than to be able to ask for your laundry to be done.

Learning a language must always be related to how much time is spent in using it each and everyday notes are rarely written in the new language and if the researcher has a partner then their conversation reverts to their natal language as soon as they get back to wherever they are living. However there are additional handicaps in the transfer of knowledge from one person to another. Whether the researcher is a national or non-national there are always additional changes in understandings as information is received from informants; what did the informant intend to say or avoid saying and what was accepted into the minds of these researchers.

So whatever the efforts put into becoming linguistically competent and the extent of involvement within a community, it seems unlikely that non-nationals can become fluent in proportion to the time that they spend in the field. Grammar can be learnt externally for the major languages but there is no published material available for many languages and even less for linguistic codes which may have no one available who is competent to teach them. Even national researchers may take some time to gain competence in the local dialects in which they have to conduct conversations with the people from whom they hope to gain information.

6. Difficulties over what language is being used

We have established that no known community has a single language which all its members use all the time and that is apart from the fluctuations which occur in every natal language depending on who is speaking to who about what and such specialised uses as are used in addressing spirits, important authorities and the poetry of songs. Even in states which have a lingua franca for official correspondence and legislation there are other lingua franca which have a limited but nevertheless almost universal use in restricted fields. KiSwahili as a commercial lingua franca is used from the East African coast up to Zaire and remains the military lingua franca in all three East African states, just as Urdu is the military language of India.

Any contemporary city contains minorities whose members speak one language at home and with friends but use the dominant language in business, school and with the local authorities. Calcutta, Accra and Kampala will each have dozens of languages in use at any one time. In a Kampala hosing estate (Parkin,1971) kiSwahili was used by people wanting to show brotherhood, Luganda by non-Luganda when wanting to show deference to the dominant community, English when speakers wanted to show their own status and a tribal dialect to show exclusiveness. In contemporary Africa and Asia there can be few children who are not to some extent multilingual for special purposes. So social scientists by learning or using a single language may be underlining a special viewpoint or keeping themselves within a particular social environment.

7. The social scientist and levels of translation

In any fieldwork situation there are continuous processes of translation going on from the simplest conversations with passers-by to that of the language used in the final report before publication under the direction of the editor. Even if what is known by the social scientist comes from their observations or even from the photographs taken and recordings made, notes are taken in the primary language of the researcher and then translated again into the language used in print. In the commonest situation, the observer and the observed talk face to face and the former abbreviates this into noting down, usually afterwards, what was relevant to their research and in terms of their own understandings.

However often the major process is the translation from one language to another and in fact into an internationally recognised language of which the commonest is English from a lesser language of much more local use. Even if the researcher and the researched share the same natal language which is the official language or lingua franca of a large nation, there will usually be the need for a translation into an internationally acceptable language in which notes are taken or which are latter translated into that language. In this process the national has the immediate advantage of speaking and reading the main local language; they have no time consuming handicap of having to learn a language of which they had no previous understandings. This learning probably takes at least six months in which the required language only is used through out the day. Then there may be a breakthrough in which the learner suddenly realises that they are speaking the new language without going through any process of translating in their heads. It is probably not a situation in which the researchers are formally taught but one in which they seek to make their lives initially more endurable and then more useful in the terms of what they are trying to do.

The natal speaker of the language to be used in the research has the advantages of knowing more of what is involved in the language and its social uses but they will still have the problems of learning the linguistic codes of the people and indeed their specialised ones for a fuller understanding of their social and psychological behaviour. Learning to use a language for an understanding of the everyday activities of those being researched is not and is never likely to be adequate for understanding of their metaphysical ideas whether this is of a Nuer leopard skin chief or a Hindu *saddhu*. Fishing and forms of spirit possession involve very different and complex linguistic understandings which will be as different as the speech patterns of lawyers, librarians and labourers. An average Sukuma will have a vocabulary of some five hundred plants of use to them in their daily lives and it will be just as complex as that of the national lawyer's language about the legal processes in which they are involved. So we see that whatever the language or languages involved in fieldwork, there are always more than one process of translation from what may have happened to what is finally taken as the 'official' viewpoint of the social scientists involved; mutual processes of filtering in and filtering out information through the minds of the researcher and the researched.

8. The turning of information into notes

Even if there is a shared natal language and the researcher and the researched are of equivalent status and the information is likely to be recorded in that language, it will be edging towards the linguistic code of professionals. But always the language of the informant will have been altered to conform to what they consider to be the understandings of the person to whom they are passing on information. If the researcher has learnt the language from the start it will always be used in the simplest forms possible; it seems very unlikely that the learner of Sukuma after six months intensive study will be using its eighteen tenses (Batibo,1985) correctly and even less likely that their use of its twelve tones will match those of a natal speaker. There will be a certain crudity in their mutual use and understandings of that language. Asian languages with their long literary history are even less likely to be adequately learnt in such short periods.

This is the stage in which information is beginning to be sorted into categories which make professional sense; the boxing in of information in ways that do not parallel the social understandings of the people from whom the information has been obtained. This will be the linguistic code of the social scientists which they do not use in their private lives. It is the end product of a process by which the language spoken by informants is turned into something else altogether and losing in the process much of the nuances of their own lives. Probably a specialised form of speaking to outsiders even before notes are taken.

Unless a tape recorder is used and in fact listened to again as researchers are often taking notes, and these are always a simplified and shortened version of what they have heard in the same way as descriptions of a ceremony which by its very nature must be complex, is going to be a short précis of all that has gone on; the writers will always be recording what is interesting to them and not necessarily what was going on in the understandings of the participants.

9. The fixing of information in writing

So we have researchers diligently filling notebook after notebook with records of what they have seen and heard and in this process what was once socially fluid becomes rigid and at once potentially independent of the person who once put these understandings onto paper. These writings are just understandings not reproductions of any exactness of what went on because once it is on paper even with the most illustrious of writers, it is no longer alive. As the researcher writes, what has been written becomes part of the professionalization of information.

The language of reports, articles and books based on research is not often one that can be easily read by the people about whom it is written nor would informants recognise much of what has been written about them if it had been translated back into their own languages from the texts in which it had become professionally embedded even if it is the researchers natal language. The final research writing is usually in a dominant international language needed for its professional use in universities, the governmental or non-governmental institutions sponsoring or the intellectual market at which the researchers are aiming. It is not in a form which is readily understood by outsiders to intellectual and bureaucratic ways of processing thought.

When documents are being considered written in one of the Romance western languages, importance and breaks in the sense of what is being written is shown by capital letters. Many other languages such as Arabic and Chinese do not have capital letters and there are many social and religious complications which come from this (Tanner,1979). There are great differences between writing about Jesus as the Son or son of God or indeed over putting the word God in print at all.

However perfect the researchers may be in understanding what is said to them and what they say to others, putting this into writing is always going to be an imperfect translation abbreviated in its extent and subject to handicaps which will have prevented it from being a quasi-photographic record of what may and indeed may not have taken place once the echoes of spoken language or seen event have disappeared (Chafe,1994).

10. Languages as the bearers of untranslatable biases

Social scientists whatever the length and extent of their professional learning are born into particular cultures and with this comes linguistic understandings which are not the result of any teaching process; it is absorbed. This is the way I express myself, it is natural even if intellectually I know that this is nonsense; one language can no more be natural than any other. There may be affinities in thinking patterns within certain language groups such as within the Romance language group, but in seems unlikely that there can be any sharing in the structures of thought between languages which have no shared structures at all such as between Burmese, Bengali and the Bantu group of which the last has speakers of similarly structured languages from Kenya to South Africa.

Indo-European languages have exclusive personal pronouns that carry with them no social categorization other than their connection with persons or groups. Other languages are more complicated than this since they carry specific social categorizations from which they cannot be separated (Suzuki,1978). The irreducible 'I' in Western thinking and social practices makes it a difficult inter-relational barrier in Japanese where it is always relational (Smith, R,1983). In Burmese personal pronouns refer to the relation of the speaker to those spoken to and the Western liberal social scientist would be uncomfortable using a superior set when talking in general and for a woman to use the inferior ones in talking to men of the same age or older than herself.

The use of the major languages has inevitable problems as in Japanese 'it is utterly impossible to form a sentence without commenting on the relationship between oneself and one's interlocutor. This language teaches one that a human being is always and inevitably involved in a multiplicity of social relationships (Kondo,1990:31). This is just one example in which by using one of the major languages, the social scientists would be conveying social facts beyond that which they are intending to convey. The national user of their own languages would know what their language was intending to convey and of course in doing so restricting the use of their research results to those in the same language group or the limited group of social scientists who have acquired a similar standard of linguistic competence.

Then there is the question of the symbols used to express languages and in which English has the advantage of having 26 only whereas Mandarin Chinese has well over forty thousand different symbols. To read a popular newspaper in English requires a knowledge of just these 26 and to read at the same level in Mandarin probably requires a knowledge of at least two thousand.

Translation in terms of complicated Western social science hypotheses is probably an impossibility, hence the point that Eastern and Western scientists are only comfortable in the shared use of mathematical and chemical symbols which have no social component. It is also the case that Chinese or Japanese concepts cannot be translated because of the paucity of the English language.

Translation is always difficult if not impossible when the language to be translated does not have the same ideas as that of the receiving language. Social scientists researching religion have particular difficulties when they use pejorative words such as 'magic' or 'witchdoctor' or when the language does not have an indefinite future tense. In general whatever words social scientists use from their own languages, their minds will be carry the meanings of their own culture and tend to interpret them in those usages in the cultures which they are studying. They are sceptics by profession and they can have no instinctive understandings of the peoples in other cultures use of magic anymore than an Englishman or a Bengali woman can understand the full implications of the winning 'haiku' in national Japanese poetry competitions.

11. The dominance of major languages and translation

Social science researchers are under a professional obligation to write up their fieldwork and to gain progressive credit in their careers as well as to fulfil their own personal ideas of fulfilment. The issue must often be the languages in which this process is inscribed and there is a triple stage process of translation. The language or linguistic code in which the fieldwork was carried out, the national or natal language in which the researchers think and finally, the language used in their reports and publications. If language is the venicle within which thought is expressed, then the researchers are the hinges round which their fieldwork is expressed; this is bound to be a filter of some inexactness in which these translation processes have varying influences.

For the national researcher the language in which their fieldwork is executed is likely to belong to the same language group as their natal language and cover a range of shared understandings and practices. Nevertheless the final work will usually be in one of the internationally dominant languages which is usually not one natally learnt by these researchers or one in which they have been professionally trained. These are important linguistic hurdles which are made more difficult to pass by the dominance of these international languages in which most professional publication is carried out.

For the non-national researcher the hurdles are different as there is little possibility of any shared linguistic and social understandings. They probably feel that the most important understandings can be more adequately and professionally expressed in a major international language in which English leads, followed by other less important but still dominant European ones. More importantly the wording used in the use of European languages becomes what the original field sources are accepted as saying although the discontinuities between the languages used would make it impossible for this to be so; they have been semantically dominated out of consideration to what may actually have been meant (Asad,1986). In addition the process of translation into print will systematize the thinking of people who are essentially if not entirely non-literate (Douglas,1966:91).

If Arab, Chinese, Japanese or Russian social scientists publish their results in their own languages, they are not likely to receive the international recognition which their intellectual content might justify; the script alone must restrict their readership to their own professional colleagues. An Indian writer has summarised these issues in the preface to one of his novels written in English (Rao,1938) 'The telling (of the story in this novel) has not been easy. One has to convey in a language that is not one's own the spirit that is one's own. One has to consider the various shades or omissions of a certain thought-movement that looks maltreated in an alien language. I use the word 'alien' yet English is not really an alien language to us. It is the language of our intellectual make-up like Sanskrit or Persian was before—but not our emotional make-up. We are all instinctively bilingual, many of us writing in our own language and in English. We cannot write like the English. We should not. We cannot write only as Indians. We have grown to look at the large world as part of us. Our method of expression therefore has to be a dialect which will some day prove to be as distinctive and colourful as the Irish or American. Time alone will justify it'.

Chapter 10

Some Conclusions on The Role of The Social Science Researcher

1. A necessary and influential inquisitiveness

Social science fieldwork is the exposure of intellectual outsiders whether nationals or non-nationals to the realities of social life other than their own which results in a personal broadening of individual understandings. This may be a worthy aim but perhaps not worth the expense of time and effort unless it is privately paid for; it is too personal.

If the fieldwork in developing societies is in anyway supported by public funds or foundations dedicated to the public interest, then there should be some close correlation to immediate public needs and understandings. To evaluate subsidised social science fieldwork as contributing new knowledge to any general understandings of human behaviour is a far too general an aim to be justified when so much of contemporary social needs remain without attention from researchers. To push back the frontiers of knowledge into anything anywhere at any time might appear to be a selfish preoccupation of no immediate discernable public value.

There is a need for relevance and there appear to be two processes by which social science could become more related to contemporary needs. Firstly, that research should be related to matters of current concern for the betterment of the lives of people and their descendants. Secondly, this research and its methodologies should link this social relevance to the understandings of the people themselves as far as is reasonably possible; there seems little point in producing research results in forms which can only be understood and appreciated by a professional elite. Every community at whatever level of economic and social development contains an intellectual elite who are well able to understand much of what is being deduced about their society. Methodologies which exclude this form of social relevance by basing themselves on scientific prototypes may not relate easily to the more immediate aims of social betterment.

On the one hand there is long-term investigative work with an almost journalistic approach to social reality as shown in work relating to the human suffering caused by the Partition of India (Butalia,1998) and more recent events such as the study of Bombay (Mehta,2004) and the artisans of Banaras (Kumar,1988). Whatever language these studies may be written in they would be appreciated by those interested in those matters; there should be a sharing of intellectual understanding and humane concern between the researchers and the researched.

At the other extreme there are studies expressed in statistical terms of information under circumscribed situations which show no more than the betting odds on such behaviour reoccurring in parallel situations which are unlikely to reoccur anyway. It may well be the case that necessarily inaccurate impressions expressed quantitatively are likely to be more influential than qualitative impressionistic accuracy because that is the form with which bureaucracies are more at home; it is both their defensive and offensive currency.

How can social science become more influential because currently in all societies it is only minimally involved when the conditions highlighted by public opinion and the political needs power-holders coincide? We have to start with our own basic professional understandings that there are very few if any facts which are not transitory and circumscribed. Unusually a whole academic profession has grown up not so much on the possible value of these transitory inaccuracies but in the hope of being able to achieve accuracy and thus to predict particular types of human behaviour. It evolved in the manipulative traditions of the hard sciences without accepting the limitations and restrictions of the data which they acquired with such care; at best the patchwork nature of whatever accuracies that have been obtained.

Whatever the reasons, social scientists from developing societies are under-represented on the editorial boards of many international journals and as sole authors of articles published in the major international journals (Saxena *et al*, 2003). One analyst classified this as a bias against the diseases of poverty (Horton,2003). This is particularly noticeable in medical articles related to tropical diseases in which only 5 per cent of exclusive authors came from societies classified as having low human development (Kelser *et al*,2004).

The social sciences have something in the nature of a choice in which direction they should develop in order to be more influential. It does seem that the scientific channel of influence is rarely likely to be available because the information is available too late and to refer to small enclaves in major societies while the human objects of research are in a constant process of changing their behaviour. Finally, the inevitable biases of the researchers are part of the information collected.

The attitude of those in power to the results of social science research are likely to be a combination of what their supporting bureaucrats consider important enough to be given to their political employers in the form of short memos and the power-holder's needs from short-term political expediency. Perhaps also there may be an attitude that avoids any suggestion that what is happening is unavoidable. The most important aspect of influence is to catch the attention of these power holders by combining research results and their presentation so that they are within the wider range of public opinion needs rather than just getting the approval of professional colleagues who remain on the fringe of decision making or influencing public opinion.

The difficulty of producing research results which might influence decision making is that this power in any society is diffused. Much of the developing world in which a major proportion of research is done, have oligarchic systems of government whether based on military coups, one party systems, fraudulent ballots or dictatorships; they are in fact almost always nepotic inward looking systems based on tribe, clan, caste, extended families or those who have gone through similar training systems whether educational, religious or military.

The impulse to change does not come from the rational discussion of policy at least partially aimed at the public good. In such situations there is little difference to 18th century England in which the political system was described as an oligarchy mitigated by riot. Collective farming in China, Russia and the Pol Pot regime in Indo-China were not the result of carefully researched policy changes any more than that of the villagisation mass movement in Tanzania with its consequences of serious declines in agricultural production. These massive changes were the result of applying the rationality of economics and factory production to highly variable and varying human behaviour outside such restricted environments.

The monumentally impressive study of contemporary Bombay (Mehta,2004) by an author who should be considered a 'half and half' researcher, puts into exciting prose more information about that city than could be covered by many social scientists. Qualitative results but the information therein is most certainly already known in general details by those in power and by implication most of those who are experiencing this particular way of living.

Of course, it is true what Gilbert White, the distinguished English naturalist who wrote in such detail about his own rural environment that 'the more concentrated the observations, the more accurate will be the recordings'. But this means at the same time that they will be less influential since such data only refers to a few people and will soon have been just part of history if anyone could be bothered to read about it. It lacks the push and stimulus of immediacy.

Most politically inspired major changes start without the social sciences being involved in the planning and in estimating the consequences. Fieldwork concerning decisions which have already been made, make social scientists unwilling partners in political processes for which research is not well designed.

2. The restrictions from terminology

The research worker in social science fieldwork is tied into the professional definitions of those three words. The word 'science' is tied to the pragmatics of cause and effect created by work on inorganic materials, 'social' by any standards is both individual and environmentally variable but the most important word is 'field' which is a concept of area.

Now contemporary science deals with veracities of lineal relationships about which there can be little doubt as to what facts are put in and what facts are taken out. There is little allowance for uncontrolled variation. By keeping the word in their job specification social scientists have defined themselves into producing results which are acceptable according to the canons of hard science. Human variations being what they are, social scientists are perhaps always unable to comply with their own professionalized expectations.

We have to accept that social factors can never be tied to exact definitions of any type and is incapable of producing any irrefutable conclusions. Any social situation is subject to immense and constant variations in the number and characteristics of those involved. There are in fact no uniformities which can be presupposed to exist in human social behaviour and when some uniformities appear to have been found, this may have been achieved by excluding from this many factors because there are no social science methodologies that can cope lineally with multiple factors.

We have to accept that the characteristics of the researchers are of constant importance as they see and hear in individual ways along the basic divide between nationals and non-nationals. They may try to exclude themselves from their own observations but they are always the variable centre but constant influence on what is seen, reported on and worked over in final reports. Then the 'field' in which their observations are made is substantially varied on a moment to moment basis so that there is a constantly changing amalgam of these three individualities.

It seems that the basic problem may be the lineal demands of the scientific mentality requiring one fact to influence its successors in an identifiable way. No methodology has been developed satisfactorily for relating multiple factors. It has to be accepted that the number of potential factors is innumerable and that quantitative studies are always narrowed abstractions. An individual sees the situations in which they are involved as researchers or as the researched, in an endless series of instant pictures; the wholenesses for which Parallel Distributive Processing is endeavouring to develop into workable methodologies.

Anthropologists have always had this workable concept of a field in which they work which has only the vaguest of definitions as to who, what, where and when are involved. Researchers who accept this qualitative approach to their work for which a scientific approach is usually going to be a fruitless endeavour and in which nationals are more likely to have an initial effective understanding as a basis for their efforts are more likely to produce data on which others can similarly imaginatively work.

3. The pattern of selectivity

No matter whether the researchers are nationals or non-nationals of whatever social and demographic characteristics, there will always be a pattern of selectivity in what they see and assess and in what they consciously at the time and retrospectively avoid seeing and perhaps unconsciously turn towards or away from in their research. No one however well trained sees all parts or persons in a community in which they are working on equal terms. Indeed in crude terms we are no more capable of assessing parts of a society in equal terms than we are of accepting equality in our own lives; inequality is part of our socio-psychological make-up however unacceptable this may be in overt political terms.

Every researcher not only has hang-ups which they will circle round if not positively avoid but positive preferences which come from their own backgrounds and research experiences. It seems likely that these are often ethical if not moral choices such as preferences for but not too close an involvement with the poor and envy as much as dislike of the well-to-do. The writers of a text book on interviewing stated that they had an irreducible preference for pretty women in fur coats and there can be no doubt at all that there are similar biases in all cultures and in all researchers.

More is known and has been written about witchcraft, an almost universal preoccupation of anthropologists and which at most can never be described as a daily activity. In comparison very little is known about the social system in every community by which they dispose of their human wastes; not only is there a distaste for such a subject despite the fact that it is universal and inevitably frequent but in its assumed crudity; it is not likely to be seen as career enhancing unless such research is seen as part of the wider topic of public health.

We are all brought up to accept some reticence in cross-sexual interactions in which there are distressful masculine and feminine patterns of public and private conversation. The researchers and their informants work within one-to-one relationships in which the ideas and practices of their separate cultures are bound to be under social and psychological strain. It is usually male researchers talking to men but rarely women except at a professional level of assumed collegiality. Women researchers are often given a quasi-male status and who are thus able to have both male and female informants but that is not the same as treating sexuality in an equally open way.

To have an abstract interesting and an understanding of commonplace behaviour which is essentially private, is not the same thing as bringing it out of this private domain into the public world of interpersonal discussion in the presence of note-books and tape-recorders. One has only to look at the range of human behaviour and the distribution of social science information to recognise that there is and always has been a pattern of selectivity. This is not only governed by personal choice within their own cultural parameters but also by the ease and difficulty off potential and actual social access to what they consider researching. Most people spend their days both in public and private according to the social acceptability of individualism, but proportionately we know very little about private activities for which we have to rely on what we are told which can never have the same validity as observed activity. Whether the national or non-national researcher is better able to intrude for that is what it is, into the affairs of others with whom by definition they will always be within some category of intruding strangers; they must always be inside such simple categorisation and remain with the personal abilities of individual researchers with their personal aptitudes for getting information and their diplomatic persistence but it will always be selective.

4. The problem of condensation

It is not so much that people are individualistic in much of their behaviour but that their group involvements are so situational that it is rarely possible to make assumptions about the similarities of such behaviour except in very general terms such as very cold weather makes riots unlikely; in fact many of these conclusions verge on what common sense would suggest as likely. Thus there is the problem that while social science has to accept that human behaviour is rarely similar enough to be grouped with any realistic hope of overall accuracy, it is almost invariably presented in categories as if they were similar, unless of course it is presented in qualitative terms in which numbers are not used as illustrations of perceived realities. We have already discussed that categorisation works well enough in some primary legal senses and in their extreme ranges but that the grey areas in which these categories merge are too commonplace and extensive to be ignored. Social scientists condense primarily in order to present their findings to others in the same profession and they do not do so in consultation with those who have been researched who only occasionally see behaviour in such abstract terms uninfluenced by personalised situational factors.

If there is any such condensation in the thinking of people at any stage of their social, political, religious and economic involvements, it is in their attitudes to people outside their personal knowledge who are grouped together by some assumed attributes for the conveniences of commonplace biases. This occurred so frequently in all societies as to be almost a constant as when the Sukuma, the English or Biharis refer to the attributes of the Maasai, the Scots and Tamils in terms which have little coincidence with social realities or indeed in the ways that most people refer to those who are demographically different to themselves by reason of age, sex or ethnic appearance.

It is easy to assume that some condensations have value as in the proportion of men to women in Musoma and Ngara areas of Tanzania which was so low as to make monogamy virtually impossible in the former and in the latter making women carry out many of the role requirements of both sexes. But to what extent do such condensations provide accurate and usable data which must be related to the age of the women, the number, sex and ages of their children, the viable and supportive nature of their extended families and the pragmatic nature of their relationships in terms of property holdings which make unions worth maintaining or indeed initiating? Widows cannot be inherited if there are no brothers to inherit them nor can mother's brothers have responsibilities for their sisters if there are none in existence.

How in fact is it possible when the data provides no obvious groupings? In a survey of Tanzanian attitudes to schools run by Christian missions (Tanner and Morgan,1976) for which a sample of 5000 were questioned from which the answers were grouped into 300 categories, but which might just as well have been six hundred and it was assumed that these were both primary and singular opinions. Nothing was known as to why each individual opinion was held or indeed whether these people ever thought in such singular rather than multiple terms. As an exercise in social accuracy it only produced answers in the form required by those making the analysis who in fact wanted to produce a simplification in what they should have known was much more complex and situational; a combination perhaps of the dominance of Western social science theory and non-national researchers.

It is worth considering the possibility that this tendency to condense data at least in part comes from the sense of superiority felt by social scientists towards those they are researching. This is shown by the compulsive way in which behaviour is condensed into patterns presumably with some idea that their future behaviour can be estimated and perhaps socially and politically manipulated for the greater good of society. We can perhaps assume that there is some collusion between social science researchers who wish to have their data in some marketable form and their principle informants who are probably stating no more than quasi-idealistic conceptions of the ways in which their society is or should be ordering itself.

Any condensations of data must be decisively inaccurate as for instance in relating high indulgence of infants to extended, polygynous, nuclear and single mother families (Whiting,1961). What constitutes high indulgence and birth order for a start and are not all families extended to some extent in emergencies and single mothers may be living with their own mothers so that there would be a tendency towards the 'normal' indulgences of alternate generations.

5. Psychological fieldwork

In sociology the terminology and practices can be moved from the language and practices of one culture to that of another and how this is done depends upon the personality and understandings of national and non-national fieldworkers and what they undertake. Most of those studied can understand in their own languages such concepts as mother-in-law avoidance joking relationships, family segmentation and the unity of alternate generations and the reasons for them. In going through the index of Evans-Pritchard's study of Azande religious practices, the Tanzanian Sukuma elders consulted were quite able to state that they did this but not that. This is easy for such people as these terms are based on concrete behaviour in particular circumstances and they are not based on abstract thought.

The major problems for both national and non-national researchers is that this way of looking at human behaviour, the bases from which social scientists work, has been provided by European and north American thinkers. Fieldwork and the writings by thinkers from other cultures came into the arena of research much later and may not yet have developed theories from different and quite easily understood base-line conceptions.

Psychology within and without a particular culture is much more difficult to transfer from one system of understandings to another perhaps because it is more complex for a variety of theoretical, personal and practical reasons. The base-lines are overwhelmingly Western. A recent summation of the possibilities of integrating contemporary psychoanalysis and cognitive science surveying contemporary models (Westen,2001) makes no mention that there may be non-Western models applicable to much larger populations than the total Western one.

Psychology is worked out in print in one of the major languages in a professional code used among the international psychological profession fraternity which would be extremely difficult to translate from the language in which it is printed into another one outside that particular linguistic group. Abstract ideas for which there can never be cross-culturally acceptable definitions are almost impossible to translate from Chinese and Japanese into German, French or English. Regarding the possible transfer into English of Chinese philosophical concepts Richards wrote (Richards, 1953:150) 'we have here indeed what may very probably be the most complex type of event yet produced in the evolution of the cosmos'.

Much of the data is supported by inferences for which correlations vary widely in their plausibility. It would be extremely difficult to translate schizophrenia for a group of non-literate Sukuma intellectuals as they would suggest such behaviour among them is the relatively common spirit possession of successful diviners and there is just no need for another definition by an outsider when they already have the diagnosis and the cures which work well enough.

Over much depends on the researcher-researched interrelationships and the extent which this is benign. In comparison to discussing mother-in-law avoidance across cultural boundaries which is relatively easy but forms of mental normality and abnormality in looking at spirit possession usually start with some assumption about normality which do not include the possibility that such people are possessed as a reality which is so common globally as to verge on normality; perhaps it is necessary to accept that spirits do exist in this case.

All these difficulties can be summed up by the fact that it is possible to discuss cognitive patters in particular cultures but that how they work out is always both individual and situational. Rationally we should surely have to accept that two Toureg men even if they were identical twins, are inevitably dissimilar in how they react to the social realities surrounding them; why should they be possessed by similar spirits and who can possibly state that the spirits are similar.

There are indeed more important difficulties that divide the national and non-national in psychological research more than in the other social sciences in that the current base lines are European and date at most from the Enlightenment. Eastern cultures are based on ideas recorded in writing that probably antedate the former by a millennium at least. By and large Western thinking is that these concepts such as the Chinese thinking about Yin and Yang are at best archaic and related to the practices of those in fringe cults. We may be forgetting scientific fashions (Kuhn,1962) and the varying faiths which we have in Freud, Jung, Adler and Klein. Thus non-national researchers approaching the social science understandings of Chinese, Indian and Japanese start from a barrier of deep-seated cultural biases. It may not be any part of their training in psychology to question in these circumstances the foundations of their training.

National researchers who are working largely within their own cultural environment but not so much speaking as thinking within these same paradigms as those whom they are researching, will have no process of translation going on in their minds and no great leap over any cultural divide. They are aware already that the bases of their thinking which ultimately came to be expressed in writing, are already available in Pali, Sanskrit, classical Arabic and the complexities of classical Chinese with its forty-four thousand different ideograms expressing different words. Moreover Western social science theoreticians manoeuvre around individualism. The individual is so important in Western thought and practice that it has taken on the practical characteristics of a moral and necessary 'good' and that their group involvements can rightly be considered a varying handicap on its proper development.

We can conclude that national researchers working in their home environment can certainly carry out valuable psychological work, but that in the Western world it has a short-term perspectives. In the non-Western world national researchers have a head-start within the intricacies of their own complexities which begin from different philosophical understandings. Moreover they have the advantages of being able to work with and to develop very ancient psychological paradigms for which there is ample written background. The non-national cannot compete with what is provided by their own relatively recent written paradigms.

6. The layers of involvement in social science research

In comparing national and non-national researchers in the context of developing societies, this ends up as an assessment of social science research by Westerners of non-Western societies. As we have seen few non-Westerners researched Western societies, except in the context of white Americans working on American Indians, while there are few if any non-Western societies which have not been visited by Westerners for research purposes. It would of course be incorrect to see this in racist terms, but research by Westerners in non-Western societies is almost always a necessary step for academic advancement; their use for research could be seen as an export which does not figure in the balance of payments.

We can assume in the absence of any conclusive evidence to the contrary that every society has a proportion of intellectuals not to be confused with educated literates, capable of fulfilling the demands of social science research. So can we not assume that if the United Kingdom has 'x' number of researchers, the United States might well have '4x' and China with a population of over a billion might well have '16x'. This is just to tentatively suggest that in marketing terms that Asia rather more than Africa is the field in which social science fieldwork may well develop. But of course the numbers of social science practitioners does not depend on the factor of intelligence in isolation, but on other prevalent issues principally cost and socio-political access.

It seems likely that the majority of social scientists publishing in a major international language about research in developing nations are non-nationals. These people can be described as primary since they instigated the research projects and arranged for the financial support. However in a growing number of projects particularly those that are more narrowly scientific there are trained and active nationals as secondary researchers co-equals in authorship with the primary ones.

Beneath these qualified and aspiring national and non-national predefined leaders of any research project there are large numbers of tertiary nationals without whom no research anywhere can be carried out. Even if we accept that these people were helping primary and secondary researchers for entirely personal ends, they remain the building blocks on which research is constructed. More particularly if we discredit them for their economic and other motives, in what ways are they different to the primary and secondary researchers who are not exactly working for disinterested motives?

We know little about these essential people since in the acknowledgements prefacing publications they are often grouped rather than given names. It would seem reasonable that anyone who spends more than a few hours in providing negative or positive information for a researcher should be acknowledged somewhere as a contributor whether they have been paid or not. There are usually fulsome acknowledgements to other academics who may have spend far less time on reading a draft than some distant elder or specialist who may have spent days in discussions with these researchers.

Very few non-nationals can afford to have another non-national in the field as a research assistant unless they are long-term partners who in effect work for nothing as was the case with Colin Turnbull and his companion in his Uganda work. There have been certain partnerships in which both worked on interrelated projects such as the Bohannans' and Margaret Mead's second marriage but still there are the rings of tertiary helpers without which they would have achieved little. Very little social science research has involved the primary researchers finding their own data without the help of secondary and tertiary nationals.

7. Differences in patterns of thinking and of history

We have already looked at the spectrum of bias which is built into human behaviour by our biological and psychological abilities and from the ways in which we are socialised. In the generally accepted sense it is not 'good' to be biased; it is seen by relatively unknown others as a social and psychological failing and it is not assessed as a form of necessary conformity.

It seems that this is a more individual way of looking at what goes on and does not cover cultural aspects of how we evaluate our own behaviour and that of others. After all the definition of bias (Oxford English Dictionary,1956:1,175/6) is to swerve from the right line and having a bias is to be unfavourably inclined which covers both individuality as well as cultural predefinitions of what is appropriate. However, a later edition of the Oxford Dictionary of English (Oxford Dictionary of English,2003:160) defines bias as 'inclination or prejudice for or against one person or group, especially in a way considered to be unfair'. So bias is deviation from something which is defined as 'right' or 'fair', moral adjectives which have immensely variable meanings so a bias in one social environment might not be one in an adjacent situation. The distaste of the social scientist for bias must come in part from their assumption that there are facts to be unearthed about human behaviour and that bias is a deflection from this.

It would seem that what facts there are in social behaviour if they are not physiologically based and in some way inorganic, are more the result of bureaucracy than part of 'normal' social interrelationships. According to statutory law in most societies people are married or not as a result of set procedures and getting a signed certificate to that effect. We can assume that nothing else on the form constitute facts in the same category of exactness.

It may well be that individuals are almost always implicitly concerned with their day to day affairs and another explicitly formulated world inherited from the past and uncritically absorbed (Gramsci,1971:337) in which social scientists are over concerned with the latter on the assumption that it dominates the former. When we come to consider whether bias has any overall meanings rather than being a purely personal tool modifying the possibilities of finding exact realities, we should be considering whether there are any overarching conceptual realities which channel understandings in particular ways independent of more individual social circumstances. It may be more valuable to have this as an idea to explore than to reduce the realities of biases to such factors as age and sex.

Let us examine 'bias' in the context of Gramsci's concept of 'common-sense'. Everyone in every social situation has views and behaviour which are amalgams of common-sense in which this is in accordance with time, place and persons with whom they are associating for that particular environment in which there are certain persistent themes which correlate with social and psychological distancing. Thus in the defined sense of bias varying from a norm, there is the snag in attaching a pejorative meaning to that word, in that norms never can have or ever have had any unilateral constancy except in some abstract ideological sense. In this approach then bias has to be assessed as so individualistic as to be out of agreement with those around them in the social circle of whoever makes the statement and thus are indeed difficult to assess except in almost pathological terms. In this the views of a hermit, paedophile or door to door salesman are not biased in any immediate social environmental sense although clearly biased from the points of view of other levels of social and psychological understandings.

There would seem to be therefore three levels of common sense understandings and related behaviour. Firstly, that of the individual and their immediate social inter-relationships; primary common-sense. We have to get on with the people with whom we live in the same apartment, house, Hindu joint household or group of huts within a single fence. Then secondly, common-sense understandings related to the community in which individuals live and finally, tertiary ones relating to the overall culture in which individuals and communities are contained. Thus the ideas of bias should be seen in relationship to these three levels of common-sense. The community which is out of functional agreement with the overall culture such as most if not all religious communities and the individuals and their families who are out of common-sense order with both their community and culture, such as criminal groups.

So bias in the sense of a non-conformist viewpoint would appear to be a constant in all social situations, but that of both national and non-national researchers have an additional form of bias because in the vast majority of fieldwork situations where they are detached from the social situations in which bias and common-sense operate together in socially acceptable ways; they are not by definition at home in primary, secondary or tertiary environments. They are isolated not only in the general sense from the three levels of common-sense to which they have been habituated in their own lives but are expected professionally to detach themselves from these ways of thinking and behaving which can only be a theoretical possibility. So there is a fourth professional level of normative-cum-common-sense behaviour based on the social impossibility of scientific detachment. This is perhaps the basic and most persistent of role conflicts for researchers in the requirement of objectivity which is paralleled by the persistence of social as much as psychological common-sense attachments at all three levels.

All researchers are burdened by their own liminal and subliminal understandings which are inserted into the same three tiers of common-sense norms. Even the national researcher is only going to be attuned to the cultural levels of common-sense but at the individual and communal level they may well be total strangers in terms of their understandings of and acceptance of primary and secondary common-sense norms. The abatement of this can only be by the physiological similarities of ethnic appearance, age and sex and possibly biological abnormalities. We have to conclude that in these four senses of common-sense both national and non-national researchers are persistently and comprehensively biased as individuals.

Instead of bias being assessed as ethically questionable and professionally unacceptable both national and non-national researchers should initiate their fieldwork with a fundamental understanding that they are seeing their work through biased perspectives. These characteristics are minimally at least ten which are accepted as records of channelling thought and behaviour in particular ways; age, sex, religion, education, ethnicity, location, status, occupation, marital status and biological factors limiting participation. These are exterior biases which the researchers bring with them into their work whether they are aware of their extent or not. These will not directly reflect the three levels of biased normalities as they become detached from the social environments in which they normally function. Then facing them are the same inevitable biases in reverse which exist in all informants which will similarly show their primary individual normative common-sense biases and their secondary and tertiary ones according to what information they are providing.

8. The uses of bias as common-sense normality

We have suggested that bias in both national and non-national researchers, their adherence to common-sense social norms at three levels of normality and with a fourth level related to professional ideology, means that they have their minds and behaviour split four ways by the requirements of fieldwork. Consciousness of this normative division would initially prevent researchers from being too tied into the social realities of any close personal ties with strangers; the type of Goffmann involvement with primary relationships. Researchers may tend to cling to the social settings into which they are first introduced; it is difficult enough to get into a strange community in the first place without shifting to another one once the elements of domestic support have been organised. On the other hand entering fieldwork with an adequate understanding of its cultural elements at the tertiary level from reading published works leaves them with an inadequate understanding and acceptance of the reality that few people are predominantly governed by the standards of their cultures. They adhere to a culture and often express it in ideal terms in talking to outsiders, but they habitually break its rules.

The non-national researchers with their systematic categorization of social behaviour may tend to see the inconsistencies and contradictions as part of their professional logic. The national researchers sharing some of their understandings with their fellow nationals are more likely to appreciate the come and go nature of social and psychological affiliations. Their fellow nationals see no inconsistencies with beliefs in certain things at the tertiary level which is practiced in a different way at the secondary level and which form no part of the practical aspects of life at the primary one and indeed in a reversed ascending scale of social and psychological commitment.

The social science researcher may accept this for the people they are involved with and indeed on reflection see it in their own behaviour and beliefs; the schizoid nature of religious belief and practice. What complicates the researcher's approach is their proto-scientific need to make their conclusions logical; they accept role conflict and its solution as logical. They find it difficult to accept the normality of holding contradictory beliefs for which there are no solutions and which are retained at the same time as different forms of normality. The social science researcher, particularly non-national ones should learn that illogical thinking is part of the normality of common-sense behaviour and thought. We should accept the normality of 'swerving from the right line' which is imposed by both culture and professionalism because an overall right line cannot exist in any social life and psychological understandings.

9. The distinctions between national and non-national researchers

Like all categorisations the distinctions between national and non-national social science researchers is based on examples at the extremes of what they represent; Malinowski as a non-national working in the Trobriand Islands (Malinowski,1935) at one extreme and Kumar as a national working in Banaras (Kumar,1992) at the other. There is no confusion as to under which heading these two distinguished researchers should be classified. Although the distinction may be quite clear when the differences are between a Bengali researching Banaras or Brussels and a Belgian doing the same thing for Brussels or Banaras, this is to simplify the social environments in which these categories can be applied. There can be few communities now anywhere in the world in which it might be even theoretically possible; all researchers even in their own countries are likely to be to a large extent internal expatriates by reason of language, education and social understandings of what is normal behaviour. Cities these days will have not so much tribal enclaves which have always existed such as the Jewish ghettoes in European cities prior to second Word War and the current Chinese areas in Liverpool and London but in the increasing separation of Hindus and Muslims in Indian cities. All contemporary cities whether in Africa, Asia or in the Western world now contain enormous numbers of people from divergent backgrounds so that in the context of research the categories of national and non-national in the conditions of fieldwork are virtually useless.

An investigative journalist (Hall,2005) or should he in reality be described as a social scientist, went to live for a year in the East End of London in an area dominated by Bangladeshi immigrants. When this Englishman tried to rent a room from a Shia Muslim letting agent he was told'you are obviously educated and this is not a place for a person like you. Everyone here is criminal. He describes his neighbours as Sikhs, Jews, Albanians and Afghans and indeed from any other category which can be thought of which must be the same today in Mumbai, Calcutta, Accra and New York. His experiences and sensitivity duplicate Jack London's account of the same area in 1903 in his 'The People of the Abyss' (London. 1903). In such a situation the category of national as opposed to that of non-national must be irrelevant when in that environment he was as much a non-national as he would have been living in a Mumbai slum.

Most researchers whether national or non-national combine some of the characteristics of both while the 'half and half's' are at least committed to trying to practice from both view points at the same time. Both are committed to finding out as much as they can without bias and in the main they have studied from the same text as and along the same theoretical lines; they have no difficulty in communicating with each other providing that they both use the same major language which means that much Asian research is not available in the West and *vice versa.*

Both types of researchers are attempting to carry out forms of fieldwork which are in practice impossible. The non-nationals cannot in fact ever detach themselves from what they are observing in order to make their observations scientific and the nationals can never become so much a part of those they are working with as to be indistinguishable. All researchers in whatever ways they categorize themselves or are categorized, are caught in the dilemma of being inevitably and permanently outsiders while at the same time wanting to be insiders; the human clashing with the proto-scientific. Then of course there is always the difficulty that the non-national researchers usually look different, dress differently and talk differently; they are not as efficient as plain clothes policemen in becoming inconspicuous.

Both forms of research whether assessed in terms of researchers or methodology, have profits and liabilities in the persisting conditions of developing societies in which the social science research is an almost necessary part of attempts to accelerate development. Whether or not this is an economic or a social rather than a politically justified necessity is not a part of the social science remit.; they are committed to making known possible elements in human behaviour which have wide-spread significance and of some immediate advantage to all those involved in human betterment.

10. The dominance of demographic factors

It is obvious enough that researchers are what they look like about which nothing can be done and that these obvious characteristics restrict what they are able and not able to do in the context of their attempts to approach their projects in neutrality. The first of these is the sex of the researcher which automatically splits what they are able to do into male and female segments with knowledge of the opposite sex being obtained indirectly or from public participatory observations. Equality of the sexes if such a thing exists at all anywhere may not extend much beyond academic ideology in day to day practices.

The second restriction comes from age since the majority of researchers would be classified in the community in which they are working as immature adults with little understanding or experience of the 'real' world unless they have achieved status by marriage and in the case of women, motherhood. The older the researcher the greater access they may have to those who hold power in a community or institution. It is unreasonable to think that older people will automatically like being asked questions by immature 'outsiders'.

Then there is also the plain fact of ethnicity; that all researchers look as if they belonged to a specific ethnic group with the whole set of stereotypes which are attached to being classified as a north as opposed to a south Indian or Chinese, Bantu, Nilotic, San or Nilo-Hamite. This classification may aid inclusion in some communities but is inevitably an alienating factor in working outside ethnic boundaries.

11. The researcher as a superior outsider

There is also the basic fact that the national as much as the non-national is always an outsider unless an individual undertakes covert research with all its ethical and personal difficulties or is working within a purely family dimension study of their own communities but still with the restrictions of age, sex and to whom they are related and by what degree of closeness. There is always the shared bias from being seriously educated and it is probably inevitable that the over-educated in terms of years spent in training makes them see those who have not been through the same prolonged process as being somewhat lacking in the elements of cleverness and understandings which they have achieved by a combination of birth and effort.

12. The classification of nationals

Whereas non-national researchers are easily classified into a single of group of people who have a passport and usually a visa and permission to do work on a specific topic in a specific area in order to enter the country, classifying nationals is far from being so simple and they cannot easily be lumped together by an identity of interests or nationality.

This way of classifying researchers in a mass can only apply to small nations or sub-divisions of a nation which have distinct social and political if not religious identities; a Balinese or Bengali working on their own island, city or province might be assessed in a unitary sense. But a Javanese researcher working in Sumatra or a Punjabi researching in Pallavaram south of Madras would share very little other than nationality. They would have nothing in common except possibly the national language of their bureaucracies, and even shared nationality is not a serious factor in most people's lives except when there are external threats which override more local and individual interests.

It seems nationals may have certain assets for research projects such as easier access to both topic and area, knowledge of how officialdom works and the ability to make adjustments within their bureaucracies. Beyond this in many other aspects they are internal expatriates sharing many difficulties with external ones being assessed by those they are researching as foreigners. They have to learn a local language and they are usually outside the range of any ethnic togetherness and of course they may be many hundreds of kilometres from their natal areas where they might have had some built-in advantages.

13. The debatable existence of any unitary knowledge in cultural thinking

There is the divisive nature of social science training not so much that it divides the researchers from the researched, but that its format is Western created and most non-Westerners are trained along these same lines. This is an off-shoot of the European Enlightenment which is taught as if it was the only way in which human behaviour and societies can be understood. A pattern of belief that is based on the assumption that assessments of human behaviour have to be done in dichotomous lineal and sequential terms of 'yes' or 'no' and that individuals cannot be two 'opposing' things at the same time.

Eastern thought might well think that it is not quite so clear given their dislike of philosophical and theological definitions and of any isolation of primary reasons as sufficiently adequate for behaviour when it may well have been initiated in far more distant factors. A general feeling that multiple simultaneous influences are better able to be understood as creating and influencing human behaviour which Western thought is beginning to investigate under the title of Parallel Distributive Processing. It seems likely that nationals from cultures with many centuries of independent thought will develop social science frameworks of thinking which are not so closely confined to Western models. Further that this will develop along the lines of nationals preferring qualitative rather than quantitative Western non-national methods.

14. The non-national researcher: an overview

The primary factor in the possibility of non-nationals doing fieldwork in another country must be political. Newly independent nations do not like nationals from the western nations doing research which may reflect on their inability to develop as quickly as their political hopes may have framed so decisively; they want to know in advance what is going to be researched and researchers who are let in have time limited visas. Academic freedom is subject to considerable circumspections and they may be sure that they are under observation by the police.

Whole countries, areas within countries and subjects may be closed off from social science research either because non-nationals will not be let in or because their own countries advise against going there. The atmosphere of restriction and covert supervision may not encourage researchers to even attempt such work. At best their research is constricted within short periods of work in the field and the difficulties of continued contacts within the communities with which they have worked-up personal knowledge and a circle of personally committed informants. So much of non-national research programmes are one off periods of fieldwork within a particular framework of getting something done within the time allowed which might in a number of cases be described as academic tourism; they are in a hurry. Even if this is partially true it has to be assessed in terms of the researchers' own career ambitions within their own countries and do not involve any long term commitment to the countries which have provided the information on which these same careers are based. They cannot plan on any assumption that they will be coming back again and again to this one community so that their research cannot incorporate change in persons, families and institutions. Much research is produced in terms of a static moment in time, which is socially unreal.

However, they do bring into the field and into the academic contacts that they may make in the countries in which they are short-term guests, viewpoints from another culture and another set of academic understandings which cannot be based on the reading of books, so that such international and possibly ongoing interrelationships are a valuable stimulus which comes from their differences. The non-national is also expensive to keep in the field and the return on substantial grants may be a low percentage in terms of the outlay which involves fares and medical support in emergencies; they have backups against misfortune. They are also often without the same amount of supervision which they might experience as support in working their own countries.

They are on the other hand not only well funded and independent of local power structures which may give them a biased view of the realities of local life but they will always be materially and very visibly better off than the people with whom they are working and thus make them the object of much petitioning for and far from disinterested offers of help. On the whole they are younger rather than older and so their research work does not benefit from previous fieldwork and of course they take a long time to get acclimatised to the conditions under which they will have to work, excessive heat, damp, debatable hygiene and lack of privacy as well as the almost total absence of easy associations with people of their own culture. Many will not have previously experienced being so totally alone in terms of the sociability to which they have been accustomed in the university environments in which they have spent their last few years. There is of course the advantage that they are not so likely to have family commitments which handicap the time that the older researcher can be actively working away from home.

The most important handicap is that of language in which the non-national unless he or she is a 'half and half' immigrant, is unlikely to reach bilingual competence in the time allowed by their visa even allowing for preparatory work. So that possibly more than a year of the field trip will be spent in getting a fairly common knowledge of the language but not one in which they might approach the bilingual ability necessary to be able to identify and understand cultural nuances. This means that a substantial part of the grant is spent in learning back-up which does not have much to do with a complex research topic and this means that usually the researchers have a handicapped dependence on interpreters who themselves are not really bilingual apart from the reality of their own biases. While non-nationals may not necessarily get ill they are almost always going to have problems which come from being forced to eat strange food because there is no other and cooked moreover under conditions which they may not appreciate. They will often have to live under conditions which are markedly different to that they would have experienced in their own countries. Non-nationals are also not likely to have local contacts which make their connections into the community to be researched which cuts down on initial introduction times. Of course, there will be local academic support but they are not likely to pass onto strangers their own contacts which are often very personal.

Of course all academics whether they admit to themselves or not are likely to have a sense of intellectual superiority to the people whom they are researching but in the case of non-nationals coming from western highly developed societies this is likely to be allied to having a sense of cultural superiority as well. Perhaps a more general failing of non-nationals is their assumption of superiority regardless of age and sex in which they approach local people expecting cooperation when in terms of the local culture they are socially inferior to the people from whom they are expecting cooperation. This in part comes from the tendency to imagine that lower material complexity in many societies than they would expect to find in their own, is correlated to an assumption than these societies are similarly simpler altogether rather than just as complex as their own.

As non-nationals they will not have been socialised into ways of getting round the difficulties of local bureaucracies or understanding the intricacies of the local power structures and how it might be manipulated. As newcomers they are more likely to be biased towards those who become their earliest informants to whom they are drawn by misunderstanding over the nature of friendship which is not based on abstract goodness of heart but on reciprocity.

Since they have not grown up in those communities into which they are now placed for research they are more likely to have illusions about their capability to have abstract understandings. They do however have the advantage that they can accumulate information which might endanger nationals and in leaving the country at the end of their allowed fieldwork period this can be publicized in safety. There is one other disadvantage in that most non-nationals come from secular intellectual backgrounds and often assess the religious activities of these communities in terms of their intellectual frailty rather than as a valuable part of their individual and collective lives. As we will see later the long-term advantage of non-national researchers to a developing nation lies in the stimulus they can provide and in the building up of local in stitutions.

15. The national researcher : an overview

The primary restraint on nationals doing social science research work in their own country must be political; they have to live with what they publish and once they get a reputation for a wide range of political involvements, their names will keep on coming up in computer searches. Developing societies with unstable political regimes and the recourse to violence as solutions to their difficulties do not like their own nationals researching into areas in which their failure to fulfil political promises are made public by their own intellectuals rather than their political opponents. Politicians who have risen quickly to power are usually suspicious of their intellectual superiors who they suspect are also contemptuous of them. Academic freedom is an ideal which is maintained by a considerable amount of self-censorship about what is researched and printed. They will certainly be on the books of the 'political' police who have the job of reporting on subversion, if national academics work consistently in such sensitive areas as voting fraud, corruption in high places and inefficiency.

Whole countries, areas within countries and subjects may be closed off to social science research because nationals will not be let in unless they already live there such as the high Himalaya border areas of China and India which are of social research interest because of polyandry and the effects of altitude on human behaviour. However nationals are in a better position to do research which would be barred to non-nationals because they are better able to work their way round official barriers and to wait for opportunities to do what they want rather than being tied to the impatience of their own career time-tables.

Their research is not so constricted to short periods of work in the field which are hard to rearrange for further visits from abroad; they can come and go relatively easily to the research areas in which they have developed interests as in the Bhasin's interest in the Himalayan region and Kumar's in Banaras; national researchers are often no more than a car or train journey away from their field specialisations. There will be a build up of personal knowledge and their own circle of informants who are personally committed to them. So national research is not necessarily one-off periods of fieldwork within a particular framework of visas and finance and getting something down on paper before the time there runs out; they need not be in a hurry and they can thus observe change as it goes on rather than taking a more static view of social processes.

Their commitment to research is related to careers in their own countries and can involve long-term commitments to communities on which their careers have become based; a much more personal development of relationships than could be possible for non-nationals who might correspond with some informants when either of them wanted something. However they do bring into the field their own long term local academic contacts as well as viewpoints which are part of their own cultural understandings from socialisation and partly from academic training within the culture in which they will continue to work. It may be difficult for them to break free of the social categories to which they belong whether of tribe or caste but, of course, they have the major advantage of looking much like most of the people with whom they are likely to do research

The national researchers are less expensive to keep in the field and much research can be done within their salaries and work commitments by using their own students and the opportunities for research which start from just outside the gates of their university. Small grants can show a higher return of completed research because they can be more closely supervised. The national researchers will also have their own familial backups against misfortune while they may not necessarily have better health than non-nationals. However, they are certainly less likely to, suffer from the results of eating strange food, and to live under conditions which are markedly different to what they are used to without the options of being able to organise something better.

On the other hand although they are not well funded which must restrict costs other than that of talking and walking, and they are better able to be independent of local power structures so avoiding a biased view of the realities of local life. However they will always be materially and visibly better off than the people with whom they may be studying and subject to endless importuning by people thinking that they have influence in other places or have plentiful sources of money. Nationals are better able to work their way round the petitions of people asking for help without giving offence.

On the whole national researchers may be older rather than younger and so their research work does benefit from previous fieldwork trips and of course they do not have the prolonged period of acclimatisation which is inevitable for non-national researchers. They will not suffer from the social isolation of non-nationals since they will usually be able to find acceptable associations with people of their own culture and similar status. They may well have the disadvantage of family commitments around which it may be somewhat easier to organise their research as was the case with Kumar's ongoing Banaras work in which partners or spouses can be supportive.

The most important asset of the national researcher must be language in which they are likely to be bilingual although of course they may have to learn local linguistic codes but these will usually be within the major national languages in which they are already fluent. So less time will be spent in getting a socially competent knowledge in which they would be able to understand, identify and comply with cultural nuances. This means that less time has to be spent in learning the cultural back-up which is part of any complex topic, and that national researchers are less handicapped by dependency on interpreters who themselves are never as bilingual as they make themselves out to be, apart from the constant difficulty of their biases.

Contemporary developing societies probably have as much dependency on networks of influence based on villages, families, classificatory kinship, in-laws, occupations as well as sub-caste or class as what may have been the case historically. So, there is always the possibility of national researchers having some tenuous connections with the community in which they want to work which may be usefully developed. Non-nationals have some ethical difficulties over manipulating influence which in these developing societies may be the only way to get things moving to the advantage of their research.

Just as non-nationals they may be unable to disguise their sense of intellectual and cultural superiority to those they are researching and we have seen how in some cases researchers have adopted an approach that informants have an obligation to provide information and indeed are expecting cooperation from people who are politically inferior to the class of the researchers; this unwise approach comes from the nationals being unable to adopt a role which is independent of pride in their own families and education which the non-nationals would be unlikely to make. Nationals should be better able to work within expected cultural obligations. As nationals they will have been socialised into ways of getting round the difficulties of local bureaucracies as well as understanding the intricacies of local power structures. They are less likely to be tied into those who had become their earliest informants because they have a better understanding of the nature of friendship as being based on reciprocity not abstract idealism.

Since they have grown up within the broader culture in which they are researching, they are less likely to have illusions about their capability to understand their work in terms of abstract theories detached from any understanding of the macro-structures into which they have been socialised. But they do have the handicap that they cannot accumulate socially dangerous information and then publish details from the safety of another country. They also may not have the disadvantage of coming from secular intellectual backgrounds and thus they are likely to asses religious activities as a valuable but not out of date intellectually restricting part of individual and communal life. As we will see later the long-term advantages of national researchers to a developing society lies in the stimulus they can provide and the supportive environment which they can provide for non-national research efforts.

16. Unitary thinking and Eastern social science

The distinction between Eastern and Western thinking is clear enough in some of the earliest philosophical and theological writings and since it predates social science evaluations, it is surely necessary to accept that there are long-standing differences which are the foundations of subsequent thinking. A Chinese thinker Xu Dashou wrote (Gernet.1985;208) 'the barbarians say I believe in eternal life. It is easy to see from this that they have understood absolutely nothing about the meaning of the word 'life'. Another Chinese thinker Huang Zhou (Gernet.1985;205) also wrote that 'the mode of action of Heaven and Earth can be summed up in a word; it is not double'. We may question whether this thinker meant that the upper case should be used when it does not exist in Chinese. It would seem therefore that Eastern thinking feels itself superior to the more recent Western patterns of thought and moreover their materialism undercuts their claim for some equality and that (M)man and (S)spirit cannot be viewed in dichotomous terms.

The early Jesuits spent a great amount of time trying to make Chinese literati understand that the natural soul was of a substance radically different from that of the body and inanimate things and that the soul was the exclusive privilege of mankind. They however got nowhere and the Chinese thinkers concluded that the Jesuits were materialists since they deprived the universe of indivisible form, turning it into brute matter directed from outside and lacking the spontaneous intelligence that all creatures display (Gernet,1985:146,203). It is not our intention to initiate a discussion of these philosophical differences but they nevertheless have substance.

Some commentators would go so far as to state categorically that the social sciences in their research have to drawback from an assumption that it can imitate the natural sciences. Its open nature, constant variations and innumerable influencing factors prevents any possible foundation for accuracy with its absence of definitions and the porous nature of its physical and psychological data. Karl

Popper stated (Banton,1964:99) 'the triumph of social anthropology is the triumph of pseudo-observational, pseudo-descriptive and pseudo-inductive generalising methodology and above all marking the triumph of a pretended objectivity and hence an imitation of the methods of natural science'. If this is so, then Eastern thought which has none of the logical consistencies of Thomistic thinking, is in a better initial position to study human behaviour outside any hope of imitating natural science which it would not be attempting to copy in the first place.

Above these methodological issues national researchers who have from birth a general knowledge of their own large scale complex societies with long literary and philosophical traditions would mean that within the thinking of Chinese, Indian and Japanese intellectuals, there are deep elements of these almost coercive understandings. These would be impossible for non-national researchers to acquire and certainly not as far as the implementation of these ideas in the practicalities of their fieldwork (Nakane,1982).

17. Cooperative efforts in research

Cooperation between researchers within one academic culture has always potential personal problems since the topic has the background of individual careers which would be particularly difficult with researchers of equal age and experience. It may well work best when an obviously senior person leads and the junior benefits from attachment to a star who does not see the other as a potential rival. What then of cooperation between nationals and non-nationals which might well have advantages to all concerned since their career patterns are within their own academic cultures and they do not see their joint work in competitive terms. Cooperation would have significant advantages. The non-national would get access to fields of enquiry which would otherwise be closed to them or would at least take significantly longer to work themselves into. An immediate benefit would be that they would not have to spend so long in language learning. For the national there would be the advantages of better access to research funding and at the very least there would be the benefits of the support and encouragement from an interested partner with viewpoints and expertise from another cultural viewpoint.

A case of fruitful cooperation was the study of urban pollution in parts of Cairo and a village for which funds were provided by a Canadian source (Mehanna and Hopkins,2000). This study researched people's attitudes to garbage and sewage in their streets as well as air and water quality and noise, involving the opinions of both men and women and participant observation. The difficulties that a non-national would have had in doing such a study do not have to be listed and the predominance of the national researcher and sensitivity to his role is shown in the order of authors' names. There can be no doubt that both authors and their institutions and the national culture within which the work was done benefited. A valid example of research work that could be done by national and non-national cooperation would be the commercial work of women (Perry,2002). There would be a sense of shared identity between such women researchers as well as a shared need to illustrate the ways in which women have achieved some degree of economic independence.

There are two main areas for cooperative research efforts. Topics which start with people in various circumstances and which may or may not be correlatable to laws and environmental situations. Then areas of research which start with the fact that such a law, situation or international convention exists, for example, the existence of a new mass disease, the installation of automation or the wearing out of a natural resource to which the behaviour of the people involved can be correlated.

In the first situation there are always problems related to the role of each researcher and there can be no assumptions that a Chinese, Indian, Peruvian or American researcher would be approaching the topic in the same way; the interrelationships are too personal and occur in different social and psychological environments. In the alternative research scenario, social scientists start with facts that exist about which there can be little disagreement and the work starts from an agreed foundation. Thus the Pakistani or Ukrainian researcher would be able to state from their viewpoints what would be the consequences of these facts. In both these situations the research process would benefit from the mixture of nationals and non-nationals as well as differences in age and sex and prevent exclusive viewpoints closing off alternatives.

One to one relationships in a research project can work themselves out within the privacy of such associations but the difficulties of a group or committee approach to a topic are more complex and more public. Personal interrelationships come into prominence which have little directly to do with the research which has nothing much to do with the fieldwork problems which come from different backgrounds. Time is taken up with discussions about approaches which may have a tenuous application to particular situations and of course costs rise with the numbers involved. Social scientists are too individualistic to make cooperation an easy approach in such situations.

18. The way forward in international research cooperation

We must first set aside, as far as possible, issues of nationalism since the flow of researchers is from the developed societies with relatively stable and surplus economics as far as the backing of research goes, with their surplus of social scientists looking for fieldwork opportunities and experience outside their existing professional and personal understandings. The number of non-nationals seeking to do research in these developed states is almost always from similar states and few if any come from developing nations except as a by-product of requirements for an already subsidised higher degree.

The flow of non-national researchers then is from the developed nations into developing ones with new ideas of national dignity and widespread thinking that they are being manipulated generally in the interests of Western society. Thus there are covert feelings of resentment and overt bureaucratic restrictions which govern both their entry into these countries and what they are permitted to research, which is probably not a feature of cross-university association. So nationalism is a factor which non-nationals have to circumvent and cope with in and around their fieldwork.

We have seen that both national and non-national social science researchers have both advantages and disadvantages which mirror each other, but on balance the nationals ones have the most advantages but are handicapped by shortage of financial backup and library support even with the availability of computers which are indicative of data sources but rarely supply enough information to be useful for researchers. Nevertheless despite these inevitable differences both have the same general objectives to widen the understandings of different parts of the one society for the primary benefit of its people, while still each working for the furthering of their own careers.

In the simple sense of keeping this national distinction, there would appear to be few advantages beyond purely personal academic selfishness. The distinction is restrictive as it does not enable the best use of financial and personal resources for what must surely be considered as the objective of all research, the betterment of a particular segment of mankind. If we take the example of a research project conducted by two Americans by birth and one Indo-American on the availability of health care in Udaipur, Rajasthan (Banerjee *et al*,2004),we find an immense amount of information about the widespread symptoms of disease and that 41per cent of those claiming to be doctors had no medical qualification. Whatever help this team received from Indian academics was not shown in joint authorship of the report. So we have some thorough research work done by non-nationals who then departed taking with them their expertise, having dumped their report into the laps of the provincial government and the local health authorities to react in any way that they see as politically relevant and economically affordable. There was apparently no contribution to the building up of any local research institution.

Surely a rule of thumb for research in countries which have limited budgets for research, would be to have most if not all research initiated on the basis of some equality in senior staff numbers between national and non-nationals. This would enable a flow of research funds into developing nations while the equality of researchers would be that both types would have access to stimuli from the thinking and cultural understandings of their opposite numbers. This would be particularly valuable for non-nationals so that they are exposed to other forms of methodologies and thinking from their professional colleagues. They would also have easier access to fields of research by the help of their local colleagues who would know their way round the labyrinths of local bureaucracies.

At the same time non-nationals would gain from the cultural understandings of national researchers with whom they are likely to be closely associated and by this sort of interactive association they will be avoiding the rather persistent local opinion that they are getting their own personal profits from whatever research they do. This cooperation also gets over the age divide with the younger non-national benefiting from the older nationals advice over the problems of status differences in their contacts. Indeed advice generally of how best to fit into the communities and culture which must be strange to outsiders, thus shortening the acclimatisation period.

If there is language learning it seems likely that the national partner will be able to advise on the appropriate linguistic code for the project, short-cuts to the more appropriate and difficult aspects of a language rather than what might be called the touristic vocabulary. It is perhaps more useful to be able to enquire about the intricacies of spirit possession and ritual and mother-in-law avoidances than to be able to ask how cooking is done.

With such cooperation the national researcher benefits from assured financial support and access to wider academic resources. Just knowing that the people in other cultures think differently is not the same as encountering this in personal relationships in which the points of difference and agreement can be talked out in encounters which may last some time and which may well continue over a series of research projects and field trips and some years of collegiality. The main difficulty over this type of long term collaboration is that university training and more particularly post graduate work is an isolating process. Cooperation in research is a learning process and is particularly so in fieldwork when there are few others nearby to offset the problems of close and contrary associations.

What kind of cooperation can be created and developed between researchers of different nationalities and different protogenetic patterns of thinking, but shared professional aims and understandings? An important preliminary consideration must be some recognition and understanding of the professional dignity of the nationals involved which may not be given due consideration when much of the findings and material resources may come from non-national sources.

What are the advantages of cooperation with a non-national for national researchers? Any cooperation is likely to be stop and go so that there are not the complications of joint work between nationals who might not be able to avoid each other socially when there are professional differences of opinion about their joint work. Secondly, the non-nationals are likely to belong to or be connected with universities or institutions which are well-funded, stable, not subject to direct political pressures and which will usually look favourably on this form of international cooperation.

The national researchers have access to research facilities and academic backup which are provided as a matter of course for fieldworkers whom they support. Above all the non-nationals with whom they are cooperating come from different cultures and the covert and overt biases which they carry with them should not be seen just as regrettable aspects of Western and non-Western socialisation. The bases for a constantly active and stimulating dialectic are between different and valid points of view. Above all these non-nationals whatever their achieved social and ascribed physiological status, provide automatic differences with which the nationals have to cope rather than simply classifying them into a rag bag of unacceptable views based as much as anything on political prejudices.

What does the non-national gain from cooperation with specific nationals. There is an immediate compulsory social nearness to a national which in the field would require months if not years to acquire, and the immediate access to the national's knowledge of likely behaviour and plausible explanations of the background to much of the seemingly unusual behaviour of the people with whom they are working. For the non-national the nationals provide innumerable short-cuts which would otherwise have wasted much scarce time.

If the cooperation in a joint project exceeds a simple twinning and the research is in fact run by a committee composed of equal numbers of nationals and non-nationals, then there are likely to be cultural difficulties. The Westerners there would need finite decisions to be reached by vote as quickly as possible which does not fit easily into the practices of other cultures in which people talk out difficulties into compromises rather more than clinical solutions.

Most social scientists aim to help the societies in which they work and since most nationals and non-nationals work in developing societies, it is the form of research most appropriate to these conditions with which we should conclude. This is low cost applied rather than theoretical fieldwork about matters which involve cross boundary issues and inter-cultural cooperation.

Throughout this study it has become clear that national social scientists do not have the resources to carry out expensive time-consuming research without funding, while the non-nationals working in isolation are perhaps in danger of becoming a focus of animosity of those who do not have the advantages of isolation and abstraction from the realities of living in poor countries with uncertain and corrupt political systems.

The way forward must surely be to increase the amount of research with funding depending on the necessity of cooperation so that both nationals and non-nationals benefit from working together. The non-national should see that a major part of their contribution is towards the building up in those countries of the institutional capacity to do research and thus building up an equality of productive and published research which is justified by their intellectual potential.

Bibliography

(The Numbers in bold faces following each reference give the text page on which the article or book is cited)

Aarts, F. and Aarts, J. (1982). *English syntactic structures: Functions and categories in sentence analysis*. Pergamon Press, Oxford. ***381, 383.***

Ablon, J. (1977). *Field method in working with middle class Americans*. Human Organisation. 36; 69-72.

Abrahams, R. (1981). *The Nyamwezi today*. Cambridge University Press, Cambridge, ***337***.

——, (1994). *Witchcraft in contemporary Tanzania*. African Studies Centre.

Abu-Lughod, L. (1988). 'Field-work of a dutiful daughter' in Altorki, S. and C.F. El-Solh (eds). *Arab women in the field. Studying your own society*. Syracuse University Press. Syracuse. 139-161. ***290.***

Agar, M. (1996). *The professional stranger*. Academic Press, London, ***66, 67, 70, 71.***

Aguilar, J.L. (1981). 'Insider research. An ethnography of a debate'. In D.A. Messerschmidt. (ed). *An anthropologist at home in North America*. Cambridge University Press. Cambridge. 15-26.

Almy, S.W. (1977). 'Anthropologists and development agencies'. *American Anthropologist*. 79(2);280-292. ***294.***

Altorki, S. (1982). 'The anthropologist in the field. A case of 'indigenous' anthropology from Saudi Arabia'. In H. Fahim.(ed). *Indigenous anthropology in non-Western countries*. Carolina University Press, Durham, North Carolina. 167-175. ***4, 245, 275-6.***

Amadiume, I. (1993). 'The mouth that spoke a falsehood will later speak the truth. Going home to the field in Eastern Nigeria'. In D. Bell et al. (eds). *Gendered fields: Women, men and ethnography*. Routledge. London, ***339.***

Amin, S. (1995). *Event, metaphor, memory*: *Chauri Chaura*.1922-1992'. University of California Press. Berkeley, ***205, 383.***

Anand, M.R. (1986) *Untouchable*. Penguin, London, ***156.***

——(1994). *Coolie*. Penguin— London.

Anderson, N. *et al*.(2004). 'National Sectional study of view on sexual violence and risk of HIV infection and AIDS among South African school pupils'. *British Medical J*. 328. ***144, 151.***

Appell, G.W. (1978). *Ethical dilemmas in anthropological enquiry*: *A case book*. Cross-Roads Press. 'Valtham, Mass. ***94.***

Arensburg, C.M. and Kimball, S.T. (1940). *Family and community in Ireland*. Harvard University Press, Cambridge, Mass. 213.

Asad, T. (1986). 'The concept of cultural translation in British social anthropology'. In J. Clifford & G.E. Marcus.(eds). *Writing culture. The poetics and politics of ethnography*. University of California Press. Berkeley. 141-164. ***393.***

——, (1993). *Genealogies of religion*. John Hopkins University Press. Baltimore. ***35, 284.***

——, (1998). Introduction to Asad, T.(ed). *Anthropology and the Colonial encounter*. Humanity Books. Amherst, New York. 9-19. ***44.***

Ashkenazi, M.A.(1997). 'Informant networks and their anthropologists'. *Human Organization*.56(4).471-478. ***332.***

Augost, O. (2005). 'One final victim of the Rape of Nanking'. *The Times*.17 March, London.

Baker, T.L. (1998). *Doing social research*. McGraw-Hill, New York.***185.***

Bannerjea, M. (2000). *The Pathan unarmed*. Oxford University Press. Karachi. ***116, 237.***

Bannerjee, A. *et al*.(2004). *Health care delivery in Rajasthan*. Massachusetts Institute of Technology, Cambridge Massachusetts. ***46, 430.***

Banton, M.(1964). 'Anthropological perspectives in sociology'. *British J. Sociology*. 15;95-112.***10, 426.***

Bartlett, F.C. (1995). *'Remembering': A study in experimental and social psychology*. Cambridge University Press, Cambridge. ***178-9.***

Barua, I. and Devi, A.(2004). 'Women markets in Manipur. An anthro-historical perspective'. *J. Human Ecology*.15(2);129-133. ***346.***

Batibo, H. (1985). 'Le keSukuma. Langue Bantude Tanzania'. Editions Recherche sur les Civilisations. Paris. ***389.***

Baviskar, B.S. (1979). 'Walking on the edge of factionalism. An industrial co-operative in rural Maharashtra'. In M.N. Srinivas *et al.*(eds). *The field worker and the field*. Oxford University Press. Delhi.154-200. ***128.***

Beals, A.R. (1970). 'Gopalpur.1958-60'. In G.D. Spindler.(ed). *Being an anthropologist. Fieldwork in eleven cultures*. Holt, Rinehart and Winston. New York. 32-57. ***325.***

Beattie, J. (1965). *Understanding an African kingdom*. Holt, Rinehart and Winston. New York.

Bell, R.M. (1950). 'The Maji Maji rebellion in Liwale district'. Tanganytika Notes and Records. 28;38-57. ***7.***

Bellwinkel, M. (1979). 'Objective appreciation through subjective involvement. A slum in Kanpur' in M.N. Srinivas *et al.*(eds). *The fieldworker and the field*. Oxford University Press. Delhi. 141-151. ***49, 184.***

Bera, S. (2004). 'Food and nutrition of the Tibetan women in India'. *The Anthropologist*. 6(3);175-180. ***97.***

Bernard, H.R. *et al.*(1984). 'The problem of informant accuracy. The validity of retrospective data'. *Annual Review of Anthropology*.13;495-517. ***178.***

Bernardi, L.B. (1959). *The Mugwe. A failing prophet. A study of a religious and public dignitary of the Meru of Kenya*'. Oxford University Press. London. ***62.***

Bernstein, B. (1974). 'Sociology and the sociology of education'. In J. Rex. (ed). *Approaches to Sociology*. Routledge and Kegan Paul, London. 153-170. ***283.***

——,(1975). *Class, codes and control*. Routledge and Kegan Paul, London, ***317.***

Berreman, G.D.(1962). 'Behind many masks'. Society for Applied Anthropology Monograph.4. Ithaca. New York. ***361.***

——, (1989). Chipko; a movement to save the Himalayan environment and people' in C. Borden.(ed). *Contemporary Indian tradition*. Smithsonian Institute, Washington, DC. ***293.***

Beteille, A.(1965). *Caste, class and power*. University of California, Berkeley. ***88, 113, 361***.

Bettelheim,B.(1986). *The informed heart*. Penguin, Harmondsworth. ***304***.

Bhasin, N.V. (1988). *Himalayan Ecology, transhumance and socials organisation. Gaddis of Himachal Pradesh*. Kamla-Raj, Delhi. ***422.***

——,(1996).*Transhumants of Himalayas. Changpas of Jammu and Kashmir and Bhutias of Sikkim*. Kamla-Raj, Delhi.

——,(1999). *Tribals of Ladakh. Ecology, human settlement and health*. Kamla-Raj. Delhi.

Blackwood, B. (1935). *Both sides of the Buka passage. An ethnographic study of social, sexual and economic questions in the Solomon Islands*. Clarendon Press. Oxford.

Blanchard, K. (1977). 'The expanded responsibilities of long-term informant relationships'. *Human Organisation*.36;66-69. ***162.***

Bochner, S. *et al*.(1975). 'Introduction'. in R.W. Brislow *et al*.(eds). *Cross-cultural perspectives on learning*. Wiley. New York. ***150.***

Bourgois, P.(2002). 'The violence of moral binaries'. *Ethnography*.3(1);221-231. ***354.***

Bohannan, P. (1981). 'Unseen community; the natural history of a research project'. In D.A. Messerschmidt.(ed). *Anthropologists at home in North America*. Cambridge University Press. Cambridge. 29-45. ***288.***

Booth,C. (1889). *Life and labour of the people of London'. Williams* and Norgate. London. ***310.***

Bowen, L. (1964). *Return to laughter*. Doubleday. New York. ***199, 325, 341.***

Braroe, N.W. and G.L. Hicks. (1967). 'Observations on the mystique of anthropology'. *Sociological Quarterly*.8;173-186. ***229.***

Brass, T. (2003). 'The sabotage of anthropology and the anthropologist as saboteur'. *J. Anthropological Society of Oxford.* 180-186. ***25.***

Briggs, J. (1970). 'Kapluna's daughter'. In P. Golde.(ed). *Women in the field. Anthropological experiences.* Aldine, Chicago. 19-44. ***183, 185, 219, 231, 372.***

——,(1992). 'Mazes of memory. How a child and a culture create each other' in W.A. Corsaro and P.J. Miller.(eds). *Interpretive approaches to children's socialization.* Jossey-Bass. San Francisco. 25-50.

*British Medical J.*328; 1459. (Editor.2004). 'Editorial on the nature of scientific interests'.

Browne, K.E.(2001). 'Female entrepreneurship in the Caribbean. A multinational pilot investigation of gender and work'. *Human Organisation* 60(2). ***326-342.***

Buechler, H.C. (1969). 'The social position of the ethnographer in the field' in F. Henry and S. Saberwal. (eds). *Stress and response in fieldwork.* Holt, Rinehart and Winston. New York, 7-19. ***164, 273, 361.***

Burton, R.F. (1893/1964). Memorial edition. 'Personal narrative of a pilgrimage to Al-Medina and Meccah'. Dover. New York. ***83.***

Burton-Jones, N.G. and M.J. Konner. (1973). 'Sex differences in behaviour of London and Bushmen children' in R.P. Michael and J.H. Crook.(eds). *Comparative ecology and behaviour of primates.* Academic Press. London, 689-751. ***311.***

Busza, J. *et al.*(2004). 'Trafficking and health'. *British Medical J.*328;1369-1371. ***68.***

Butalia, U. (1998). *The other side of silence. Voices from the Partition of India.* Penguin. Delhi. ***165, 177, 396.***

Castenada, C. (1968). *The teachings of Don Juan. A Yaqui way of knowledge.* University of California Press. Berkeley. ***9, 153, 244.***

Cernea, M. (1982). 'Indigenous anthropologists and development oriented research' in Fahim, H.(ed). *Indigenous anthropology in non-western countries*. Carolina Academic Press. Durham, 121-137. ***203, 255, 282***.

Chafe, W. (1994).*Discourse, consciousness and time. The flow and displacement of conscious experience in speaking and writing*. University of Chicago Press. Chicago. ***390.***

Chagnon, N.A. (1983). *Yanomamo. The fierce people*. Holt, Rinehart and Winston. New York.

Chakavarti, A. (1979). 'Experiences of an encapsulated observer. A village in Rajasthan' in M.N. Srinivas *et al.*(eds). *The fieldworker and the field*. Oxford University Press. Delhi. 38-57. ***78, 115, 136.***

Chang, I. (1998). 'The rape of Nanking. The forgotten Holocaust of World War Two'. Basic Books. New York. ***304.***

Chaudhuri, N.C. (1951). *The autobiography of an unknown Indian*. University of California Press, Berkeley. ***226.***

——, (1959). *A passage to England*. Macmillan. London. ***26.***

——, (1967). *Intellectual India*. Associated Publishing House. New Delhi. ***156.***

——, (1976). *Culture in a vanity bag*. Bombay, ***156.***

Cheater, A.P. (1987). 'The anthropologist as citizen; the defracted self' in D.C. Rubin.(ed). *Autobiographical memory*. Oxford University Press. Oxford. 164-179. ***47, 371.***

Cheesmond, A.K. and Fenwick, A. (1981). 'Human excretion behaviour in a schistosomiasis endemic area of the Gezira, Sudan'. J. *Tropical Medicine and Hygiene*.84;101. ***160.***

Chen, N. *et al*. (1995). 'Social functions and adjustments in Chinese children; a longitudinal study'. *Developmental Psychology*.31;531-539. ***251.***

Chilungu, J.W. (1976). 'Issues in the ethics of research methods. An interpretation of the Anglo-American perspective'. *Current Anthropology*.17(3);457-481. ***220.***

Chrisman, N.J. et al.(1999). 'Community partnership research with the Yakama Indian Nation.' *Human Organisation*.58(2).134-141. ***253.***

Clifford, J.(1982). 'Power and dialogue in ethnography. Marcel Griaule's initiation' in G.W. Stocking.(ed). *Observers observed. Essays in the ethnographic fieldwork.* University of Wisconsin Press, Madison, Wisconsin. ***111, 121.***

——, (1986). 'Introduction; partial truths'. In J. Clifford and G. Marcus.(eds). *Writing culture. The poetics and politics of ethnography.* California University Press. Berkeley. 1-26. ***181.***

Cohen, Y.A. (1982). 'Studying a nation state anthropologically'. In H. Fahim.(ed). *Indigenous anthropology in non-Western countries.* Carolina Academic Press, Durham. North Carolina. 193-201. ***143, 282.***

Colson, C. (1982). 'Anthropological dilemma in the late twentieth century' in H. Fahim.(ed). *Indigenous anthropology in non-Western countries.* Carolina Academic Press. Durham. 253-262.

Condon, R. (1962). *The Manchurian Candidate.* McGraw-Hill. New York. **109.**

Conklin, B.A. (1995). 'These are our own bodies, this was our custom. Mortuary cannibalism in an Amazonian society'. *American Anthropologist.*22(1);75-101. ***182, 329.***

Cory, H. (1953). *Sukuma Law and Custom*'. Oxford University Press. London. ***42, 137, 379.***

Cory, H. & Hartnoll, M.M. (1945). *The customary law of the Haya tribe*'. International African Institute. London. ***42.***

Cuisenior, J. (1982). 'Indigenous anthropology and cultural heritage' in H. Fahim.(ed). *Indigenous anthropology in non-Western countries.* Carolina University Press. Durham. 61-71. 267.

Cushing, F.H. (1978). Zuni. Selected writings. J. Green. (ed.). University of Nebraska Press. ***383.***

Daily Telegraph Newspaper. (March 3rd 2003). London.

Dambisya, Y.M. (2004). 'The fate and career destinations of doctors who qualified at Uganda's Makerere Medical School in1984;a retrospective cohort study'. *British Medical J.*329;600-601. ***56.***

Deepa, K.S. *et al.* (2004). 'Seasonal variations in nutritional states of adolescent girls'. J. *Human Ecology.*15(2);147-151. ***133.***

Deng, F.M. (1971). *Tradition and Modernisation. A challenge for law among the Dinka of the Sudan*. Yale University Press, New Haven. ***255.***

——, (1972). *The Dinka of the Sudan*. Holt, Rinehart and Winston. New York.

Desai, A. (1980). Clear light of day. Baumgartner. Bombay. ***156.***

Devisch, R. (2005). 'The villagisation of Kinshasa as people's means to domesticate rampaging violence' in N.S. Kastfelt.(ed). *Religion and African civil wars*. Hurst. London. 102-118.

Dharwadker, V. (2003). 'The historical formation of Indian-English literature'. In S. Pollock.(ed). *Literary cultures in history. Reconstruction from South Asia*. University of California Press. Berkeley, 199-267. ***234.***

Diamond, J. (2005). *Collapse. How societies choose to fail or survive*. Penguin. Harmondsworth.

Diamond, S. (1964). 'Nigerian discovery. The politics of fieldwork'. In Vidich, A.J. *et al*.(eds).*Reflections on community studies*. Wiley. New York, 119-154. ***211, 271-2.***

Diver, R. and Patel, D.(2002). *Cinema, India. The visual culture of Hindi film*. Reaktion Books. London. ***156.***

Douglas, J. (1972). *Research on deviance*. Random House, New York.

Douglas, M. (1966). *Purity and danger*. Routledge & Kegan Paul. London. ***199, 393.***

Dragadze, T. (1987). 'Field-work at home-the USSR'. In A. Jackson.(ed). *Anthropology at home*. Tavistock. London.154-163. ***35.***

Drangert, J.O. (1993). 'Who cares about water, Household water development in Sukumaland, Tanzania'. Linkoping Studies in Arts and Science. 85. Linkoping, ***79, 174, 229.***

Dua, V. (1979). 'A woman's encounter with Arya Samaj and Untouchables' in M.N. Srinivas *et al*.(eds). *The fieldworker and the field*. Oxford University Press. Delhi. 115-126. ***91, 115, 357, 381.***

Dube, S.C. (1973). *Social science in changing society*. Lucknow University Press. Lucknow. ***40, 255.***

Dworkin, R. (1976). 'Taking rights seriously'. Duckworth. London. ***43.***

Economist, The. (1904).'Counting people. Where is everybody'. 27 November;35. London.

Elwin, V. (1947). *The Murria and their ghotul.* Oxford. University Press. Bombay.

Engle, P.L. *et al.*(1997). 'Smoke exposure of women and young children in highland Guatemala; prediction and recall accuracy'. *Human Organization*. 56(4).408-417. ***178.***

Englund, H.(1996). 'Witchcraft, modernity and the person. The morality of accumulation in Central Malawi'. *Critique of Anthropology*.16;257-79. ***84, 376.***

Englund, H. and J. Leach.(2000). 'Ethnography and the meta-narrative of modernity'. *Current Anthropology*.41(2);225-248. ***374.***

Esmail, A.(2004). 'The prejudices of good people'. *British Medical J*.328;1448-9. ***114.***

Evans-Pritchard, E.E. (1940). *The Nuer*. Oxford University Press. Oxford. ***2, 41, 43-4, 45, 54, 74, 78, 225.***

——, (1956). 'Nuer Religion'. Oxford University Press. Oxford. ***110, 401.***

Fahim, H. (1977). 'Foreign and indigenous anthropology; the perspective of an Egyptian anthropologist'. *Human Organisation*. 36; 80-86.

——(1979). 'Field research in a Nubian village; the experience of an Egyptian anthropologist' in G.M. Foster et al. (eds). Long-term field research in social anthropology. Academic Press. New York. ***287, 362.***

——, (1982). 'Communication among anthropologists across non-Western countries' in H. Fahim.(ed). *Indigenous anthropology in non-Western countries*. Carolina Academic Press. Durham, North Carolina. 138-151. 40, 127.

Faris, J.C. (1998). 'Pax Britannica and the Sudan; S.F. Nadel'. in T. Asad.(ed). *Anthropology and the Colonial encounter.* Humanity Press, Amherst, New York, 153-170. ***41.***

Fei Hsiao Tong. (1939). 'Peasant life in China' in B. Pasternak. *A conversation with. Current Anthropology*. 29(4). 637-662. ***248.***

Feldman, K.D. (1981). 'Anthropology under contract; two examples from Alaska'. In D.A. Messerschmidt.(ed). *An anthropologist at home in North America.* Cambridge University Press. Cambridge. 223-237. ***291.***

Finlinson, H.A. *et al.*(1999). 'Access to sterile syringes by injection drug users in Puerto Rico'. *Human Organisation* 58(2).201-211. ***253.***

Firth, R.(1969). 'Forward'. in J Beattie and J. Middleton.(eds). *Spirit mediumship and society in Africa.* Routledge and Kegan Paul, London. 1-XIV. ***153.***

Firth, S. (1972). 'From wife to anthropologist' in S.T. Kimball and J.B. Watson.(eds). *Cross-cultural boundaries. The anthropological experience.* Chandler. San Francisco. 10-32. ***189, 230, 232, 343.***

Fisher, J.F. (1996). 'An interview with Dor Bahadur Bista'. *Current Anthropology.* 37(2);349-356. ***168, 287.***

Fleuhr-Lobban, C. (1994). 'Informed consent in anthropological research. We are not exempt'. *Human Organisation.* 53(1).1-10. ***99.***

Freeman, D. (1983). Margaret Mead and Samoa. The making and unmaking of an anthropological myth'. Harvard University Press. Cambridge, Mass. ***51, 139, 164, 330.***

French, R. (1997). 'The National Native Title Tribunal and the Native Title Act. Agendas for change'. In G. Meyers.(ed). 'Implementing the Native Title Act. The next step. Facilitating negotiated agreements. Selected discussion papers of the National Native Title Tribunal.1996'. National Native Title Tribunal. Canberra. ***298.***

Fuller, C.J.(2003). *The renewal of the priesthood. Modernity and traditionalism in a South Indian temple.* Princeton University Press. New Jersey. ***80, 92.***

Gans, H.J.(1982). 'The participant observer as a human being. Observations on the personal aspects of field work'. In R.G. Burgess.(ed). *Field research. A source book and field manual.* Allen and Unwin. London. 53-61. ***212, 326.***

Gardiner, M. (1984). *Footprints on Malekula. A memoir of Bernard Deacon.* Salamander Press, Edinburgh.

Gellner, E. (1970). 'Concepts and society' in B.R. Wilson.(ed). *Rationality*. Blackwell. Oxford. 18-49. **366.**

Geertz, C. (1973). 'Linguistic etiquette' in J. Pride, and J. Holmes.(eds). *Sociolinguistics*. Penguin. Harmondsworth. 169-79. ***74, 274.***

Gernet, J. (1985). China and the Christian impact. A conflict of cultures'. Cambridge University Press. Cambridge. ***426.***

Glazier, S.D. (1993). 'Responding to the anthropologist. When the Spiritual Baptists of Trinidad read what I wrote about them'. in C.B. Bretell.(ed). *When they read what we write*. Bregin & Garvey. Westport, Conn.36-48. ***111, 115.***

Goffman, E. (1969). *The presentation of self in everyday life*. Penguin. Harmondsworth. ***54, 210.***

Gold, A.G. (1988). *Fruitful journeys. The ways of Rajasthani S. pilgrims*. University of California. Berkeley.

Gold, R.L. (1969). 'Roles in sociological observation' in G.S. McCall and J.L. Simmons. (eds). *Issues in participant observation*. Addison-Wesley. Reading, Mass. ***182.***

Golde, P. (1986). Introduction. in P. Golde.(ed). *Women in the field*. Anthropological experience. Aldine. Chicago. 75-85. ***341.***

Goyle, A. *et al.* (2004). 'Nutrient intake of children residing in squatter settlements on pavements and along road sides in Jaipur city'. *J. Human Ecology*.15(2);143-146. ***133.***

Gramasci, A. (1971). *Selections from The Prison Notebooks*. Lawrence and Wishart, London. ***410.***

Griaule, M. (1965). *Conversation with Ogotemeli*. Oxford University Press. London. ***121.***

Green, R.H. (1981). 'Magendo in the political economy of Uganda; pathology, parallel system or dominant sub-mode of production'. Discussion Paper 164. Institute of Development Studies. University of Sussex. ***228, 266, 289.***

Grinker, R.R. (2000).*In the arms of Africa. The life of Colin C. Turnbull*. St Martin's Press. New York.

Gronewold, S.(1972). 'Did Frank Hamilton Cushing go native'. In S.T. Kimball and J.B. Watson.(eds). *Cross-cultural boundaries. The anthropological experience*. Chandler. San Francisco. 33-50. ***187, 213, 218.***

Gronhaug, R. (1974).'Micro-macro relations'. PhD mss. University of Bergen. Norway.

Gulliver, P. (1957). Interim Report on land and population in the Arusha chiefdom. Tanzania National Archives. Dar es Salaam. ***43, 373.***

Guppy, N. (1958). *The Wai Wai. Through the forests north of the Amazon.* Murray. London. ***349.***

Gupta, K.A.(1979). 'Travails of a woman fieldworker. A small town in Uttar Pradesh' in M.N. Srinivas et al.(eds). *The fieldworker and the field.* Oxford University Press. Delhi. 103-114. ***91, 345.***

Gwaltney, J.L. (1981). 'Commonsense and science; urban core black observations'. In D.A. Messerschmidt. (ed). *An anthropologist at home in North America.* Cambridge University Press. Cambridge. 46-61. ***176, 285.***

Haimendorf, F. von C. *The Sherpas of Nepal. Buddhist Highlanders.* University of California Press. Berkley.

Hall,. T. (2005). *Salaam Brick Lane. A year in the Near East End.* Murray. London. ***414.***

Handler, R. (1993). 'Fieldwork in Quebec, scholarly reviews and anthropological dialogue' in C.B. Brittell.(ed). *When they read what we write.* Bergen & Garvey. Westport, Conn. 67-74. ***143.***

Hashmi, J. (1998). 'Banished' in Memon, M.U.(ed). *An epic unwritten.* Penguin. Delhi. 191-194. ***157.***

Hau'Ofa, E. (1975). 'Anthropology and Pacific Islanders'. *Oceania.*45(4);283-89. ***118.***

——, (1991). 'Anthropology at home. A South Pacific Island's experience'. In Fahim, H.(ed). *Indigenous anthropology in non-western countries.* Carolina Academic Press. Durham, North Carolina. 213-222. ***312.***

Hearn, Lafcadio *alias* Koizumi Yakumo. (1894). *Glimpses of unfamiliar Japan.* ***383.***

Hennigh, L. (1981). 'The anthropologist as key informant inside a rural Oregon town' in D.A. Messerschmidt.(ed). *The anthropologist at home in North America.* Cambridge University Press. Cambridge. 121-132. ***25, 163, 193, 271, 273-4, 276, 370.***

Hessler, P. (2001). *River Town. Two years on the Yangtse*. John Murray. London. ***35, 227-8.***

Hill. S.F *et al.*(2004). 'Mortality among 'never smokers' living with smokers; two cohort studies.1981-4 and 1996-9'. *British Medical J.*328;988-9. ***12.***

Hollander, A.N.J. (1967). 'Social description. The problem of reliability and validity'. In D.G. Jongmans and P.C.W. Gutkind.(eds). *Anthropologists in the field*. Humanities Press. New York.1-34. ***112.***

Horton, R. (2003). 'Medical journals; evidence of bias against the diseases of poverty'. *Lancet*.361;712-3. ***397.***

Horwitz, R.P. (1993). 'Just stories of ethnographic authority'. In C.B. Brettel.(ed). *When they read what we write. The politics of ethnography*. Bergin and Garvey. Westport, Conn. ***287.***

Hotchkiss, J.C. (1967). 'Children and conduct in a Ladino community of Chiapas, Mexico'. *American Anthropologist*.69;711-718. ***92.***

Howell, N. (1990). 'Surviving fieldwork. A report of the Advisory Panel on Health and Safety in fieldwork'. American Anthropological Association Publication 26. Washington, DC. ***341-2***.

Hsu, F. (1975). 'Intercultural understanding; genuine and specious'. *Anthropology and Education Quarterly*.8;202-9. ***108***.

Hua, C. (2001). *A society without fathers or husbands. The Na of China*. Zone Books. New York. ***250-1, 318.***

Huang Zhou. (1985). Quoted in J. Gernet. *China and the Christian impact. A conflict of cultures*. Cambridge University Press. Cambridge. 205.

Hughes, E.C. (1960). 'Introduction; the place of fieldwork in social science'. In B. Junker. (ed).*Fieldwork. An introduction to the social sciences*. Chicago University Press. Chicago, ***264.***

Humphreys, L. (1970). *The tea-room trade. Impersonal sex in public places*. Aldine. New York. ***182.***

Jahoda, G. (1969). 'Scientific training and the persistence of traditional beliefs among West African University students'. *Nature*. 220;1365. ***375.***

Jayaraman, R. (1979). 'Problems of entry; Sri Lanka, Zambia and Australia' in M.N. Srinivas *et al.*(eds). *The fieldworker and the field.* Oxford University Press. Delhi 266-275. ***58.***

Johanson, D. and Edey, M.(1981). *Lucy. The beginnings of mankind.* Warner Books. New York. ***90.***

Jones, D.J. (1970). 'Towards a native anthropology'. *Human Organisation.* 29;251-59.

Jones, S.L. (1994). 'A constructive relationship for religion with the science and profession of psychology'. *American Psychologist* 49:184-199. ***110.***

Jorgensen, D. (1981). 'Correspondence'. *Man(ns).*16(3);470-2. ***24.***

Josephidies, L.(1997). 'Representing the anthropologist's predicament'. In A. James *et al.*(eds). *After writing culture. Epistemology and praxis in contemporary anthropology.* Routledge. London. 16-33. ***245.***

Joshi, P.C. (1979). 'Fieldwork experience relived and reconsidered' in M.N. Srinivas et al.(eds). *The fieldworker and the field.* Oxford University Press, Delhi. 73-99. ***132, 142, 183, 358.***

Kadt, E de. (1970). *Catholic radicals in Brazil.* Oxford University Press. New York. ***25.***

Kashoki, M.E.(1982). 'Indigenous scholarship in African universities. The human factor'. In H. Fahim.(ed). *Indigenous anthropology in non-Western countries.* Carolina University Press. Durham, North Carolina. 35-51. ***39.***

Kelser, J. *et al.*(2004). 'Representation of authors and editors from countries with different human development indexes in the leading literature on tropical medicine; survey of current evidence'. *British Medical J.*328;1229-1232. ***397.***

Kendrick, M.A. (1974). 'Viral hepatitis in American missionaries abroad'. *J. Infectious Diseases.*129;227-229. ***372.***

Keskitalo, A.I. (1976). 'Research as an inter-ethic relation'. *Acta Borealis.*1;15-42. ***256.***

Kimball, S.T. (1972). 'Learning a new culture' in S.T. Kimball and J.B. Watson.(eds) *Crossing cultural boundaries. The anthropological experience.* Chandler. San Francisco. 182-192. ***213.***

Knutsson, K.E. and Selinus, R. (1970). 'Fasting in Ethiopia' *American J. Clinical Nutrition*.23;956-9. ***133.***

Kochar, V.K. (1977). 'Sanitation and Culture. II. Behavioural aspects of the disposal of excreta in a rural West Bengal region'. *Indian J. Preventative and Social Medicine*. 8;142-151. ***251.***

——,(1978). 'Human factors in hookworm transmission in a rural West Bengal region'. *Indian J. Preventative and Social Medicine*. 9(3). ***251.***

Kochar, V.K. *et al*. (1976). 'Human factors in the regulation of parasitic infections. Cultural ecology of hookworm population in rural West Bengal' in F.X. Grollig and H.B. Haley.(eds). *Medical Anthropology*. Mouton. The Hague. 287-312. ***251.***

Koentjaraningrat. (1982). 'Anthropology in developing countries'. In Fahim, H.(ed). *Indigenous anthropology in non-Western countries*. Carolina Academic Press. Durham. North Carolina. 176-192. ***111, 274.***

Kondo, D.K. (1990). *Crafting selves. Power, gender and discourses of identity in a Japanese workplace*. University of Chicago Press. Chicago. ***197, 236, 240, 241, 391.***

Kuhn, T.S. (1962). *The structure of scientific revolutions*. Chicago University Press. Chicago. ***158, 307, 406.***

Kumar, N. (1988). *The artisans of Banaras. Popular culture and identity*. Princeton University Press, Princeton. New Jersey.

——,(1992). 'Friends, brothers and informants'. Fieldwork memories of Banaras. University of California, Berkeley. ***51, 76, 78, 83, 104, 314, 322, 326, 362, 373, 396, 414, 420, 424.***

——,(2001). 'Children and the Partition'. In S. Kaul. (ed). *The partitions of memory. The afterlife of the division of India*. Permanent Black. Delhi.269-301. ***265.***

Ladd, J. (1957). *The structure of a moral code. A philosophical analysis of ethical discourse applied to the ethics of the Navaho Indians*. Harvard University Press. Cambridge, Mass. ***124,149.***

Le Fort, R. (1966). *The teaching of Gurdjieff*. Gollancz. London.

Laing, R.D. (1968). *The politics of experience*. Penguin. Harmondsworth. ***137.***

Laitin, D.D. (1986). *Hegemony and culture. Politics and religious change among the Yoruba.* Chicago University Press. Chicago, ***106, 369.***

Lal, D. (1995). *Unintended consequences.* MIT Press. Cambridge, Mass.

——,(2004). *In praise of empires. Globalization and order.* Palgrave. New York. ***236, 243.***

Landis, R. (1970). 'A woman anthropologist in Brazil'. In P. Golde.(ed). *Women in the field. Anthropological experience.* 12, 4. Aldine, Chicago. 118-139. ***340, 341.***

Lattimore, O. (1950). *Ordeal by slander.* Atalantic-Little Brown. New York. ***56.***

Lefort, R. (1966). *The teachings of Gurdjireff.* Amazon. London, ***153.***

Levi-Strauss, C. (1966). 'Anthropology; its achievements and future'. *Current Anthropology.*7;124-7. ***111.***

Lewis, D.(1973). 'Anthropology and colonialism'. *Current Anthropology.*14;581-602. ***207.***

Lewis, O. (1959). *Five families. Basic* Books. New York. ***98, 112.***

——,(1951). *Life in a Mexican village. Tepoztlan restudied.* Illinois University Press. Urbana. ***98, 230.***

Lienhardt, G. (1979). 'Review of F.M. Deng's Africans of two 10 worlds'. Ethnic and Racial Studies.2;516-519. ***239.***

London, J. (1903). *The people of the abyss.* Project Gutenberg. New York. ***414.***

Lugalla, J.L.P. (1997). 'Development, change and poverty in the informal sector during the era of structural adjustment in Tanzania'. *Canadian J. of African Studies.*31(3);424-451. ***252.***

Machlup, F. (1956). 'The inferiority complex of the social sciences' in M.H. Sennholz.(ed). *On freedom and free enterprise.* Van Nostrand. Princeton.161-172. ***124.***

Macintyre, A. (1984). *After virtue. A study of moral theory.* University of Notre Dame Press. ***43.***

Macionis, J. (1989). *Sociology.* Prentice Hall. Englewood Cliffs. ***17.***

Madan, T.N. (1975). 'On living intimately with strangers'. In A. Beteille & T.N. Madan.(eds). *Encounters and experience.* Vikas. Delhi. 131-156. ***56, 239.***

——,(1982). 'Indigenous anthropology in non-Western countries; an overview' in H. Fahim.(ed). *Indigenous anthropology in non-Western countries*. Carolina Academic Press. Durham. 7, 263-268. ***288.***

Majumdar, D.N. (1962), *Himalayan polyandry : structure, function and culture change: a field study of Jaunsar Bawar.* Asia Publishing House. New York. *310*

Malhotra, P. and Prasad, B.G.(1966). 'A study of morbidity among children under five years of age in an urban area of Delhi'. *Indian J. Medical Research.*54;285-314. ***199.***

Malinowski, B. (1938). Introduction. J. Kenyatta. *Facing Mount Kenya*. Secker and Warburg. London. ***45, 258, 414.***

——,(1961). *The Argonauts of the Western Pacific*. Dutton. New York. ***272.***

——,(1967). *A diary in the strict sense of the term.* Harcourt, Brace and World. New York. ***50, 105, 212.***

Mao Tse Tung. (1969) *Collected Works*. ***49, 118.***

Marcus, G.E. (1986). 'Contemporary problems of ethnography in the modern world system'. in J. Clifford & G.E. Marcus.(eds). *Writing culture. The poetics and politics of ethnography*. University of California Press. Berkeley. 165-193. ***143.***

Marmot, M.G. (2004). 'Evidence based policy or policy based evidence'. *J. Epidemiology and Community Health*. 58. ***365.***

Marsh, D.R. *et al.*(2004). 'The power of positive deviance'. *British Medical J.*329;1177-1179. ***170.***

Mascarenhas-Keyes, S. (1988). 'The native anthropologist; constraints and strategies in research'. In A. Jackson.(ed). *Anthropology at home*. Tavistock. London. 180-195. ***31, 242, 338-9.***

Mayaram, S. (1996). 'Speech, silence and the makers of Partition violence in Mewat' in S. Amin and D. Chakrabarty.(eds). *Subaltern Studies IX*. Oxford University Press. Delhi. 126-163. ***304, 379.***

Mayer, A.C.(1975). 'Becoming a professional observer'. In A. Beteille and T.N. Madan. (eds). *Encounter and Experience.* Vikas. Delhi. 27-41.

McQuay, R.M.(1967). 'Parasitologic studies in a group of furloughed missionaries'. *American J. Tropical Medicine*.16;154-166. ***344, 372.***

Mead, M.(1928). *Coming of age in Samoa.* Morrow Quill, New York. ***139, 296, 330, 333-4.***

——(1956). *New lives for old. Cultural transformation - Manus. 1928-1953.* Morrow. New York. ***76.***

——,(1972). 'Fieldwork in high cultures' in S.T. Kimball and J.B. Watson. (eds). *Crossing cultural boundaries. The anthropological experience.* Chandler. San Francisco. 120-132. ***65, 66, 123, 190, 197, 203, 315.***

Mehanna, S & N.S. Hopkins.(2000). 'Managing personal health in a polluted urban common'. *J. Social Science*.4(1);111-116. ***428.***

Mehta, S.(2004). *Maximum city, Bombay lost and found.* Headline. London. ***76, 267, 354-5, 356, 396, 398.***

Meron, R.K. (1972). 'Insiders and Outsiders. A chapter in the sociology of knowledge'. In *Varieties of political expression in sociology.* University of Chicago Press, Chicago. 9-47. 7.

Messenger, J.(1989). *Inis Beag revisited. The anthropologist as observant participator.* Sheffield Publishing. Salem, Wisconsin. ***98, 102.***

Messerschmidt, D.A. (1981). 'Constraints in government research; the anthropologist in a rural school district'. In Messerschmidt, D.A.(ed). *The anthropologist at home in North America.* Cambridge University Press. Cambridge. ***376-7.***

Miller, D.B. (1976). 'A genius like feeling for the social stuff'. *South Asia*.6;112-20. ***11.***

Milne, L. (1923). *The home of an Eastern clan. A study of the Palaungs of the Shan States.* Oxford University Press. Oxford. ***7, 45.***

Milne, L. and W.W. Cochrane. (1910). *Shans at home.* John Murray. London.

Minocha, A.A. (1979). 'Varied roles in the field. A hospital in Delhi' in M.N. Srinivas et al.(eds). *The field worker and the field.* Oxford University Press. Delhi. 201-215. ***70.***

Morton, K. and Shaw, B.(2003). 'Shifting sands. Applied anthropology in Native Title litigation'. *Anthropology in Action.*10(2);50-57. ***298.***

Mott, L.R.B. (1982). 'Indigenous anthropology and Brazilian Indians' in H. Fahim.(ed). *Indigenous anthropology in non-western countries.* Carolina Academic Press, Durham. 112-117. ***203.***

Mukherjee, R. (1976). 'The value base of social anthropology. The context of India in particular'. *Current Anthropology.*17;71-95. ***40.***

Mwilala, A.J. (2000). 'Resource use and management in communal grasslands by pastoralists in Tanzania'. *J. Social Sciences.*4(1);65-70.

Myrdal, G. (1944). *An American Dilemma. The Negro Problem and American Democracy.* Harper and Brothers. New York. ***65, 155.***

Naderi, N.A.(1982). 'Some considerations regarding anthropological dilemmas' in Fahim, H.(ed). *Indigenous anthropology in non-Western countries.* Carolina Academic Press. Durham, North Carolina, 242-249. ***252, 274.***

Nadel, S.F. (1947).*The Nuba. An anthropological study of the hill tribes of Kordofan.* Oxford University Press, Oxford. ***41, 43.***

Naidu, R.(1980). 'The impact of processions in the city of Hyderabad' in Chatterjee. P. and Jeganathan P.(eds). *Community, gender and violence.* Hurst. London. 204-207. ***354.***

Naipaul, V.S. (1964). *An area of darkness.* Vintage. New York. ***156.***

——,(1977). *India; a wounded civilization.* Vintage, New York. ***156.***

——,(1990). *India; a million mutinies now.* Penguin, New York. ***156.***

Nakane, C.(1982). 'The effect of cultural tradition on anthropologists' in H. Fahim.(ed). *Indigenous anthropology in non-Western countries.* Carolina University Press. Durham, North Carolina. 52-60. ***427.***

Nakhleh, K. (1979). 'On being a native anthropologist'. In G. Huizer and B. Mannheim. (eds). *The politics of anthropology. From colonialism and sexism. Toward a view from below*. Mouton. The Hague. 343-35. ***339.***

Nash, D. (1963). 'The ethnologist as stranger. An essay on the sociology of knowledge'. *South-Western J. Anthropology*. 19;149-167.

Nash, D. and Wintrop, R.(1972). 'The emergence of self-consciousness in ethnography'. *Current Anthropology*. 13;527-42. ***242.***

Neitz, M.J. (1987). *Charisma and community. A study of religious commitment within the Charismatic Renewal*. Transactions. New Brunswick

Nida, E.A. (1964). *Towards a science of translating*. Orbis. Maryknoll. ***379.***

Ogbu, J.U. (1974). *The next generation; an ethnography of education in an urban neighbourhood*. Academic Press. New York, ***309.***

Okely, J. (1987). 'Anthropology & autobiography. Participatory experience and embodied knowledge'. In J. Okely & H. Callaway.(eds). *Anthropology and Autobiography*. Routledge. London. ***34.***

Ole-Lengisugi, N.(1996). 'Livestock health care. Maasai pastoralists; experience in the Arusha region'. Proceedings, Conference on Indigenous Knowledge and Sustainable Development. Sokoine University of Agriculture. Morogoro, Tanzania.

Ornstein, R.E. (1972). *The psychology of the unconscious*. Viking Press, New York. ***151.***

Ortiz, C.B. (1982). 'Anthropology in the Puerto Rican colonial context; analysis and projection' in H. Fahim,(ed). *Indigenous anthropology in non-Western countries*. Carolina Academic Press. Durham. 97-111. ***257.***

Orwell, G. (1946). *Animal farm*. Harcourt Brace. New York.

Owens, M. and Owens, D. (1984). *Cry of the Kalahari*. Houghton Mifflin. Boston. ***372.***

Owusu, M. (1978). 'The ethnography of Africa; the usefulness of the useless'. *American Anthropologist*.80;310-334. ***46.***

Oxford Dictionary of English. (2003). Oxford University Press. Oxford.

Oxford English Dictionary. (1956). Oxford University Press. Oxford.

Pandey, T.N. (1979). 'The anthropologist-informant relationship. The Navajo and Zuni in America and the Tharu in India'. In M.N. Srinivas.(ed). *The fieldworker and the field.* Oxford University Press. Delhi. 246-265. ***4, 98, 223.***

Parkin, D.J. (1971). 'Language choice in two Kampala housing estates' in W.H. Whiteley (ed). *Language use and social change. Problems of multilingualism with special reference to Eastern Africa.* Oxford University Press. Oxford. 347-363. ***387.***

Pasternak, B. (1989). 'A conversation with Fei Xiaotong'. *Current Anthropology.* 29(4); 637-662. ***56, 61, 263.***

Patton, M.Q. (1990). *Qualitative evaluation and research methods*'. Sage. Newbury Park. ***175.***

Patwardhan, S. (1979). 'Making my way through caste images. Untouchables in Poona' in M.N. Srinivas *et al.*(eds). *The fieldworker and the field.* Oxford University Press. Delhi.152-150. ***107.***

Pels, P. (1999). 'Professions of duplicity. A prehistory of ethical codes in anthropology'. *Current Anthropology.*40(2);101-136. ***42, 94, 279-80.***

Pelto, P.J.(1970). *Anthropological research. The structure of enquiry.* Harper and Row. New York. ***92.***

Perry, D. (2002). 'Micro-credit and women money lenders. The shifting terrain of credit in rural Senegal'. *Human Organisation.*61(1). ***428.***

Petersen, R. (1982). 'Some ethical questions in connection with research activity in an asymmetrical ethnic situation' in H. Fahim.(ed). *Indigenous anthropology in non-Western countries.* Carolina Academic Press. Durham.223-249. ***303.***

Powdermaker, H. (1967). *Stranger and Friends.* Nation, New York.

Ramaswamy, E.A.(1979). 'Being where the action was. Trade unions in Coimbatore'. In M.N. Srinivas *et al.*(eds). *The fieldworker and the field*. Oxford University Press. Delhi. 173-183.

Rao, R. (1938). *Preface to 'Kanthapura'*. New Directions. New York, 393.

Rapport, N. (1996). 'Cultural form and individual meanings'. In J van Bremen et al.(eds). *Horizons of understanding*. Leiden University Press. Leiden. ***136.***

Redfield, R. (1950). *Chan Kom. The village that chose progress*. Chicago University Press. Chicago. ***65, 79, 80, 203, 229.***

Reymond, L. (1958). *My life with a Brahmin family*. Rider. London, ***221.***

Richards. I.A. (1953). 'Towards a theory of translating'. In A.F. Wright.(ed). *Studies in Chinese thought*. Chicago University Press. Chicago,157-164. ***405.***

Richardson, I. (1959). *The role of tone in the structure of Sukuma*. School of Oriental and African Studies. London. ***319.***

Rici, J.A. *et al*.(1995). 'Assessing the validity of informant recall; results of a time use pilot study in peri-urban Egypt'. *Human Organization*.54(3).304-308. ***178.***

Roberts, D.F. and Tanner, R.E.S. (1963). 'The effects of parity on birth-weight etc in a Bantu sample'. *British J. Preventive and Social Medicine*. 17(4). 209-218. ***52.***

Rosenhan, D.(1973). 'On being sane in an insane place'. Science;179.

Rushdie, S.(1980). *Midnight's children*. Avon. New York. ***156.***

——,(1988). *Satanic Verses*. Viking. New York. ***158.***

Saberwal, S. (1969). 'Rapport and resistance among the Embu of Central Kenya'. In S. Saberwal and F. Henry.(eds). *Stress and response in fieldwork*. Holt, Rinehart and Winston. New York, 47-62. ***320.***

Saberwal, S. and F. Henry. (1969). Introduction to S. Saberwal and F. Henry.(eds). *Stress and response in fieldwork*. Holt, Rinehart and Winston. New York, ***307.***

Sanday, P.R.(1986). *Divine hunger. Cannibalism as a cultural system*. Cambridge University Press. New York. ***329.***

Saran, A.K. (1958). 'India'. in J.S. Roucek.(ed). *Contemporary Sociology*. Philosophical Library. New York. 1013-34. ***60.***

Sargent, R.P. *et al.*(2004). 'Reduced insistence of admission for myo-cardial infarction associated with public smoking ban; before and after study'. *British Medical J.*328;9779. ***12.***

Saxena, S. *et al.*(2003). *How international are the editorial boards of leading psychiatry journals*? Lancet.361;609. ***397.***

Sayaranan, R. (1979). 'Problems of Entry. Sri Lanka, Zambia and Australia' in Srinivas. M.N (ed). *The fieldworker in the field.* Oxford University Press. Delhi. 246-265. ***58.***

Scheper-Huges, N. (1982). *Saints, Scholars and Schizophrenics. Mental illness in rural Ireland.* University of California Press Berkeley. ***101-2, 116.***

Schwartz, M.S. and Schwartz. C.G.(1955). 'Problems in participant observation'. *American J. of Sociology*.60;343-53.

Scott, J.(1973). 'Black science and nation building'. In J.A. Ladner.(ed). *The death of white sociology.* Vintage Books. New York. 289-309. ***48.***

Scott, J.C. (1969). 'The analysis of corruption in developing nations'. *Comparative Studies in Society and History.*11;315-41. ***266.***

Scotton, C.M. & Zhu, W. (1983). 'Tongzhi in China; language change and its conversational consequences'. *Language in Society.*12(4);477-94.

Seligman, C.G. (1950). *The races of Africa.* Oxford University Press. Oxford.

Sen, A.K. (1992). *Inequality re-examined.* Clarendon Press. Oxford.

——, (1999). *Development and freedom.* Oxford University Press. Oxford. ***243.***

Serematakis, C.N. (1991). *The last word. Women, death and divination in inner Mani.* University of Chicago. Chicago. ***121, 244-5.***

Serpell, R. (1982). *Culture's influence on behaviour.* Methuen. London. ***48.***

Sethi, S. and M.G. Seligman. (1993). 'Optimism and fundamentalism'. *Psychological Science*, 4(4). 256-9. ***114.***

Shah, A.M. (1979). 'Studying the present and the past. A village in Gujarat' in M.N. Srinivas et al.(eds). *The fieldworker and the field.* Oxford University Press. Delhi.29-37. ***115, 169.***

Shami, S. (1988). 'Studying your own. The complexities of a shared culture'. in S. Altorki and El-Solh. C.F.(eds). *Arab women in the field. Studying your own society*. Syracuse University Press. Syracuse.115-138. ***179.***

Sharma, N. *et al.*(2004). 'Maternal beliefs of Dogra mothers of Jammu and their children's perceived competence pre-school'. *J. Human Ecology*.15(2);153-156. ***346***.

Sheehan, E.A. (1993). 'The student of culture and the ethnography of Irish intellectuals'. In C.B. Britten.(ed). *When they read what we write*. Bergin and Garvey. Westport, Conn. 75-89. ***71, 102, 139, 303, 324.***

Sheth, N.R. (1979). 'Trial and error in a new field. A factory in Gujarat' in M.N. Srinivas *et al.*(eds). *The fieldworker and the field*. Oxford University Press. Delhi. 163-172. ***354.***

Shokeid, M. (1997). 'Negotiating multiple viewpoints. The cook, the native, the publisher and the ethnographer'. *Current Anthropology*. 38(4); 631-645. ***140.***

Simmel, G. (1971). 'The problem of sociology'. In D. Levine.(ed). *On individuality and social forms*. Chicago University Press. Chicago.23-40. ***136, 137.***

Singh, S.K. (2004). 'Road accident analysis: A case study of Patna city'. *Urban Transport J*.3(2);60-75. ***130.***

Sivakuma, C. (1979). 'An apprenticeship marked by fiustration. College students in Mysore' in M.N. Srinivas *et al* (eds). *The field worker and the field*. Oxford University Press. Delhi. ***216-227.***

Skinner, J. (1997). 'Doing anthropology as action or in-action'. *Anthropology in Action*. 4(2);13-16. ***19.***

——,(2002). 'Considering ethics. From classroom teaching to research committee vetting'. *Anthropology in Action*.9(3).25-27. ***78.***

Smith, N. (1954). *Baba of Kano. A woman of the Moslem Hausa*. Faber. London. ***89, 205.***

Smith, R.J. (1983). *Japanese society. Tradition, self and the social order*. Cambridge University Press. Cambridge. ***237, 391.***

Sorbo, G.M. (1982). 'Anthropology at home and abroad. A discussion of epistemology and ethical issues'. in H.

Fahim.(ed). *Indigenous anthropology in non-Western countries.* Carolina Academic Press. Durham, North Carolina. 152-163. ***248.***

Srinivas, M.N. (1952). *Religion and society among the Coorgs of South India.* Asia Publishing House, London. ***200.***

—,(1979).The field worker and the field. A village in Karnataka'. In Srinivas, M.N. *et al.*(eds). *The fieldworker and the field.* Oxford University Press. Delhi. ***115, 136, 161, 249, 310, 324, 340.***

Stavenhargen, R. (1971). 'Decolonizing applied social science'. *Human Organisation.* 30(4). ***208.***

Stewart, S. (1994). *Crimes of writing. Problems in the containment of representation.* Duke University Press. Durham, North Carolina. ***93.***

Strathern, A. (1981). *Field work and theory in social anthropology.* University College Press. London. ***337.***

Suzuki, D.T. (1958). *The Zen doctrines of the mind.* Rider. London, ***153.***

Suzuki, T. (1978). *Japanese and the Japanese. Words in context.* Trans. S Miura. Kodansha. Tokyo. ***237, 391.***

Szwed, J.F. (1969). 'An American anthropological dilemma. The politics of Afro-American culture'. In D. Hymes.(ed). *Reinventing anthropology.* Pantheon. New York, 153-81. ***248.***

Tannenbaum, N. (2000). 'Protest, tree ordination and the changing context of political ritual'. *Ethnology.*29(2).109-127. ***293.***

Tanner, R.E.S. (1956). 'A preliminary enquiry into Sukuma diet in the Lake Province, Tanganyika'. *East African Medical J.*33(8). 305-324. ***146.***

——,(1960). 'Majadiliano. Kitabu cha kuelimisha wafanya kazi'. International Labour Office. Geneva.

——,(1964). 'Conflict within small European communities in Tanganyika'. *Human Organisation.*23(4);319-327. ***4, 89.***

——,(1966a). 'Drug addiction in East Africa;', *International J. of the Addictions.*1(1).9-29. ***89.***

——,(1966b). 'Cattle theft in Musoma,1958-60'. Tanganyika Notes and Records.65; 31-42.

——,(1966c). 'European leadership in small communities in Tanganyika prior to Independence'. *Race*.7(3);289-302. ***4, 89.***

——,(1970a). 'An East African prison'. Scandinavian Institute of African Studies. Uppsala. ***34, 161, 167, 304.***

——,(1970). 'Three studies in East African criminology'. Scandinavian Institute of African Studies. Uppsala. ***76, 119, 130.***

——,(1978). 'Rumour and the Buganda emergency'. *J. Modern African Studies*.16;329-338. ***52, 205.***

——,(1979). 'Word and Spirit in contemporary African religious practice and thought'. *J. Religion in Africa*.9(2);123-135. ***390.***

——,(1983). 'Law and Order Statistics in the Third World'. Institute of Development Studies. University of Sussex. ***130.***

——,(1993). 'East African ethical ideas and translation ; a bilateral process'. *Anthropos*. 88;29-37. ***96, 98, 145, 379.***

—, (1996). 'Some reflections on a foreigner's use of Swahili'. *Nordic J. African Studies*. 5(1).107-117.

——,(2001). 'Informants and Researchers; cooperation, self-interest and reciprocity. An examination of changing relationships'. *J. Social Sciences*. 5(4);279-290. ***34.***

—— (2003). 'Postcards. A neglected source of anthropological data'. *The Anthropologist.* 6(3). 215-218. ***120.***

Tanner, R.E.S. and F. Kamoga.(1968). *Mtawala. Masimulizi ya Machiavelli*. East African Publishing House. Nairobi. ***224.***

Tanner, R.E.S. and Morgan, J. (1979). *Opinion Survey of attitudes to Church and education in Tanzania.* East African Publishing House. Nairobi. ***403.***

Trant, N. (1975). 'Interview with Nathaniel Tarn' Boundaries. 24(1);1-34. ***225.***

Trice, H.M. (1970). 'The 'outsider' role in field study'. In J. Filstead.(ed). *Qualitative methodology. First hand involvement with the social world.* Markham. Chicago. 72-82. ***221.***

Tre, H.M. (1970). 'The 'outsider' role in field study' in W.J. Filstead.(ed). 'Qualitative methodology. First hand involvement with the social world'. Markham. Chicago.72-82.

Turnbull, C. (1973). 'The Mountain People'. Simon and Schuster, 8, New York. ***93, 198, 327.***

Turner, E. (1996). The hands feel it. Healing and spirit presence among a northern Alaskan people. Northern Illinois University Press. Dekalb.

——,(2000). Interview with M. Engelke. *Current Anthropology*. 41; 843-52. ***315.***

Turner, V. (1967). *The forest of symbols*. Aspects of Ndembu ritual. Cornell University Press. Ithaca, ***119, 124, 201, 335.***

——,(1970). 'Introduction' Colonialism in *Africa.1870-1960*. L.H. Gann & P. Duignan. (eds). Cambridge University Press. Cambridge.1-5. ***44.***

Uberoi, J.P.S. (1968). 'Science and Swaraj'. *Contributions to Indian Sociology*.2;119-123. ***60***.

Ueda, S.(1993). 'Sixteen ways of saying 'no' in Japanese. Patterns of communication in and out of Japan'. ICU Communication Department, Tokyo.

Unni, K.R. (1979). 'On tracks and traits in my fieldwork' in M.N. Srinivas et al.(eds). *The fieldworker and the field*. Oxford University Press. Delhi. 58-72. ***115, 164.***

Valencia, A. *et al*.(1990). 'A Yaqui point of view on Yaqui ceremonies and anthropologists' in R. Schechner and W. Appel.(eds). *By means of performance. Intercultural Studies on Theatre and Ritual*, Cambridge University Press. Cambridge. 96-106.

Vansina, J. (1985). *Oral tradition as history*. James Currey. London. ***174***.

Varadachar, B.D. (1979). 'The bottom view up. Some cognitive categories' in M.N. Srinivas *et al*.(eds). *The fieldworker and the field*. Oxford University Press. Delhi. 127-140. ***15, 117, 141, 348, 359.***

Vidich, A.J. & Benshan, J. (1968). *Small town in mass society. Class, power and religion in a rural community*'. Princeton University Press. Princeton, New Jersey. ***98, 143.***

Wallace, A.F.C. (1961). 'The psychic unity of human groups'. In B. Kaplan. (ed). *Studying personality cross-culturally*. Harper and Row. New York. ***136.***

——,(1962).'Culture and cognition'. *Science*. 135. 129-163. ***136.***

Wallerstein, I. (1977). 'The tasks of historical social science. An editorial'. *Binghamton Review*.1;3-7. ***38.***

Warman, A. (1982). 'Indigenous thought' in H. Fahim.(ed). *Indigenous anthropology in non-Western countries*. Carolina University Press. Durham. ***248.***

Weber, M. (1978). 'Economy anu Society'. An outline of interpretive sociology. University of California Press. Berkeley. ***335.***

Weil, A. (1972). *The natural mind. A new way of looking at drugs and the higher consciousness*. Houghton Mifflin. Boston.

Weil, S. (1995). 'It is futile to trust in man. Methodological difficulties in studying non-mainstream populations with reference to the Ethiopian Jews in Israel'. *Human Organisation*.54(1). ***153, 168, 332.***

Weiss, B.(1998). 'Electric vampire. Haya rumour of the commodified body' in M. Lambek and N. Strathern.(eds). *Body and Person. Comparative perspectives from Africa and Melanesia*. Cambridge University Press. Cambridge.172-194. ***375.***

Werthein, W.F. (1972). 'Counter-insurgency research at the turn of the century. Snouck Hurgronje and the Acheh war'. *Sociologischogios*.19;320-28. ***88.***

Westen, D. (2001). 'Beyond binary oppositions in psychological anthropology; integrating psychoanalysis and cognitive science'. In C.C. Morris and H.F. Mathews.(eds). *The psychology of cultural experience*. Cambridge University Press. Cambridge. 21-47. ***405.***

Whitehead, A.N. (1953). *Dialogues*. Cambridge University Press. Cambridge. ***151.***

Whiting, J.W.M. (1967). 'Sorcery, sin and the superego. A cross-cultural study of some mechanisms of social control' in C.S. Ford.(ed). *Cross-cultural approaches. Readings in comparative research*. Human Relations Area Files Press. New Haven, Conn. ***404.***

Whyte, W.F. (1981). Street Corner Society. Chicago University Press. Chicago. ***369.***

——,(1995). 'Encounters with participatory action research'. *Qualitative Sociology*.18;289-300.

Wijsen, F. (2001). 'When elephants fight the grass gets hurt. Muslim-Christian relationships in upcountry Tanzania'. In F.

Wijsen and P. Nissen.(eds). *Mission is a must.* Rodkopi. Amsterdam. ***309.***

Wintrob, R.M.(1969). 'An inward focus. A consideration of psychological stress in field work'. in F. Henry and S. Saberwal. (eds). *Stress and response in field work.* Holt, Rinehart and Winston. New York. 63-76. ***159.***

Wijeyewardene, G.E.T. (1961). 'Some aspects of village solidarity in Ki-Swahili speaking coastal communities of Kenya and Tanganyika'. PhD thesis. Cambridge University. ***309.***

Wolcott, H.F. (1981). 'Home and away. Personal contrasts in ethnographic style'. In D.A. Messerschmidt.(ed.) *Anthropologists at home in North Americas.* Cambridge University Press. Cambridge. 255-265. ***79, 191, 259, 261-2.***

Wood, M. (1995). 'White skins, 'Chinese' and 'real people'. Some spatial transformations of the Western Province, Papua New Guinea'. *Oceania.*66;23-50. ***374.***

Xu-Dashou. (1985) quoted in Gernet, J. China and the Christian *impact. A conflict of cultures.* Cambridge University Press. Cambridge. 208.

Yang, M.M.C. (1972) 'How a Chinese village was written up' in S.T. Kimball and J.B. Watson. (eds). *Crossing cultural boundaries. The anthropological experience.* Chandler. San Francisco. 63-73. ***47.***

Zelditch, M. (1970). 'Some methodological problems in field studies'. In S.T. Kimball and J.B. Watson. (eds). *Cross-cultural boundaries. The anthropological experience.* Chandler. San Francisco. 217-231. ***159.***

Zhong Wei Chen. (2004). 'Biography of a Chinese micro-surgeon'. *British Medical J.*328;1019.

Zimmer, R. (1981). 'Observer participation and consultancy; research in urban food cooperatives'. in D.A. Messerschimdt. (ed). *Anthropologists at home in North America.* Cambridge University Press, Cambridge. 62-76.

INDEX